WHAT EVERY CHEMICAL TECHNOLOGIST WANTS TO KNOW ABOUT...

Volume I

EMULSIFIERS AND WETTING AGENTS

Compiled by

Michael and Irene Ash

Chemical Publishing Co., Inc.
New York, N.Y.

Printed in the United States of America

PREFACE

This reference book is the first volume in the set of books entitled WHAT EVERY CHEMICAL TECHNOLOGIST WANTS TO KNOW ABOUT . . . SERIES. This compendium serves a unique function for those involved in the chemical industry—it provides the necessary information for making the decision as to which trademark chemical product is most suitable for a particular application.

The chemicals included in this first book of the series have their major function as emulsifiers, wetting agents, and detergents, however, complete cross-referencing is provided for the multiple functions of all the chemicals.

The first section which is the major portion of each volume contains the most common generic name of the chemicals as the main entry. All these generic entries are in alphabetical order. Synonyms for these chemicals are then listed. The CTFA name appears alongside the appropriate generic name. The structural and/or molecular formula of the chemical is listed whenever possible. The generic chemical is sold under various tradenames and these are listed here in alphabetical order for ease of reference along with their manufacturer in parentheses. The *Category* subheading lists all the possible functions that the chemical can serve. Because of differences in form, activity, etc., individual tradenames of the generic chemical are used in particular applications more frequently. These are delineated in the *Applications* section. The differences in properties, toxicity/handling, storage/handling, and standard packaging are specified in the subsequent sections wherever distinguishing characteristics are known.

The second section of the volume TRADENAME PRODUCTS AND GENERIC EQUIVALENTS helps the user who only knows a chemical by one tradename to locate its main entry in section 1. The user can look up this tradename in this section of the book and be referred to the appropriate, main-entry, generic chemical name.

The third section GENERIC CHEMICAL SYNONYMS AND CROSS REFERENCES provides a way of locating the main entries by knowing only one of the synonyms. If the generic chemical is not in the volume, it will refer you to the volume in which it is contained.

The fourth section TRADENAME PRODUCT MANUFACTURERS lists the full addresses of the companies that manufacture or distribute the tradename products found in the first section.

The following is a list of the six volumes that comprise this series:

Volume I	Emulsifiers and Wetting Agents
Volume II	Dispersants, Solvents and Solubilizers
Volume III	Plasticizers, Stabilizers and Thickeners
Volume IV	Conditioners, Emollients and Lubricants
Volume V	Resins
Volume VI	Polymers and Plastics

This series has been made possible through long hours of research and compilation and the dedication and tireless efforts of Roberta Dakan who helped make this distinctive series possible. Our appreciation is extended to all the chemical manufacturers and distributors who supplied the technical information.

M. and I. Ash

NOTE

The information contained in this series is accurate to the best of
our knowledge; however, no liability will be assumed by the
publisher for the correctness or comprehensiveness of such
information. The determination of the suitability of any of the
products for prospective use is the responsibility of the user. It
is herewith recommended that those who plan to use any of the
products referenced seek the manufacturer's instructions for the
handling of that particular chemical.

OTHER BOOKS BY MICHAEL AND IRENE ASH

A Formulary of Paints and Other Coatings, Volumes I and II
A Formulary of Detergents and Other Cleaning Agents
A Formulary of Adhesives and Sealants
A Formulary of Cosmetic Preparations
The Thesaurus of Chemical Products, Volumes I and II
Encyclopedia of Industrial Chemical Additives, Volumes I–IV
Encyclopedia of Surfactants, Volumes I–IV
Encyclopedia of Plastics, Polymers and Resins, Volumes I–III

ABBREVIATIONS

@ ....at
anhyd. ....anhydrous
APHA ....American Public Health Association
approx. ....approximately
aq. ....aqueous
ASTM ....American Society for Testing and Materials
avg. ....average
B.P. ....boiling point
Btu ....British thermal unit
C ....degrees Centigrade
CAS ....Chemical Abstracts Service
cc ....cubic centimeter(s)
CC ....closed cup
cm ....centimeter(s)
cm³ ....cubic centimeter(s)
COC ....Cleveland Open Cup
compd. ....compound, compounded
conc. ....concentrated, concentration
cP, cps ....centipoise
cs, cSt ....centistokes
CTFA ....Cosmetic, Toiletry and Fragrance Association
DEA ....diethanolamine
disp ....dispersible, dispersion
dist ....distilled
DOT ....Department of Transportation
DW ....distilled water
EO ....ethylene oxide
equiv. ....equivalent
F ....degrees Fahrenheit
F.P. ....freezing point
FDA ....Food and Drug Administration
ft³ ....cubic foot, cubic feet
g ....gram(s)
gal ....gallon(s)
HLB ....hydrophile-lipophile balance
insol. ....insoluble
IPA ....isopropyl alcohol
kg ....kilogram(s)
l, L ....liter(s)
lb ....pound(s)
M.P. ....melting point
M.W. ....molecular weight
max ....maximum
MEA ....monoethanolamine
MEK ....methyl ethyl ketone
mfg. ....manufacture
MIBK ....methyl isobutyl ketone
min ....minute(s)
min. ....mineral, minimum
MIPA ....monoisopropanolamine

misc.	miscible
ml	milliliter(s)
mm	millimeter(s)
NF	National Formulary
no.	number
o/w	oil-in-water
OC	open crucible
PEG	polyethylene glycol
pH	hydrogen-ion concentration
pkgs	packages
PMCC	Pensky Marten closed cup
POE	polyoxyethylene, polyoxyethylated
PPG	polypropylene glycol
pt.	point
R&B	Ring & Ball
RD	Recognized Disclosure
ref.	refractive
rpm	revolutions per minute
R.T.	room temperature
s	second(s)
sol.	soluble, solubility
sol'n.	solution
sp.gr.	specific gravity
SS	stainless steel
std.	standard
SUS	Saybolt Universal seconds
TCC	Taggart closed cup
TEA	triethanolamine
tech.	technical
temp.	temperature
theoret.	theoretical
TLV	threshold limit value
TOC	Taggart open cup
UL	Underwriter's Laboratory
USP	United States Pharmacopoeia
uv, UV	ultraviolet
veg	vegetable
visc.	viscosity, viscous
w/o	water-in-oil
wt	weight
$\approx$	approximately equal to
<	less than
>	greater than
$\leq$	less than or equal to
$\geq$	greater than or equal to

TABLE OF CONTENTS

Ammonium lauryl ether sulfate

SYNONYMS:
Ammonium laureth sulfate (CTFA)
Poly (oxy-1,2-ethanediyl), α-sulfo-ω-(dodecyloxy)-, ammonium salt

STRUCTURE:

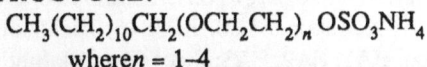

$$CH_3(CH_2)_{10}CH_2(OCH_2CH_2)_n OSO_3NH_4$$
where n = 1–4

CAS No.:
32612-48-9 (generic)

TRADENAME EQUIVALENTS:
Alkasurf EA-60 [Alkaril]
Carsonol ALES-4, SES-A [Carson]
Cedepal SA-406 [Domtar]
Conco Sulfate 216, WEM, WM [Continental]
Cycloryl MA 330, MA 360 [Cyclo]
Drewpon EKZ [Drew Produtos]
Empicol EAB [Marchon]
Manro ALES 27 [Manro]
Maprofix LES-60A, MB, MBO [Onyx]
Nutrapon AL60, KF3846 [Clough]
Polystep B-11 [Stepan Europe]
Richonol S-1300, S-1300C [Richardson]
Sactipon 2OA [Lever Industriel]
Sactol 2OA [Lever Industriel]
Sipon 201-20, EA, EA2, EAY [Alcolac]
Standapol EA1, EA2, EA3 [Henkel]
Steol CA-460 [Stepan]
Sterling ES600 [Canada Packers]
Sulfotex OT [Henkel]

1

Ammonium lauryl ether sulfate (cont'd.)

TRADENAME EQUIVALENTS *(cont'd.):*
  Texapon EA-1, EA-2, EA-3, EA-40 [Henkel/Canada]
  Texapon NA [Henkel Argentina]
  Tylorol A [Thomas Triantaphyllou S.A.]
  Ungerol AM3-60 [Unger Fabrikker AS]
  Witcolate S1300C [Witco/Organics]

CATEGORY:
  Detergent, wetting agent, surfactant, base, cleansing agent, degreasing agent, dispersant, emulsifier, flash foamer, foam stabilizer, foaming agent, penetrant, solubilizer

APPLICATIONS:
  Automobile cleaners: car shampoo (Carsonol SES-A; Cedepal SA-406; Sulfotex OT); tire cleaners (Conco Sulfate 216)
  Bath products (Alkasurf EA-60; Carsonol SES-A; Conco Sulfate 216; Empicol EAB; Manro ALES 27; Nutrapon AL60, KF3846; Sipon 201-20, EA, EA2, EAY; Standapol EA1, EA2, EA3; Sterling ES600; Texapon EA-2, EA-3, EA-40; Tylorol A)
  Cleansers: (Standapol EA1, EA2)
  Cosmetic industry preparations: (Conco Sulfate WM; Empicol EAB; Nutrapon KF3846; Sipon EA, EAY; Witcolate S1300C); shampoos (Alkasurf EA-60; Carsonol SES-A; Cedepal SA-406; Cycloryl MA 330; Drewpon EKZ; Empicol EAB; Manro ALES-27; Maprofix LES-60A; Nutrapon AL60, KF3846; Richonol S-1300C; Sipon 201-20, EAY; Standapol EA1, EA2, EA3; Steol CA-460; Sterling ES600; Texapon EA-1, EA-2, EA-3; Tylorol A; Ungerol AM3-60)
  Degreasers: (Cycloryl MA330, MA360; Sterling ES600)
  Household detergents: (Carsonol SES-A; Conco Sulfate WM; Richonol S-1300C; Sipon 201-20; Sterling ES600; Sulfotex OT); dishwashing (Alkasurf EA-60; Cedepal SA-406; Maprofix LES-60A; Nutrapon AL60; Richonol S-1300C; Steol CA-460); laundry detergent (Alkasurf EA-60; Cycloryl MA360; Sipon 201-20, EA, EAY); light-duty cleaners (Cycloryl MA330, MA360; Nutrapon AL60); liquid detergents (Carsonol SES-A; Conco Sulfate 216; Maprofix LES-60A; Steol CA-460; Sterling ES600; Sulfotex OT; Ungerol AM3-60); powdered detergents (Sterling ES600)
  Industrial applications: dyes (Alkasurf EA-60); textile/leather processing (Conco Sulfate 216)
  Industrial cleaners: (Alkasurf EA-60; Carsonol SES-A; Steol CA-460; Sulfotex OT; Witcolate S1300C); all-purpose cleaners (Conco Sulfate 216); metal processing surfactants (Conco Sulfate 216); sanitary supply (Alkasurf EA-60); textile cleaning (Conco Sulfate 216; Steol CA-460); textile processing (Steol CA-460)
  Pet shampoos: (Sipon EA, EAY)
  Pharmaceutical applications: antiseptic products (Sipon EA, EAY); contraceptive products (Sipon EA, EAY); deodorant preparations (Sipon EA, EAY)

PROPERTIES:
Form:
  Liquid (Carsonol SES-A; Conco Sulfate 216; Conco Sulfate WM; Cycloryl MA330,

Ammonium lauryl ether sulfate *(cont'd.)*

MA-360; Drewpon EKZ; Maprofix LES-60A; Nutrapon AL60, KF3846; Richonol
S-1300C; Sactipon 2OA; Sactol 2OA; Sipon 201-20, EA, EA2; Steol CA-460;
Sterling ES600; Texapon EA-40, NA; Tylorol A; Witcolate S-1300C)
Clear liquid (Manro ALES 27)
Clear, slightly visc. liquid (Alkasurf EA-60)
Visc. liquid (Sipon EAY; Standapol EA1, EA2, EA3; Sulfotex OT; Texapon EA-1,
EA-2, EA-3)
Paste (Ungerol AM3-60)

Color:
Light amber (Alkasurf EA-60)
Pale yellow (Conco Sulfate 216; Cycloryl MA330, MA360; Manro ALES 27; Steol
CA-460; Sulfotex OT; Texapon EA-40)
Water-white (Standapol EA1, EA2, EA3; Texapon EA-1, EA-2, EA-3)
Gardner 3 (Richonol S-1300C)

Odor:
Mild (Manro ALES 27)
Mild, pleasant (Sipon EA)
Typical (Alkasurf EA-60; Conco Sulfate 216)

Composition:
23–24% active (Texapon NA)
23–25% active (Drewpon EKZ)
24–26% active (Texapon EA-2)
24.5–25.5% active (Standapol EA2)
25–27% active (Texapon EA-1)
25.5–26.5% active (Standapol EA1)
26% active (Sipon EAY)
26–28% active (Texapon EA-3)
27% active (Manro ALES 27; Sipon EA)
27–29% active (Cycloryl MA330)
57–60% active (Maprofix LES-60A)
57.5% active (Sterling ES600)
58% active (Alkasurf EA-60)
58–60% active (Cycloryl MA360; Sulfotex OT; Texapon EA-40)
59% active (Polystep B-11; Steol CA-460)

Solubility:
Sol. in water (Alkasurf EA-60; Standapol EA-3; Steol CA-460; Witcolate S1300C)

Density:
8.4 lb/gal (Carsonol SES-A)
8.54 lb/gal (Sulfotex OT)
8.6 lb/gal (Richonol S-1300C)
1.020 g/ml (Standapol EA3)

Sp.gr.:
1.016 (Steol CA-460)

Ammonium lauryl ether sulfate (cont'd.)

1.03 @ 20 C (Manro ALES 27)
1.04 @ 20 C (Alkasurf EA-60)
1.06 @ 25/20 C (Maprofix LES-60A)

Visc.:
67 cps (Steol CA-460)
300 cps (Sipon EA)
2500 cps (Standapol EA1, EA2, EA3)
3000 cps @ 20 C (Manro ALES 27)

Stability:
Stable in hard water (Steol CA-460)

Compatibility:
Compatible with anionics, cationics, most amphoterics, and many natural and synthetic gums (Standapol EA3)

Biodegradable: (Carsonol SES-A, Manro ALES 27, Steol CA-460)

pH:
6.0–7.0 (10% sol'n.) (Standapol EA-3; Sulfotex OT)
6.5–7.0 (5% sol'n.) (Texapon NA)
6.5–7.0 (10% sol'n.) (Carsonol SES-A; Standapol EA-1, EA-2)
6.5–7.5 (Richonol S-1300C)
6.7 (10% sol'n.) (Sipon EA, EAY)
7.0 (1% in dist. water) (Steol CA-460)

Cloud Pt.:
–5 C (Sipon EA)
0 C (Sipon EAY; Standapol EA2)
5 C max. (Standapol EA1, EA3)
10 C (Sulfotex OT)
19 C (Steol CA-460)
< 40 F (Alkasurf EA-60)
70 F (Richonol S-1300C)

TOXICITY/HANDLING:
Releases ammonia at pH > 7 (Sipon EA; Sipon EAY; Steol CA-460)
Irritating to skin and eyes in conc. form; although nontoxic, ingestion should be avoided (Standapol EA3; Sulfotex OT)

STORAGE/HANDLING:
Store in closed containers above 7 C (Standapol EA3; Sulfotex OT)
Combustible liquid (Sulfotex OT)

STD. PKGS.:
45 gal drums or road tankers (Manro ALES 27)
480 lb net fiber drums; bulk, tank wagons, rail cars (Standapol EA3; Sulfotex OT)

Ammonium lauryl sulfate (CTFA)

SYNONYMS:
Dodecyl ammonium sulfate
Sulfuric acid, monododecyl ester, ammonium salt

EMPIRICAL FORMULA:
$C_{12}H_{26}O_4S \cdot H_3N$

STRUCTURE:
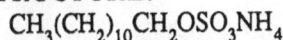
$CH_3(CH_2)_{10}CH_2OSO_3NH_4$

CAS No.:
2235-54-3

TRADENAME EQUIVALENTS:
Akyposal ALS 33 [Chem-Y]
Alkasurf ALS [Alkaril]
Avirol 200 [Henkel]
Carsonol ALS, ALS Special [Carson]
Cedepon LA-30 [Domtar/CDC]
Conco Sulfate A [Continental]
Cycloryl MA [Cyclo]
Drewpon NH, NHAC [Drew Produtos]
Emal AD-25 [Kao Corp.]
Emersal 6430 [Emery]
Empicol AL30 [Albright & Wilson/Australia]
Empicol AL30/T [Albright & Wilson]
Empicol AL70 [Albright & Wilson/Detergents]
Lakeway 101-20 [Bofors Lakeway]
Lonzol LA-300 [Lonza]
Lorol Liquid NH Sulphonated [Ronsheim & Moore]
Manro ALS30 [Manro]
Maprofix NH, NH54, NHL, NHL-22 [Onyx]
Mars AMLS [Mars]
Norfox ALS [Norman, Fox & Co.]
Nutrapon HA 3841 [Clough]
Polystep B-7 [Stepan]
Richonol AM [Richardson]
Sactipon 2A [Lever Industriel]
Sactol 2A [Lever Industriel]
Sermul EA 129 [Servo BV]
Sipon L22, L-22HV [Alcolac]
Standapol A, AMS-100 [Henkel]
Stepanol AM, AM-V [Stepan]
Sterling AM, AM-HV [Canada Packers]
Texapon A [Henkel/Canada]
Texapon A400, Special [Henkel KGaA]

Ammonium lauryl sulfate *(cont'd.)*

TRADENAME EQUIVALENTS *(cont'd.)*:
Witcolate AM [Witco/Organics]
Zoharpon LAA [Zohar Detergent]

CATEGORY:
Detergent, wetting agent, emulsifier, dispersant, foaming agent base, cleansing agent, solubilizer, surfactant

APPLICATIONS:
Automobile cleaners: car shampoo (Carsonol ALS, ALS Special; Mars AMLS; Stepanol AM, AM-V)

Bath products: bubble bath (Carsonol ALS, ALS Special; Conco Sulfate A; Cycloryl MA; Empicol AL30/T, AL70; Lakeway 101-20; Lonzol LA-300; Maprofix NH, NHL-22; Mars AMLS; Nutrapon HA 3841; Standapol A, AMS-100; Stepanol AM, AM-V; Sterling AM, AM-HV; Texapon A, Special)

Cleansers: (Standapol A, AMS-100; Texapon A); body cleansers (Conco Sulfate A; Emersal 6430); cleansing creams (Carsonol ALS, ALS Special; Mars AMLS; Stepanol AM, AM-V)

Cosmetic industry preparations: (Alkasurf ALS; Empicol AL30/T; Manro ALS30; Sipon L-22HV; Witcolate AM); cosmetic base (Empicol AL70); personal care products (Carsonol ALS, ALS Special; Lakeway 101-20; Lonzol LA-300; Stepanol AM, AM-V); shampoo base (Akyposal ALS 33; Cedepon LA-30; Zoharpon LAA); shampoos (Alkasurf ALS; Carsonol ALS, ALS Special; Conco Sulfate A; Cycloryl MA; Drewpon NH, NHAC; Emersal 6430; Empicol AL30, AL30/T, AL70; Lonzol LA-300; Lorol Liquid NH Sulphonated; Manro ALS 30; Maprofix NH, NHL-22; Mars AMLS; Nutrapon HA 3841; Richonol AM; Sipon L-22, L-22HV; Standapol A, AMS-100; Stepanol AM, AM-V; Sterling AM, AM-HV; Texapon A, Special); shaving preparations (Carsonol ALS, ALS Special; Sipon L-22; Stepanol AM, AM-V)

Farm products: insecticides (Stepanol AM, AM-V)

Food applications: fruit washing (Stepanol AM, AM-V); vegetable scrubber (Carsonol ALS, ALS Special)

Household detergents: (Alkasurf ALS; Cedepon LA-30; Conco Sulfate A; Lonzol LA-300; Sterling AM, AM-HV; Witcolate AM); carpet & upholstery shampoos (Carsonol ALS, ALS Special; Conco Sulfate A; Empicol AL30; Lonzol LA-300; Lorol Liquid NH Sulphonated; Mars AMLS; Norfox ALS; Stepanol AM, AM-V); dishwashing (Mars AMLS); laundry detergent (Emersal 6430); light-duty cleaners (Cycloryl MA; Nutrapon HA 3841); liquid detergents (Carsonol ALS, ALS Special; Stepanol AM, AM-V)

Industrial applications: construction (Avirol 200); dyes (Avirol 200; Sipon L-22); fire fighting (Avirol 200; Lakeway 101-20; Sipon L-22); industrial processing (Lakeway 101-20; Sipon L-22HV); textile/leather processing (Sipon L-22; Stepanol AM, AM-V)

Industrial cleaners (Carsonol ALS, ALS Special; Lonzol LA-300; Maprofix NHL; Mars AMLS; Stepanol AM, AM-V; Witcolate AM); metal processing surfactants

6

Ammonium lauryl sulfate *(cont'd.)*

(Stepanol AM, AM-V); railway cleaners (Stepanol AM, AM-V); sanitary supply (Mars AMLS)

Pet shampoos (Carsonol ALS, ALS Special; Sipon L-22)

Pharmaceutical applications: antiseptic soaps (Sipon L-22); vaginal preparations (Sipon L-22)

PROPERTIES:

Form:

Liquid (Akyposal ALS33; Alkasurf ALS; Avirol 200; Cedepon LA-30; Conco Sulfate A; Cycloryl MA; Drewpon NH, NHAC; Emal AD-25; Emersal 6430; Empicol AL30/T, AL70; Lakeway 101-20; Lorol Liquid NH Sulphonated; Maprofix ALS 30, NHL, NHL-22; Norfox ALS; Nutrapon HA3841; Richonol AM; Sactipon 2A; Sactol 2A; Sermul EA129; Sipon L-22HV; Texapon A; Texapon Special; Witcolate AM; Zoharpon LAA)

Clear liquid (Carsonol ALS Special; Mars AMLS; Polystep B-7; Sipon L-22)

Viscous liquid (Carsonol ALS; Lonzol LA-300; Maprofix NH; Standapol A, AMS-100; Stepanol AM; Sterling AM, AM-HV)

Viscous liquid to light gel (Stepanol AM-V)

Clear to slightly hazy, viscous liquid (Manro ALS 30)

Liquid/paste (Empicol AL30)

Color:

Light (Alkasurf ALS; Polystep B-7)

Low (Carsonol ALS)

Light amber (Avirol 200)

Light yellow (Conco Sulfate A; Cycloryl MA; Lakeway 101-2; Manro ALS 30; Mars AMLS; Stepanol AM)

Amber (Empicol AL30)

White (Lakeway 101-20)

Water-white (Standapol A, AMS-100; Texapon A)

Gardner 1 (Emersal 6430)

Gardner 2 (Richonol AM)

Odor:

Low (Carsonol ALS)

Mild (Manro ALS 30)

Pleasant (Conco Sulfate A)

Typical (Alkasurf ALS; Avirol 200)

Composition:

27% active (Alkasurf ALS)

27–28% active (Maprofix NHL-22)

27–29% active (Drewpon NH; Stepanol AM-V)

27–30% active (Lonzol LA-300; Maprofix NH, NHL)

27.5–28.5% active (Sipon L22; Standapol A, AMS-100; Texapon A)

27.7% active (Sterling AM)

28% active (Manro ALS30; Polystep B-7; Sermul EA 129; Sipon L-22HV; Sterling

7

AM-HV)
28% fatty alcohol sulfate (Avirol 200)
28–30% active (Stepanol AM)
28.5% active (Richonol AM)
29% active (Empicol AL30)
29% conc. (Lakeway 101-20)
30% active (Carsonol ALS, ALS Special; Mars AMLS)
37.5% active (Drewpon NHAC)
49–51% active (Maprofix NH54)

Solubility:
Disp. in butyl stearate (Emersal 6430)
Disp. in glycerol trioleate (Emersal 6430)
Disp. in min. oil (Emersal 6430)
Sol. in water (Alkasurf ALS; Carsonol ALS, ALS Special; Emersal 6430; Stepanol AM, AM-V; Witcolate AM)

Density:
8.4 lb/gal (Carsonol ALS, ALS Special)
8.5 lb/gal (Lakeway 101-20; Mars AMLS)
8.6 lb/gal (Emersal 6430; Richonol AM)

Sp.gr.:
1.01 @ 25/20 C (Maprofix NH, NHL-22)
1.02 (Mars AMLS)
1.02 @ 20 C (Manro ALS 30)
1.02 @ 25/20 C (Maprofix NHL)

Visc.:
600 cps max. (Mars AMLS)
1000 cps (Avirol 200; Lakeway 101-20; Sipon L-22)
1500–3000 cps (Standapol A, AMS-100)
5000 cps @ 20 C (Manro ALS 30)
6500 cps max. (Conco Sulfate A)
14,000 cs @ 20 C (Empicol AL30)
2490 cSt @ 100 F (Emersal 6430)

Stability:
Good > pH 4 (Carsonol ALS, ALS Special)
Good to anionic and nonionic surfactants, hard water, alkali, acids, lower alcohols, and electrolytes (Mars AMLS)
Product slowly degrades above pH 7 (Sipon L-22)

Storage Stability:
May become pastes or hazy liquids under cold storage; becomes clear liquid on warming (Empicol AL30)

pH:
6.0–7.0 (Richonol AM)
6.0–7.0 (10% sol'n.) (Lonzol LA-300; Manro ALS 30; Stepanol AM, AM-V)

Ammonium lauryl sulfate (cont'd.)

6.3–7.0 (10% sol'n.) (Conco Sulfate A)
6.5 ± 0.5 (5% sol'n.) (Empicol AL30)
6.5–7.0 (10% sol'n.) (Carsonol ALS, ALS Special; Standapol A, AMS-100)
6.8 (10% sol'n.) (Sipon L-22)
7.0 (10%) (Lakeway 101-20)
7.0–7.5 (10% sol'n.) (Mars AMLS)
Flash Pt.:
> 200 F (Maprofix NH, NHL, NHL-22)
Cloud Pt.:
5 C (Conco Sulfate A)
< 10 C (Avirol 200; Standapol AMS-100)
14 C (Lakeway 101-20; Sipon L-22)
> 100 C (5% saline) (Emersal 6430)
65 F (Richonol AM)
Biodegradable: (Carsonol ALS, ALS Special; Emersal 6430; Manro ALS 30; Sermul
EA129)
STD. PKGS.:
45-gal drums or road tankers (Manro ALS 30)

Ammonium myristyl ether sulfate

SYNONYMS:
Ammonium myreth sulfate (CTFA)
Poly (oxy-1,2-ethanediyl), α-sulfo-ω-(tetradecyloxy)-, ammonium salt
EMPIRICAL FORMULA:
$(C_2H_4O)_x \cdot C_{14}H_{30}O_4S \cdot H_3N$
STRUCTURE:
$CH_3(CH_2)_{12}CH_2(OCH_2CH_2)_nOSO_3NH_4$
where $n = 1–4$
CAS No.:
27731-61-9
TRADENAME EQUIVALENTS:
Cycloryl ME361 [Cyclo]
Standapol EA-40 [Henkel]
Texapon EA-40 [Henkel (Canada)]
CATEGORY:
Base, foaming agent, detergent, surfactant
APPLICATIONS:
Bath products: bubble bath (Standapol EA-40; Texapon EA-40)
Cleansers: (Cycloryl ME361; Standapol EA-40; Texapon EA-40)
Cosmetic industry preparations: shampoo base (Cycloryl ME361); shampoos (Stan-
dapol EA-40; Texapon EA-40)

9

Ammonium myristyl ether sulfate *(cont'd.)*

PROPERTIES:
Form:
    Liquid (Cycloryl ME361; Standapol EA-40; Texapon EA-40)
Color:
    Light/pale yellow (Cycloryl ME361; Texapon EA-40)
    Yellow (Standapol EA-40)
Composition:
    58–60% active (Cycloryl ME361; Standapol EA-40; Texapon EA-40)

Ammonium pareth-25 sulfate (CTFA)

SYNONYMS:
    Ammonium salt of a sulfated polyethylene glycol ether of a mixture of synthetic $C_{12\text{-}15}$ fatty alcohols
    $C_{12\text{-}15}$ alcohol ethoxysulfate, ammonium salt
STRUCTURE:
    $R(OCH_2CH_2)_n OSO_3NH_4$

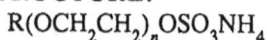

        where   R represents the $C_{12\text{-}15}$ alcohols radical and
                $n = 1\text{--}4$
CAS No.:
    RD No. 977064-04-2
TRADENAME EQUIVALENTS:
    Dobanol 25-3A/60 [Shell]
    Neodol 25-3A [Shell]
    Standapol AP-60 [Henkel]
    Witcolate AE-3 [Witco/Organics]
CATEGORY:
    Detergent, emulsifier, foaming agent
APPLICATIONS:
    Cosmetic industry preparations: personal care cleansers (Standapol AP-60); shampoos (Standapol AP-60)
    Household detergents: (Witcolate AE-3); detergent base (Witcolate AE-3); dishwashing (Neodol 25-3A); light-duty cleaners (Dobanol 25-3A/60; Neodol 25-3A; Standapol AP-60); liquid detergents (Neodol 25-3A
    Industrial cleaners: light-duty (Standapol AP-60)
    Pet shampoos (Standapol AP-60)
PROPERTIES:
Form:
    Liquid (Dobanol 25-3A/60; Witcolate AE-3)
    Clear, viscous liquid (Neodol 25-3A; Standapol AP-60)

Color:
  Light (Neodol 25-3A)
  Yellow (Standapol AP-60)
Odor:
  Mild, ethanol (Neodol 25-3A)
Composition:
  58–60% active (Standapol AP-60)
  59% active (Neodol 25-3A)
  59–61% active in aq. ethanol (Dobanol 25-3A/60)
Solubility:
  Sol. in water (Standapol AP-60; Witcolate AE-3)
Ionic Nature:
  Anionic (Dobanol 25-3A/60; Neodol 25-3A; Witcolate AE-3)
M.W.:
  432 avg. (Neodol 25-3A)
Sp.gr.:
  1.02 (Neodol 25-3A)
Density:
  1.025 g/ml (Standapol AP-60)
  1.04 kg/l (20 C) (Dobanol 25-3A/60)
  8.50 lb/gal (60 F) (Neodol 25-3A)
Visc.:
  45 cs (100 F) (Neodol 25-3A)
  60 cs (20 C) (Dobanol 25-3A/60)
  100–200 cps (Standapol AP-60)
Flash Pt.:
  26 C (Dobanol 25-3A/60)
  74 F (PMCC) (Neodol 25-3A)
Cloud Pt.:
  10 C (Dobanol 25-3A/60)
  10 C max. (Standapol AP-60)
pH:
  6.0–7.0 (10% sol'n.) (Standapol AP-60)
  6.5–8.0 (5% aq.) (Dobanol 25-3A/60)
  7.3 (Neodol 25-3A)
Biodegradable: (Neodol 25-3A; Standapol AP-60); > 95% (Dobanol 25-3A/60)
TOXICITY/HANDLING:
  Severe eye irritant, and skin irritant on prolonged contact with conc. form (Neodol 25-3A)
  Eye irritant; skin irritant on repeated/prolonged contact (Dobanol 25-3A/60)
STORAGE/HANDLING:
  Storage temp. should not exceed 40 C; vapor is inflammable; use flame-proofed

Ammonium pareth-25 sulfate *(cont'd.)*

equipment; store and handle in areas free of naked lights and sparks (Dobanol 25-3A/60)

STD. PKGS.:
218-kg mild steel drums (Dobanol 25-3A/60)

Ammonium POE (5) lauryl ether sulfate

SYNONYMS:
Ammonium laureth-5 sulfate (CTFA)
EMPIRICAL FORMULA:
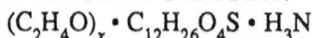
$(C_2H_4O)_x \cdot C_{12}H_{26}O_4S \cdot H_3N$
STRUCTURE:

$CH_3(CH_2)_{10}CH_2(OCH_2CH_2)_nOSO_3NH_4$
where avg. $n = 5$
CAS No.
32612-48-9 (generic)
TRADENAME EQUIVALENTS:
Carsonol ALES-5 [Carson]
CATEGORY:
Detergent
APPLICATIONS:
Bath products: bubble bath (Carsonol ALES-5)
Cosmetic industry preparations: shampoos (Carsonol ALES-5)
Household detergents: liquid detergents (Carsonol ALES-5)
PROPERTIES:
Form:
Liquid (Carsonol ALES-5)
Ionic Nature: Anionic (Carsonol ALES-5)

Ammonium POE (12) lauryl ether sulfate

SYNONYMS:
Ammonium laureth-12 sulfate (CTFA)
EMPIRICAL FORMULA:
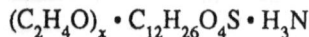
$(C_2H_4O)_x \cdot C_{12}H_{26}O_4S \cdot H_3N$
STRUCTURE:

$CH_3(CH_2)_{10}CH_2(OCH_2CH_2)_n OSO_3NH_4$
where avg. $n = 12$

Ammonium POE (12) lauryl ether sulfate *(cont'd.)*

CAS No.:
32612-48-9 (generic)
RD No.: 9770-69-08-1
TRADENAME EQUIVALENTS:
Standapol 230E Conc. [Henkel]
CATEGORY:
Detergent, foaming agent
APPLICATIONS:
Bath products (Standapol 230E Conc.)
Cleansers (Standapol 230E Conc.); body cleansers (Standapol 230E Conc.)
Cosmetic industry preparations: shampoos (Standapol 230E Conc.)
PROPERTIES:
Form:
Clear liquid (Standapol 230E Conc.)
Color:
Straw yellow (Standapol 230E Conc.)
Composition:
28–30% active (Standapol 230E Conc.)
Solubility:
Sol. in water, hydrophilic solvents (Standapol 230E Conc.)
Gel Pt.:
10 C max. (Standapol 230E Conc.)
Stability:
Stable at pH 2–7 under normal temp. conditions (Standapol 230E Conc.)
pH:
6.5–7.0 (10% sol'n.) (Standapol 230E Conc.)
Cloud Pt.:
0 C (Standapol 230 E Conc.)

Butyl oleate, sulfated

TRADENAME EQUIVALENTS:
Chemax SBO [Chemax]
Hipochem Dispersol SB [High Point]
Marvanol SBO 60% [Marlowe-Van Loan]
CATEGORY:
Detergent, wetting agent, emulsifier, dyeing assistant, leveling agent, softener
APPLICATIONS:
Industrial applications: dyes and pigments (Hipochem Dispersol SB); industrial
   processing (Marvanol SBO 60%); metalworking (Chemax SBO); textile/leather
   processing (Chemax SBO)
PROPERTIES:
Form:
Liquid (Chemax SBO; Hipochem Dispersol SB; Marvanol SBO 60%)
Color:
Amber (Hipochem Dispersol SB; Marvanol SBO 60%)
Odor:
Mild sulfate (Hipochem Dispersol SB)
Composition:
55% active (Hipochem Dispersol SB)
60% active (Marvanol SBO 60%)
Solubility:
Disp. in water at all temps. (Hipochem Dispersol SB)
Ionic Nature: Anionic
Sp.gr.:
1.01 (Hipochem Dispersol SB)
Density:
8.41 lb/gal (Hipochem Dispersol SB)
Stability:
Excellent (Hipochem Dispersol SB)
STD. PKGS.:
Bulk or drums (Hipochem Dispersol SB)

Calcium dodecylbenzene sulfonate

SYNONYMS:
Dodecylbenzene sulfonate, calcium salt
TRADENAME EQUIVALENTS:
Arylan CA [Lankro]
Conco AAS-75S [Continental]
Nansa EVM62H, EVM70 [Albright & Wilson/Detergents]
Polyfac ABS-60C [Westvaco]
Richonate CS-6021H [Richardson]
CATEGORY:
Surfactant, emulsifier, wetting agent, solubilizer
APPLICATIONS:
Degreasers (Arylan CA)
Farm products: insecticides/pesticides (Arylan CA; Conco AAS-75S; Nansa EVM62H, EVM70)
Household detergents: dishwashing (Richonate CS-6021H); light-duty cleaners (Richonate CS-6021H)
Industrial applications: dyes and pigments (Conco AAS-75S); lubricating/cutting oils (Nansa EVM62H, EVM70)
Industrial cleaners: all-purpose cleaners (Richonate CS-6021H); drycleaning compositions (Richonate CS-6021H)
PROPERTIES:
Form:
Liquid (Nansa EVM62H, EVM70; Polyfac ABS-60C)
Viscous liquid (Arylan CA; Conco AAS-75S)
Color:
Dark amber (Conco AAS-75S)
Dark (Polyfac ABS-60C)
Gardner 3 (30% solids) (Richonate CS-6021H)
Odor:
Alcoholic (Arylan CA)
Composition:
60% active (Conco AAS-75S; Polyfac ABS-60C; Richonate CS-6021H)
62% conc. (Nansa EVM62H)
67% conc. (Nansa EVM70)
70% active (Arylan CA)
Solubility:
Sol. in kerosene (Conco AAS-75S)

15

Calcium dodecylbenzene sulfonate *(cont'd.)*

Insol. in min. oil (Conco AAS-75S)
Sol. in naphtha (Conco AAS-75S)
Insol. in water (Arylan CA; Conco AAS-75S)
Sol. in xylene (Conco AAS-75S)
Ionic Nature: Anionic
Sp.gr.:
1.012 @ 20 C (Arylan CA)
1.02 (Conco AAS-75S)
1.04 (Polyfac ABS-60C)
Density:
8.7 lb/gal (Richonate CS-6021H)
Visc.:
$\approx$ 100–150 cps (Conco AAS-75S)
3025 cs @ 20 C (Arylan CA)
6000 cps (Polyfac ABS-60C)
Pour Pt.:
< 0 C (Arylan CA)
45 F (Polyfac ABS-60C)
Flash Pt.:
94 F (Abel CC) (Arylan CA)
$\approx$ 105 F (CC) (Conco AAS-75S)
133 F (PMCC) (Polyfac ABS-60C)
pH:
5–8 (1% aq.) (Arylan CA)
6.0 (1% in 1:1 water/IPA) (Polyfac ABS-60C)
6.0–7.0 (Richonate CS-6021H)
6.0–9.0 (5% in 50% aq. isopropanol) (Conco AAS-75S)
TOXICITY/HANDLING:
Avoid prolonged contact with concentrated form (Arylan CA)
STORAGE/HANDLING:
Contains flammable solvent—do not mix with strong oxidizing or reducing agents (potentially explosive); store in SS, glass, resin, or polythene-lined equipment (Arylan CA)
Combustible DOT classification (Polyfac ABS-60C)
STD. PKGS.:
55-gal drums, tank trucks, rail cars (Polyfac ABS-60C)

Calcium stearoyl lactylate (CTFA)

SYNONYMS:
Calcium stearoyl-2-lactylate
Octadecanoic acid, 2-(1-carboxyethoxy)-1-methyl-2-oxoethyl ester, calcium salt

EMPIRICAL FORMULA:
$C_{24}H_{44}O_6 \cdot \frac{1}{2}Ca$

STRUCTURE:

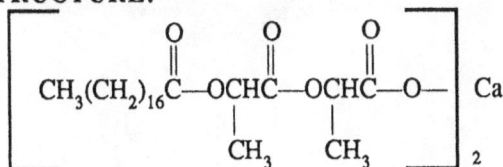

CAS No.:
5793-94-2

TRADENAME EQUIVALENTS:
Admul CSL 2007, CSL 2008 [PPF Int'l.]
Grindtek FAL 2 [Grinsted]
Lamegin CLS [Grunau]
Pationic CSL [Patco]
Radiamuls CSL 2980 [Oleofina]
Stearolac C [Paniplus]
Verv [Patco]

CATEGORY:
Emulsifier, softener

APPLICATIONS:
Cosmetic industry preparations: (Pationic CSL)
Food applications: (Admul CSL, Lamegin CSL, Radiamuls CSL, Stearolac C, Verv)

PROPERTIES:
Form:
Flake (Admul CSL, Grindtek FAL 2, Radiamuls CSL)
Powder (Admul CSL, Lamegin CSL, Pationic CSL, Radiamuls CSL, Stearolac C, Verv)
Color:
Cream (Grindtek FAL 2)
Light tan (Pationic CSL)
Composition:
30–40% lactylic acid, 2–5% calcium (Radiamuls CSL 2980)
100% conc. (Pationic CSL)
Solubility:
Sol. warm in toluene, white spirit (Grindtek FAL 2)
Partly sol. warm in peanut oil (Grindtek FAL 2)
Disp. in mineral oil, propylene glycol, IPA (Grindtek FAL 2, Pationic CSL)
Ionic Nature: Anionic

Calcium stearoyl lactylate *(cont'd.)*

Visc.:
  85 cps (2% in min. oil) (Pationic CSL)
M.P.:
  40–45 C (Pationic CSL)
HLB:
  2.0 (Grindtek FAL 2)
  5.0 (Pationic CSL)
Acid No.:
  50–65 (Pationic CSL)
Saponification No.:
  195–230 (Pationic CSL)
pH:
  4.90 (2% aq.) (Pationic CSL)
Surface Tension:
  37.00 dynes/cm (0.1% conc.) (Pationic CSL)
TOXICITY/HANDLING:
  Store in cool dry place; avoid exposure to temps. > 32 C (Pationic CSL)

Calcium sulfate *(CTFA)*

SYNONYMS:
  Anhydrite
  Calcium sulfonate
  Gypsum
  Plaster of Paris
  Sulfuric acid, calcium salt (1:1) dihydrate
STRUCTURE:
  $CaSO_4$ or $CaSO_4 \cdot 2HOH$
CAS No.
  7778-18-9; 10101-41-4
TRADENAME EQUIVALENTS:
  Alkamuls AG-CA [Alkaril]
  Emcol D 24-25 [Witco]
  Hybase C-300 [Witco/Sonneborn]
CATEGORY:
  Emulsifier, detergent, corrosion inhibitor
APPLICATIONS:
  Farm products: insecticides/pesticides (Alkamuls AG-CA)
  Industrial applications: industrial processing (Emcol D 24-25); lubricating/cutting oils
    (Hybase C-300); metalworking (Hybase C-300)

PROPERTIES:
Form:
  Liquid (Hybase C-300)
  Clear liquid (Alkamuls AG-CA)
Color:
  Amber (Alkamuls AG-CA)
  Dark brown (Hybase C-300)
Composition:
  29% active (Hybase C-300)
  < 1% water (Alkamuls AG-CA)
Solubility:
  Sol. in oil (Hybase C-300)
  Sol. in organic solvents (Hybase C-300)
Sp.gr.:
  1.13 @ 60 F (Hybase C-300)
Density:
  9.4 lb/gal (Hybase C-300)
Visc.:
  800 SUS @ 210 F (Hybase C-300)
Flash Pt.:
  380 F (Hybase C-300)
pH:
  6.0–8.0 (5% DW) (Alkamuls AG-CA)
TOXICITY/HANDLING:
  May cause eye burn (Hybase C-300)
STD. PKGS.:
  Drums, T/W, T/C (Hybase C-300)

Capric acid diethanolamide

SYNONYMS:
  N,N-Bis (2-hydroxyethyl) decanamide
  Capramide DEA (CTFA)
  Capric amide DEA
  Decanamide, N,N-bis (2-hydroxyethyl)-
  Decylic amide DEA
EMPIRICAL FORMULA:
  $C_{14}H_{29}NO_3$

Capric acid diethanolamide (cont'd.)

STRUCTURE:

$$CH_3(CH_2)_8C\!\!-\!\!N(CH_2CH_2OH)_2$$

CAS No.
136-26-5

TRADENAME EQUIVALENTS:
Alrosol C [Ciba-Geigy]
Alrosol Conc. [Ciba-Geigy]
Comperlan CD [Henkel/Canada]
Emid 6544 [Emery]
Monamid 150-CW [Mona]
Standamid CD [Henkel]

CATEGORY:
Foam booster, wetting agent, detergent, foam stabilizer, superfatting agent

APPLICATIONS:
Bath products: bubble bath (Monamid 150-CW)
Cleansers: hand cleanser (Alrosol C, Conc.; Monamid 150-CW)
Cosmetic industry preparations: (Monamid 150-CW; Standamid CD); conditioners
(Monamid 150-CW); shampoos (Alrosol C, Conc.; Monamid 150-CW); shaving
preparations (Alrosol Conc.)
Degreasers: (Alrosol Conc.)
Farm products: agricultural sprays (Monamid 150-CW)
Household detergents: (Standamid CD); carpet & upholstery shampoos (Alrosol C;
Monamid 150-CW); dishwashing (Monamid 150-CW); liquid detergents (Alrosol
Conc.); powdered detergents (Emid 6544; Monamid 150-CW)
Industrial applications: dyes and pigments (Monamid 150-CW); lubricating/cutting
oils (Monamid 150-CW); petroleum industry (Monamid 150-CW); polishes and
waxes (Monamid 150-CW); textile/leather processing (Monamid 150-CW)
Industrial cleaners: (Standamid CD); drycleaning compositions (Monamid 150-CW);
metal surfactants (Alrosol C, Conc.; Monamid 150-CW); textile cleaning (Monamid 150-CW)

PROPERTIES:
Form:
Liquid (crystallizes on aging) (Monamid 150-CW)
Clear viscous liquid (Alrosol Conc.)
Viscous liquid (Alrosol C; Comperlan CD; Standamid CD)
Color:
Light amber (Alrosol C)
Amber (Comperlan CD; Standamid CD)
Yellow (Alrosol Conc.)
GVCS-33 7 max. (Monamid 150-CW)

Capric acid diethanolamide *(cont'd.)*

Composition:
   96–98% active (Standamid CD)
   98% conc. (Comperlan CD)
   100% active (Emid 6544; Monamid 150-CW)
Solubility:
   Sol. in aromatic hydrocarbons (@ 10%) (Monamid 150-CW)
   Sol. in butyl stearate (Emid 6544)
   Sol. in chlorinated hydrocarbons (@ 10%) (Monamid 150-CW)
   Sol. in ethyl alcohol (@ 10%) (Monamid 150-CW)
   Disp. in glycerol trioleate (Emid 6544)
   Disp. in min. oil (Emid 6544)
   Sol. in natural oils and fats (@ 10%) (Monamid 150-CW)
   Sol. in perchloroethylene (Emid 6544)
   Sol. in Stoddard solvent (Emid 6544)
   Sol. in water (@ 10%) (Monamid 150-CW); disp. (Emid 6544)
Ionic Nature: Nonionic
Sp.gr.:
   0.99 (20 C) (Monamid 150-CW)
Density:
   8.25 lb/gal (Monamid 150-CW)
M.P.:
   30 C (Emid 6544)
Acid No.:
   0–2 (Monamid 150-CW)
Stability:
   Good in acid, alkaline, or neutral salts in moderate conc. (Standamid CD)
pH:
   9.0–10.0 (10% sol'n.) (Standamid CD)
   10.3–11.3 (10% sol'n.) (Monamid 150-CW)
Biodegradable: (Monamid 150-CW)
STD. PKGS.:
   450 lb net Liquipak (Monamid 150-CW)

Capryloamphodipropionate *(CTFA)*

SYNONYMS:
   Capryloamphocarboxypropionate
EMPIRICAL FORMULA:
   $C_{18}H_{34}N_2O_6 \cdot 2Na$

Capryloamphodipropionate (cont'd.)

STRUCTURE:

$$CH_3(CH_2)_6C(=O)-NH-CH_2-CH_2-N(-CH_2CH_2COONa)-CH_2CH_2OCH_2CH_2COONa$$

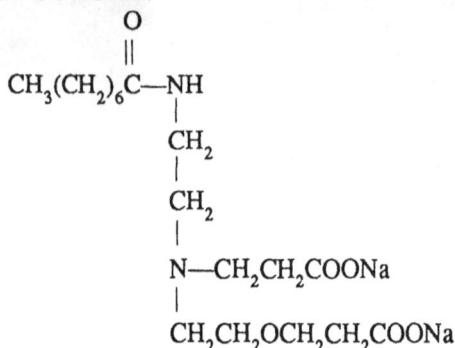

TRADENAME EQUIVALENTS:

Miranol J2M-SF Conc. [Miranol] (salt-free)

Monateric 811 [Mona]

CATEGORY:

Wetting agent, penetrant, corrosion inhibitor, detergent

APPLICATIONS:

Degreasers: (Miranol J2M-SF Conc.)

Food applications: fruit/vegetable washing (Miranol J2M-SF Conc.)

Industrial applications: polymers/polymerization (Miranol J2M-SF Conc.); rubber (Miranol J2M-SF Conc.)

Industrial cleaners: (Monateric 811); bottle cleaners (Miranol J2M-SF Conc.); textile cleaning (Miranol J2M-SF Conc.); wax strippers (Miranol J2M-SF Conc.)

PROPERTIES:

Form:

Liquid (Miranol J2M-SF Conc.; Monateric 811)

Color:

Amber (Monateric 811)

Composition:

38–40% solids (Miranol J2M-SF Conc.)

50% active. (Monateric 811)

Solubility:

Sol. in water (Monateric 811)

Ionic Nature: Amphoteric

Sp.gr.:

1.04 (Monateric 811)

Density:

8.7 lb/gal (Monateric 811)

Acid No.:

Nil (Monateric 811)

Stability:

Clear in 10% NaOH; good tolerance to high electrolyte systems (Monateric 811)

pH:
  8.8–9.3 (Miranol J2M-SF Conc.)
  10.4 (10% sol'n.) (Monateric 811)
Biodegradable: (Miranol J2M-SF Conç.)

Cetyl trimethyl ammonium p-toluene sulfonate

SYNONYMS:
  Cetrimonium tosylate (CTFA)
  Cetyl trimethyl ammonium tosylate
  1-Hexadecanaminium, N,N,N-trimethyl-, salt with 4-methylbenzenesulfonic acid
    (1:1)
  N-Hexadecyl-N,N,N-trimethyl ammonium *p*-toluene sulfonate
  N,N,N-Trimethyl-1-hexadecanaminium salt with 4-methylbenzenesulfonic acid
EMPIRICAL FORMULA:
  $C_{19}H_{42}N \cdot C_7H_7O_3S$
STRUCTURE:

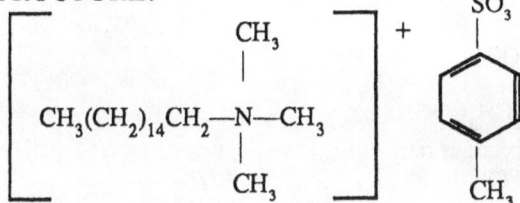

CAS No.
  138-32-9
TRADENAME EQUIVALENTS:
  Cetats [Hexcel]
CATEGORY:
  Surfactant, germicide
APPLICATIONS:
  Pharmaceutical applications: toothpaste (Cetats)
PROPERTIES:
Form:
  Free-flowing powder (Cetats)
Color:
  White (Cetats)
Odor:
  Characteristic (Cetats)
Composition:
  99% min. assay (Cetats)

Cetyl trimethyl ammonium p-toluenesulfonate (cont'd.)

Solubility:
  Sol. in alcohol (Cetats)
  Insol. in ether (Cetats)
  Insol. in ethyl acetate (Cetats)
  Insol. in MEK (Cetats)
  1% sol. in water (Cetats)
Ionic Nature: Cationic
M.W.:
  455.72 (Cetats)
Stability:
  Good heat stability (Cetats)
pH:
  5–8 (1% aq. sol'n.) (Cetats)

Cocoamphopropionate (CTFA)

STRUCTURE:

$$\underset{\displaystyle \text{RC—NH—CH}_2\text{CH}_2\text{—N—CH}_2\text{CH}_2\text{COONa}}{\overset{\displaystyle \overset{\text{O}}{\underset{\displaystyle \|}{}} \qquad\qquad\qquad \overset{\text{CH}_2\text{CH}_2\text{OH}}{\underset{\displaystyle |}{}}}{}}$$

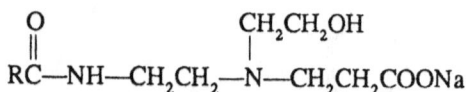

  where RCO⁻ represents the coconut fatty radical

TRADENAME EQUIVALENTS:
  Amphoterge K [Lonza]
  Cycloteric MV-SF [Cyclo]
  Mackam CSF [McIntyre]
  Miranol CM-SF Conc. [Miranol]
  Monateric CA-35% [Mona]
  Rewoteric AM-KSF [Rewo Chemische Werke]
  Schercoteric MS-SF [Scher]
CATEGORY:
  Wetting agent, detergent, foaming agent, sequestrant, emulsifier, dispersant, germicide, viscosity builder
APPLICATIONS:
  Baby products (Rewoteric AM-KSF)
  Bath products: bubble bath (Monateric CA-35%)
  Cleansers: body cleansers (Amphoterge K; Monateric CA-35%)
  Cosmetic industry preparations: (Monateric CA-35%); conditioners (Monateric CA-35%); hair dye formulations (Monateric CA-35%); hair rinses (Monateric CA-35%); shampoos (Amphoterge K; Mackam CSF; Monateric CA-35%; Schercoteric MS-SF)

Cocoamphopropionate (cont'd.)

Household detergents: (Cycloteric MV-SF; Miranol CM-SF Conc.; Monateric CA-35%); dishwashing (Amphoterge K; Schercoteric MS-SF); heavy-duty cleaner (Amphoterge K; Mackam CSF)

Industrial cleaners: (Monateric CA-35%; Rewoteric AM-KSF; Schercoteric MS-SF); all-purpose cleaners (Mackam CSF); floor cleaners (Monateric CA-35%); metal processing surfactants (Mackam CSF; Monateric CA-35%)

Pharmaceutical applications: intimate hygiene products (Rewoteric AM-KSF)

PROPERTIES:

Form:
Liquid (Amphoterge K; Mackam CSF; Monateric CA-35%; Schercoteric MS-SF)
Clear liquid (Cycloteric MV-SF; Miranol CM-SF Conc.)
Viscous liquid (Rewoteric AM-KSF)

Color:
Light amber (Miranol CM-SF Conc.)
Amber (Monateric CA-35%)

Composition:
35% active (Monateric CA-35%)
36–38% solids (Miranol CM-SF Conc.)
38% conc. (Schercoteric MS-SF)
38–40% active (Cycloteric MV-SF)
39% conc. (Mackam CSF)
40% conc. (Amphoterge K; Rewoteric AM-KSF)

Solubility:
Sol. in alcohol (Miranol CM-SF Conc.)
Sol. in water (Miranol CM-SF Conc.)

Ionic Nature:
Amphoteric (Amphoterge K; Miranol CM-SF Conc.; Monateric CA-35%; Rewoteric AM-KSF; Schercoteric MS-SF)

M.W.:
360 (Monateric CA-35%)

Sp.gr.:
1.02 (Monateric CA-35%)

Density:
8.5 lb/gal (Monateric CA-35%)

Acid No.:
50 ± 5 (Monateric CA-35%)

Stability:
Stable to high concs. of acids and alkalis; compatible with high electrolyte sol'ns. (Monateric CA-35%)

pH:
9.0–11.0 (Cycloteric MV-SF)
9.5–10.5 (Miranol CM-SF Conc.)

Cocoamphopropionate *(cont'd.)*

Surface Tension:
    29.5 dynes/cm (0.1% conc.) (Monateric CA-35%)
Biodegradable: (Miranol CM-SF Conc.; Monateric CA-35%)

Coco dimethyl amine

SYNONYMS:
    Amines, coco alkyl dimethyl
    Dimethyl cocamine (CTFA)
    Dimethyl cocoamine
    Dimethyl coconut amine
    Dimethyl coconut *t*-amine
STRUCTURE:

$$R\!-\!N\!\!<\!\!\begin{array}{l}CH_3\\CH_3\end{array}$$

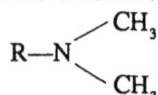

    where R represents the coconut radical
CAS No.:
    61788-93-0
TRADENAME EQUIVALENTS:
    Armeen DMCD [Armak]
    Armeen DMMCD [Armak]
Distilled:
    Kemamine T-6502D [Humko Sheffield]
    Lilamin 367D [Lilachim S.A.]
CATEGORY:
    Chemical intermediate, catalyst, emulsifier, anticorrosive, lubricant, bactericide, dispersant, extraction reagent
APPLICATIONS:
    Household detergents: surfactant raw material (Armeen DMCD, DMMCD)
    Industrial applications: petroleum industry (Kemamine T-6502D; Lilamin 367D); plastics (Kemamine T-6502D); polymers/polymerization (Lilamin 367D); printing inks (Lilamin 367D); rubber (Lilamin 367D)
PROPERTIES:
Form:
    Liquid (Armeen DMCD, DMMCD; Lilamin 367D)
Color:
    Yellow (Armeen DMCD, DMMCD)
    Gardner 1 max. (Kemamine T-6502D)
    Gardner 2 max. (Lilamin 367D)

Composition:
  95% active (Lilamin 367D)
  95% conc. (Kemamine T-6502D)
  98% active (Armeen DMCD, DMMCD)
Solubility:
  Insol. in water (Armeen DMCD, DMMCD)
Ionic Nature: Cationic
M.W.:
  ≈ 230 (Lilamin 367D)
Sp.gr.:
  0.79 (Armeen DMCD, DMMCD)
Visc.:
  3.95 cs (Armeen DMMCD)
  4.20 cs (Armeen DMCD)
F.P.:
  −19 C (Armeen DMMCD)
B.P.:
  42–150 C (3 mm Hg) (Armeen DMCD)
  90–125 C (Armeen DMMCD)
Flash Pt.:
  108 C (COC) (Armeen DMCD)
  109 C (OC) (Lilamin 367D)
  125 C (COC) (Armeen DMMCD)
Iodine No.:
  3 max. (Armeen DMCD, DMMCD)
  15 max. (Lilamin 367D)
TOXICITY/HANDLING:
  Skin irritant, severe eye irritant (Armeen DMCD, DMMCD)
STORAGE/HANDLING:
  Avoid contact with strong oxidizing agents (Armeen DMCD, DMMCD)
STD. PKGS.:
  200 L. bung-type steel durms (Armeen DMCD)

Dicoco dimethyl ammonium chloride

SYNONYMS:

Dicocodimonium chloride (CTFA)

Dimethyl dicoco ammonium chloride

Dimethyl dicoconut ammonium chloride

Quaternary ammonium compounds, dicoco alkyl dimethyl, chlorides

Quaternium-34

STRUCTURE:

$$\left[\begin{array}{c} CH_3 \\ | \\ R-N-R \\ | \\ CH_3 \end{array} \right]^+ \quad Cl^-$$

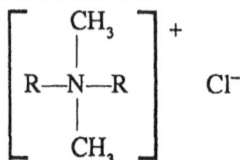

         where R represents the coconut radical

CAS No.

61789-77-3

TRADENAME EQUIVALENTS:

Accoquat 2C-75, 2C-75-H [Armstrong]

Adogen 462 [Sherex]

Arquad 2C-75 [Armak]

Jet-Quat 2C-75 [Jetco]

Kemamine Q6502C [Humko Sheffield]

M-Quat 2475 [Mazer]

CATEGORY:

Emulsifier, coupling agent, antistat, flocculating agent, foaming agent, wetting agent, dispersant, corrosion inhibitor, bactericide, germicide, softener, defoamer, coagulant

APPLICATIONS:

Automobile products: car spray wax (Accoquat 2C-75, 2C-75-H)

Cosmetic industry preparations: hair conditioners (Arquad 2C-75)

Industrial applications: asphalt (Jet Quat 2C-75); dyes/pigments (Kemamine Q6502C); ore flotation (Arquad 2C-75); petroleum industry (Adogen 462; Jet Quat 2C-75); polishes and waxes (Arquad 2C-75); textile/leather softening/processing (Arquad 2C-75; Jet Quat 2C-75; Kemamine Q6502C); water/sewage treatment (Arquad 2C-75; Kemamine Q6502C)

Industrial cleaners: metal processing surfactants (Arquad 2C-75); sanitizers/germicides (Jet Quat 2C-75; Kemamine Q6502C)

Dicoco dimethyl ammonium chloride *(cont'd.)*

PROPERTIES:
Form:
   Liquid (Accoquat 2C-75, 2C-75-H; Adogen 462; Kemamine Q6502C; M-Quat 2475)
   Semiliquid (Arquad 2C-75)
Color:
   Amber (Accoquat 2C-75, 2C-75-H)
   Gardner 4 max. (Kemamine Q6502C)
   Gardner 7 (Arquad 2C-75)
Odor:
   Mild (Accoquat 2C-75, 2C-75-H)
Composition:
   75% active (Kemamine Q6502C)
   75% active in aq. isopropanol (Arquad 2C-75)
   75% active in hexylene glycol (Accoquat 2C-75-H)
   75% active in isopropanol (Accoquat 2C-75)
   75% conc. (M-Quat 2475)
   75% quat. content (Adogen 462)
Ionic Nature: Cationic
Solubility:
   Disp. in water (Kemamine Q6502C)
M.W.:
   447 (Arquad 2C-75)
   465 (Kemamine Q6502C)
Sp.gr.:
   0.86 (Accoquat 2C-75)
   0.89 (Arquad 2C-75)
Density:
   7.2 lb/gal (Accoquat 2C-75)
Pour Pt.:
   10 F (Arquad 2C-75)
Flash Pt.:
   < 80 F (Arquad 2C-75)
HLB:
   11.4 (Arquad 2C-75)
pH:
   9 max. (5% sol'n.) (Kemamine Q6502C)
Surface Tension:
   30 dynes/cm (0.1%) (Arquad 2C-75)
TOXICITY/HANDLING:
   Skin irritant, severe eye irritant; protective clothing, goggles, gloves should be worn
      (Arquad 2C-75)
STORAGE/HANDLING:
   Flammable liquid (Accoquat 2C-75; Arquad 2C-75)

Dicoco dimethyl ammonium chloride *(cont'd.)*

STD. PKGS.:
55-gal epoxy-phenolic lined drums; bulk SS tank trucks (Arquad 2C-75)

Didecyl dimethyl ammonium chloride

SYNONYMS:
1-Decanaminium, N-decyl-N,N-dimethyl-, chloride
N-Decyl-N,N-dimethyl-1-decanaminium chloride
Didecyldimonium chloride (CTFA)
Dimethyl didecyl ammonium chloride

EMPIRICAL FORMULA:
$C_{22}H_{48}N \cdot Cl$

STRUCTURE:

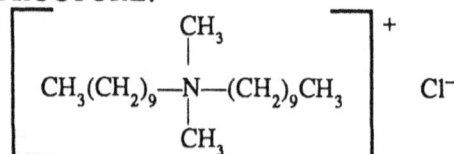

CAS No.
7173-51-5

TRADENAME EQUIVALENTS:
Bio-Dac 50-22 [Bio-Lab]
BTC 1010 [Onyx]
BTCO 1010 [Onyx]
Querton 210CL [Lilachim S.A.]

CATEGORY:
Surfactant, fungicide, germicide, antimicrobial

APPLICATIONS:
Farm products: fungicide (BTC 1010; BTCO 1010)
Household detergents: hard surface cleaner/disinfectant (BTC 1010; BTCO 1010)
Industrial cleaners: sanitizers/germicides (Bio-Dac 50-22; BTC 1010; BTCO 1010)

PROPERTIES:
Form:
Liquid (BTC 1010; BTCO 1010; Querton 210CL)
Clear liquid (Bio-Dac 50-22)
Color:
Pale yellow to water-white (Bio-Dac 50-22)
Gardner 3 max. (Querton 210CL)
Composition:
45% conc. (Querton 210CL)
50% active (Bio-Dac 50-22)

Didecyl dimethyl ammonium chloride *(cont'd.)*

50% min. quaternary (BTC 1010; BTCO 1010)
Solubility:
   Sol. in water (BTC 1010)
Ionic Nature: Cationic
M.W.:
   ≈ 360 (Querton 210CL)
Sp.gr.:
   0.89 (25/20 C) (BTC 1010; BTCO 1010)
   0.927 (20 C) (Bio-Dac 50-22)
Density:
   7.73 lb/gal (Bio-Dac 50-22)
Flash Pt.:
   86 F (BTC 1010; BTCO 1010)
   109 F (Seta) (Bio-Dac 50-22)
pH:
   6.0–9.0 (10% w/w) (Bio-Dac 50-22)

Diethanolamine lauryl sulfate

SYNONYMS:
   DEA-lauryl sulfate (CTFA)
   Lauryl sulfate, diethanolamine salt
   Sulfuric acid, monododecyl ester, compd. with 2,2'-iminodiethanol (1:1)
EMPIRICAL FORMULA:
   $C_{12}H_{26}O_4S \cdot C_4H_{11}NO_2$
STRUCTURE:
   $CH_3(CH_2)_{10}CH_2OSO_3H \cdot HN(CH_2CH_2OH)_2$
CAS No.
   143-00-0
TRADENAME EQUIVALENTS:
   Alkasurf DLS [Alkaril]
   Alkasurf WADX [Alkaril] (modified)
   Carsonol DLS [Carson]
   Conco Sulfate EP [Continental]
   Condanol DLS35 [Dutton & Reinisch]
   Cycloryl DA [Cyclo]
   Duponol EP [DuPont] (tech.)
   Empicol DA, DLS [Albright & Wilson/Marchon]
   Maprofix DLS-35 [Onyx]
   Polystep B-8 [Stepan]
   Sandoz Sulfate EP [Sandoz Color & Chem]

31

Diethanolamine lauryl sulfate *(cont'd.)*

TRADENAME EQUIVALENTS *(cont'd.):*
  Sipon LD [Alcolac]
  Standapol DEA [Henkel]
  Stepanol DEA [Stepan]
  Sterling WADE [Canada Packers]
  Texapon DEA [Henkel/Canada]

CATEGORY:
  Detergent, wetting agent, emulsifier, base, foaming agent, surfactant, visc. modifier, conditioner

APPLICATIONS:
  Automobile cleaners: (Empicol DA); car shampoo (Carsonol DLS; Stepanol DEA)
  Bath products: (Sandoz Sulfate EP); bubble bath (Carsonol DLS; Conco Sulfate EP; Empicol DA; Maprofix DLS-35; Sipon LD; Standapol DEA; Stepanol DEA; Sterling WADE; Texapon DEA)
  Cleansers: (Alkasurf WADX; Sterling WADE); cleansing creams (Carsonol DLS; Stepanol DEA); cleansing lotions (Carsonol DLS; Conco Sulfate EP); hand cleanser (Empicol DA)
  Cosmetic industry preparations: (Alkasurf DLS, WADX; Carsonol DLS; Duponol EP; Stepanol DEA); creams and lotions (Stepanol DEA); hair products (Sandoz Sulfate EP); personal care products (Sandoz Sulfate EP); shampoo base (Condanol DLS35; Cycloryl DA); shampoos (Alkasurf DLS; Carsonol DLS; Conco Sulfate EP; Duponol EP; Empicol DA, DLS; Maprofix DLS-35; Sipon LD; Standapol DEA; Stepanol DEA; Sterling WADE; Texapon DEA); shaving preparations (Carsonol DLS; Stepanol DEA)
  Farm products: insecticides/pesticides (Stepanol DEA)
  Food applications: fruit/vegetable washing (Carsonol DLS; Stepanol DEA)
  Household detergents: (Alkasurf DLS; Conco Sulfate EP); carpet & upholstery shampoos (Carsonol DLS; Conco Sulfate EP; Stepanol DEA); liquid detergents (Carsonol DLS; Stepanol DEA)
  Industrial applications: polymers/polymerization (Carsonol DLS; Polystep B-8; Stepanol DEA); textile/leather processing (Stepanol DEA)
  Industrial cleaners: (Carsonol DLS; Stepanol DEA); dairy cleaners (Stepanol DEA); metal processing surfactants (Stepanol DEA); railway cleaners (Stepanol DEA); textile cleaning (Duponol EP)
  Pet shampoos: (Carsonol DLS)
  Pharmaceutical applications: (Stepanol DEA); cleanser (Alkasurf WADX)

PROPERTIES:
Form:
  Liquid (Alkasurf DLS; Conco Sulfate EP; Condanol DLS35; Cycloryl DA; Maprofix DLS-35; Sandoz Sulfate EP; Sipon LD; Sterling WADE)
  Clear liquid (Carsonol DLS; Empicol DLS; Polystep B-8; Standapol DEA; Stepanol DEA; Texapon DEA)
  Clear to slightly hazy, visc. liquid (Alkasurf WADX)

32

Diethanolamine lauryl sulfate *(cont'd.)*

Clear, moderately visc. liquid @ 20 C (Empicol DA)
Slightly viscous liquid (Duponol EP)
Color:
Light/pale (Alkasurf DLS; Carsonol DLS; Polystep B-8)
Very pale golden (Duponol EP)
Pale straw (Alkasurf WADX)
Straw (Empicol DLS)
Light/Pale yellow (Conco Sulfate EP; Condanol DLS35; Empicol DA; Sandoz Sulfate
   EP; Standapol DEA; Stepanol DEA; Texapon DEA)
Yellow (Cycloryl DA)
Odor:
Bland (Alkasurf WADX)
Bland and clean (Duponol EP)
Mild (Carsonol DLS)
Pleasant (Conco Sulfate EP)
Slight typical (Condanol DLS35)
Typical (Alkasurf DLS)
Composition:
30% active (Alkasurf WADX)
33% active (Polystep B-8)
33–35% active (Maprofix DLS-35; Stepanol DEA)
33–36% active (Duponol EP)
34% active (Empicol DLS)
34 ± 1% active (Empicol DA)
34–36% active (Cycloryl DA)
35% active (Carsonol DLS; Sterling WADE)
35% active in water (Alkasurf DLS; Condanol DLS35)
35–37.5% active (Sandoz Sulfate EP)
36% conc. (Standapol DEA)
36–38% active (Texapon DEA)
36.0–39.9% active (Conco Sulfate EP)
40% active (Sipon LD)
Solubility:
Sol. in polar solvents with some electrolyte precipitation (Duponol EP)
Sol. in water (Alkasurf DLS, WADX; Carsonol DLS; Stepanol DEA); in all proportions (Duponol EP); readily diluted in cold water (Empicol DA)
Ionic Nature: Anionic
M.W.:
379 (Empicol DLS)
Sp.gr.:
≈ 1.0 (Condanol DLS35)
1.03 (25/20 C, Maprofix DLS-35)
1.05 (20 C) (Empicol DA)

Diethanolamine lauryl sulfate *(cont'd.)*

Density:
8.4 lb/gal (Duponol EP)
8.49 lb/gal (Carsonol DLS)
1.05 g/cc (Empicol DLS)

Visc.:
Low (Condanol DLS35)
50–150 cps (27 C) (Duponol EP)
225 cps (Sandoz Sulfate EP)
225 cps max. (Conco Sulfate EP)
500 cs max. @ 20 C (Empicol DLS)

Flash Pt.:
> 200 F (Maprofix DLS-35)

Cloud Pt.:
< 0 C (Empicol DLS)
5 C (Duponol EP)
6 C (Sandoz Sulfate EP)
6 C max. (Conco Sulfate EP)

Clear Pt.:
25 C max. (Conco Sulfate EP)

Stability:
Stable to freezing, heat (150 F), uv light, aging (Duponol EP)
Hard water stability (Conco Sulfate EP)
Good foam stability (Maprofix DLS-35)

pH:
7.0 ± 0.5 (10% aq. sol'n.) (Empicol DA)
7–8 (1% aq., Empicol DLS)
7.3–7.7 (10% sol'n.) (Stepanol DEA)
7.5–8.5 (10% sol'n.) (Carsonol DLS; Conco Sulfate EP; Sandoz Sulfate EP)

Biodegradable: (Carsonol DLS; Conco Sulfate EP)

TOXICITY/HANDLING:
Avoid prolonged contact with skin; spillages are slippery (Empicol DA)

STORAGE/HANDLING:
May cause some corrosion on iron, steel, aluminum (Duponol EP)
Excellent cold storage properties (Empicol DA)

STD. PKGS.:
Lacquered drums
200 kg net lined open-head mild steel drums (Empicol DA)

Diethylene glycol monolaurate

SYNONYMS:
Diglycol laurate
Diglycol monolaurate
Dodecanoic acid, 2-(2-hydroxyethoxy) ethyl ester
PEG-2 laurate (CTFA)
PEG 100 monolaurate
POE (2) monolaurate

EMPIRICAL FORMULA:
$C_{16}H_{32}O_4$

STRUCTURE:

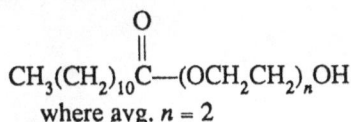

$$CH_3(CH_2)_{10}C\text{---}(OCH_2CH_2)_nOH$$
where avg. $n = 2$

CAS No.:
141-20-8

TRADENAME EQUIVALENTS:
Cithrol DGML N/E [Croda]
Hodag DGL [Hodag]
Lipo Diglycol Laurate [Lipo]
Mapeg DGLD [Mazer]
Pegosperse 100L, 100ML [Glyco]
Radiasurf 7420 [Oleofina]
Self-emulsifying grade:
Cithrol DGML S/E [Croda]
Lipo DGLS [Lipo]
Radiasurf 7421 [Oleofina]
Sole-Onic CDS [Hodag; Swift]

CATEGORY:
Emulsifier, dispersant, wetting aid, lubricant, antistat, binding agent, defoamer, opacifier, plasticizer, rust inhibitor, scouring and detergent aid, softener, thickener, w/o emulgent, spreading agent

APPLICATIONS:
Bath preparations: bath oils (Lipo DGLS)
Cosmetic industry preparations: (Radiasurf 7420, 7421; Sole-Onic CDS); creams and lotions (Lipo DGLS)
Farm products: insecticides/pesticides (Radiasurf 7420, 7421; Sole-Onic CDS)
Industrial applications: dyes and pigments (Radiasurf 7420, 7421, Sole-Onic CDS); lubricating/cutting oils (Cithrol DGML N/E, DGML S/E; Radiasurf 7420, 7421; Sole-Onic CDS); paint mfg. (Radiasurf 7420, 7421; Sole-Onic CDS); paper mfg. (Cithrol DGML N/E, DGML S/E); plastics (Radiasurf 7420, 7421, Sole-Onic CDS); polishes and waxes (Cithrol DGML N/E, DGML S/E; Radiasurf 7420, 7421; Sole-Onic CDS); printing inks (Radiasurf 7420, 7421; Sole-Onic CDS); textile/

Diethylene glycol monolaurate *(cont'd.)*

    leather processing (Cithrol DGML N/E, DGML S/E; Radiasurf 7420, 7421)
Industrial cleaners: metal processing surfactants (Cithrol DGML N/E, DGML S/E)
Pharmaceutical applications: (Radiasurf 7420, 7421, Sole-Onic CDS)

PROPERTIES:

Form:

    Liquid (Hodag DGL; Lipo DGLS; Mapeg DGLD; Pegosperse 100L, 100ML; Radiasurf 7421; Sole-Onic CDS)

    Paste (Cithrol DGML N/E, DGML S/E; Radiasurf 7420)

Color:

    White (Radiasurf 7420, 7421)

    Straw (Pegosperse 100L, 100ML)

    Yellow (Lipo DGLS)

Composition:

    100% active (Lipo DGLS)

    100% conc. (Cithrol DGML N/E, DGML S/E; Hodag DGL; Mapeg DGLD; Pegosperse 100L, 100ML)

Solubility:

    Sol. in acetone (Pegosperse 100ML); miscible (Pegosperse 100L)

    Sol. in benzene (Radiasurf 7420, 7421)

    Sol. in ethanol (Pegosperse 100L, 100ML)

    Sol. in ethyl acetate (Pegosperse 100ML); miscible (Pegosperse 100L)

    Sol. cloudy in hexane (Radiasurf 7420, 7421)

    Sol. in isopropanol (Radiasurf 7420, 7421)

    Sol. in methanol (Pegosperse 100ML); miscible (Pegosperse 100L)

    Sol. in min. oil (Pegosperse 100L); sol. cloudy (Radiasurf 7420, 7421); miscible (Pegosperse 100ML)

    Sol. in naphtha (Pegosperse 100L); partly sol. (Pegosperse 100ML)

    Sol. in toluol (Pegosperse 100L)

    Sol. in trichlorethylene (Radiasurf 7420, 7421)

    Miscible with veg. oil (Pegosperse 100L, 100ML)

    Disp. in water (Pegosperse 100L)

Ionic Nature:

    Nonionic (Cithrol DGML N/E; Hodag DGL; Mapeg DGLD; Pegosperse 100L, 100ML)

    Anionic (Cithrol DGML S/E; Sole-Onic CDS)

Sp.gr.:

    0.901 (98.9 C) (Radiasurf 7421)

    0.942 (37.8 C) (Radiasurf 7420)

    0.96 (Pegosperse 100ML)

    0.97 (Pegosperse 100L)

Visc.:

    3.60 cps (98.9 C) (Radiasurf 7421)

    15.20 cps (37.8 C) (Radiasurf 7420)

Diethylene glycol monolaurate *(cont'd.)*

Solidification Pt.:
  < 13 C (Pegosperse 100L)
  16–19 C (Pegosperse 100ML)
Flash Pt.:
  164 C (Radiasurf 7420)
  173 C (Radiasurf 7421)
Cloud Pt.:
  28.5 C (Radiasurf 7420)
  29.5 C (Radiasurf 7421)
HLB:
  5.7 avg. (Radiasurf 7420)
  6.0 ± 0.5 (Pegosperse 100ML)
  6.3 avg. (Radiasurf 7421)
  6.5 (Hodag DGL)
  7.4 ± 0.5 (Pegosperse 100L)
  8.3 (Mapeg DGLD)
  8.3 ± 1 (Lipo DGLS)
Acid No.:
  3 max. (Radiasurf 7420, 7421)
  4 max. (Lipo DGLS; Pegosperse 100L, 100ML)
Iodine No.:
  < 9 (Pegosperse 100L)
  < 10 (Pegosperse 100ML, Radiasurf 7421)
Saponification No.:
  160–170 (Lipo DGLS; Pegosperse 100L)
  180–190 (Pegosperse 100ML)
  183–193 (Radiasurf 7421)
  195–205 (Radiasurf 7420,
pH:
  3.0–6.0 (5% aq. disp.) (Pegosperse 100ML)
  8–10 (5% aq. disp.) (Pegosperse 100L)
STD. PKGS.:
  190 kg net bung drums or bulk (Radiasurf 7420, 7421)

Diethylene glycol monooleate

SYNONYMS:
  Diethylene glycol oleate
  Diglycol monooleate
  Diglycol oleate
  9-Octadecenoic acid, 2-(2-hydroxyethoxy) ethyl ester

37

Diethylene glycol monooleate *(cont'd.)*

SYNONYMS *(cont'd.):*
PEG 100 monooleate
PEG-2 oleate (CTFA)
POE (2) monooleate

EMPIRICAL FORMULA:
$C_{22}H_{42}O_4$

STRUCTURE:

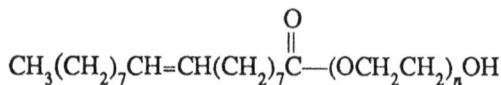

$$CH_3(CH_2)_7CH=CH(CH_2)_7\overset{\displaystyle O}{\overset{\|}{C}}-(OCH_2CH_2)_nOH$$

where avg. $n = 2$

CAS No.:
106-12-7

TRADENAME EQUIVALENTS:
Cithrol DGMO N/E [Croda]
Hodag DGO [Hodag]
Nikkol MYO-2 [Nikko]
Pegosperse 100 O [Glyco]
Radiasurf 7400 [Oleofina]
Self-emulsifying grades:
Cithrol DGMO S/E [Croda]
Witconol DOS [Witco/Organics]

CATEGORY:
Emulsifier, dispersant, wetting aid, antistat, defoamer, lubricant, opacifier, plasticizer, rust inhibitor, scouring and detergent aid, w/o emulgent

APPLICATIONS:
Cosmetic industry preparations: (Radiasurf 7400; Witconol DOS)
Farm products: insecticides/pesticides (Radiasurf 7400)
Industrial applications: (Witconol DOS); dyes and pigments (Radiasurf 7400); lubricating/cutting oils (Cithrol DGMO N/E, DGMO S/E; Hodag DGO; Radiasurf 7400); paint mfg. (Radiasurf 7400); paper mfg. (Cithrol DGMO N/E, DGMO S/E), plastics (Radiasurf 7400); polishes and waxes (Cithrol DGMO N/E, DGMO S/E; Radiasurf 7400); printing inks (Radiasurf 7400); textile/leather processing (Cithrol DGMO N/E, DGMO S/E; Radiasurf 7400)
Industrial cleaners: metal processing surfactants (Cithrol DGMO N/E, DGMO S/E)
Pharmaceutical applications: (Radiasurf 7400)

PROPERTIES:
Form:
Liquid (Cithrol DGMO N/E, DGMO S/E; Hodag DGO; Nikkol MYO-2; Pegosperse 100 O, Radiasurf 7400, Witconol DOS)
Color:
Amber (Pegosperse 100 O, Radiasurf 7400)

Diethylene glycol monooleate *(cont'd.)*

Composition:
100% conc. (Hodag DGO; Pegosperse 100 O; Witconol DOS)

Solubility:
Miscible with acetone (Pegosperse 100 O)
Sol. in benzene (@ 10%) (Radiasurf 7400)
Sol. in ethanol (Pegosperse 100 O)
Miscible with ethyl acetate (Pegosperse 100 O)
Sol. in hexane (@ 10%, cloudy) (Radiasurf 7400)
Sol. in isopropanol (@ 10%) (Radiasurf 7400)
Sol. in methanol (Pegosperse 100 O)
Sol. in min. oil (@ 10%, cloudy, Radiasurf 7400); miscible with (Pegosperse 100 O)
Sol. in naphtha (Pegosperse 100 O)
Sol. in toluol (Pegosperse 100 O)
Sol. in trichlorethylene (@ 10%) (Radiasurf 7400)
Sol. in veg. oil (@ 10%, cloudy, Radiasurf 7400); miscible with (Pegosperse 100 O)
Disp. in water (Pegosperse 100 O, Radiasurf 7400)

Ionic Nature:
Nonionic (Cithrol DGMO N/E; Hodag DGO; Nikkol MYO-2; Pegosperse 100 O; Radiasurf 7400; Witconol DOS)
Anionic (Cithrol DGMO S/E)

M.W.:
478 avg. (Radiasurf 7400)

Sp.gr.:
0.929 (37.8 C) (Radiasurf 7400)
0.93 (Pegosperse 100 O)

Visc.:
58.50 cps (37.8 C) (Radiasurf 7400)

Solidification Pt.:
< 0 C (Pegosperse 100 O)

Flash Pt.:
180 C (COC) (Radiasurf 7400)

Cloud Pt.:
–6.5 C (Radiasurf 7400)

HLB:
3.5 ± 0.5 (Pegosperse 100 O)
3.8 avg. (Radiasurf 7400)
4.5 (Nikkol MYO-2)
4.7 (Hodag DGO)

Acid No.:
4 max. (Radiasurf 7400)
80–95 (Pegosperse 100 O)

39

Diethylene glycol monooleate *(cont'd.)*

Iodine No.:
 65–75 (Radiasurf 7400)
 70–80 (Pegosperse 100 O)
Saponification No.:
 155–168 (Radiasurf 7400)
 160–175 (Pegosperse 100 O)
Ref. Index:
 1.4642 (Radiasurf 7400)
pH:
 8–9 (5% aq. disp.) (Pegosperse 100 O)
STD. PKGS.:
 190-kg net bung drums or bulk (Radiasurf 7400)

Diethylene glycol monostearate

SYNONYMS:
 Diethylene glycol stearate
 Diglycol monostearate
 Diglycol stearate
 Octadecanoic acid, 2-(2-hydroxyethoxy) ethyl ester
 PEG 100 monostearate
 PEG-2 stearate (CTFA)
 POE (2) monostearate
EMPIRICAL FORMULA:
 $C_{22}H_{44}O_4$
STRUCTURE:

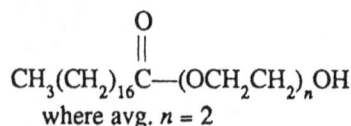

 where avg. $n = 2$
CAS No.:
 106-11-6
 9004-99-3 (generic)
TRADENAME EQUIVALENTS:
 Cithrol DGMS N/E [Croda Chem. Ltd.]
 Clindrol SDG [Clintwood]
 DMS-33 [Hefti]
 Drewmulse DGMS [Drew Produtos]
 Hodag DGS [Hodag]
 Kessco Diethylene Glycol Monostearate [Armak]
 Kessco Diglycol Stearate Neutral [Armak]

Diethylene glycol monostearate *(cont'd.)*

TRADENAME EQUIVALENTS *(cont'd.):*
  Nikkol MYS-2 [Nikko]
  Nopalcol 1-S [Diamond Shamrock]
  Pegosperse 100 S [Glyco]
  Radiasurf 7410 [Oleofina]
  Schercemol DEGMS [Scher]
  Witconol CAD [Witco/Organics]
Self-emulsifying grades:
  Cithrol DGMS S/E [Croda Chem. Ltd.]
  Grocor 5221 SE [A. Gross]
  Kessco Diglycol Stearate S.E. [Armak]
  Lipo DGS-SE [Lipo]
  Radiasurf 7411 [Oleofina]
CATEGORY:
  Emulsifier, dispersant, wetting aid, antistat, binding and thickening agent, defoamer, lubricant, opacifier, plasticizer, rust inhibitor, scouring and detergent aid, solubilizer, w/o emulgent
APPLICATIONS:
  Bath preparations: bath oils (Lipo DGS-SE)
  Cosmetic industry preparations: (DMS-33; Grocor 5221 SE; Hodag DGS; Kessco Diethylene Glycol Monostearate; Nikkol MYS-2; Nopalcol 1-S; Radiasurf 7410, 7411; Schercemol DEGMS; Witconol CAD); creams and lotions (Lipo DGS-SE); personal care products (Clindrol SDG); shampoos (Drewmulse DGMS); toiletries (Kessco Diethylene Glycol Monostearate; Witconol CAD)
  Farm products: insecticides/pesticides (Radiasurf 7410, 7411)
  Industrial applications: dyes and pigments (DMS-33; Drewmulse DGMS; Radiasurf 7410, 7411); lubricating/cutting oils (Radiasurf 7410, 7411); paint mfg. (Radiasurf 7410, 7411); plastics (Radiasurf 7410, 7411); polishes and waxes (Radiasurf 7410, 7411); printing inks (Radiasurf 7410, 7411); textile/leather processing (Grocor 5221 SE; Radiasurf 7410, 7411)
  Pharmaceutical applications: (Nikkol MYS-2; Radiasurf 7410, 7411)
PROPERTIES:
Form:
  Solid (Cithrol DGMS N/E, DGMS S/E; Clindrol SDG; Grocor 5221 SE; Hodag DGS; Nopalcol 1-S; Radiasurf 7410, 7411; Schercemol DEGMS)
  Beads (Grocor 5221 SE; Lipo DGS-SE; Pegosperse 100 S)
  Flakes (DMS-33; Drewmulse DGMS; Grocor 5221 SE; Kessco Diethylene Glycol Monostearate, Kessco Diglycol Stearate Neutral; Kessco Diglycol Stearate S.E.; Lipo DGS-SE; Nikkol MYS-2)
Color:
  White (Clindrol SDG; Grocor 5221 SE; Pegosperse 100 S; Radiasurf 7410, 7411)
  White to off-white (Lipo DGS-SE)
  White to cream (Schercemol DEGMS)

Diethylene glycol monostearate *(cont'd.)*

Composition:
85% active (Drewmulse DGMS)
98% conc. (Nopalcol 1-S)
100% active (Clindrol SDG; Grocor 5221 SE)
100% conc. (Cithrol DGMS N/E, DGMS S/E; DMS-33; Hodag DGS; Nikkol MYS-2; Pegosperse 100 S; Witconol CAD)

Solubility:
Miscible hot with acetone (Pegosperse 100 S)
Sol. in alcohols (Clindrol SDG)
Sol. in benzene (Radiasurf 7410, 7411)
Sol. in chlorinated, aromatic, and aliphatic hydrocarbons (Clindrol SDG)
Sol. hot in ethanol (Pegosperse 100 S)
Miscible hot with ethyl acetate (Pegosperse 100 S)
Sol. in glycols (Clindrol SDG)
Sol. in hexane (Radiasurf 7410, 7411)
Sol. in hydrocarbons (Grocor 5221 SE)
Sol. in isopropanol (Radiasurf 7410, 7411)
Sol. hot in methanol (Pegosperse 100 S)
Sol. in min. oil (Pegosperse 100 S; Radiasurf 7410, 7411)
Sol. hot in naphtha (Pegosperse 100 S)
Sol. hot in toluol (Pegosperse 100 S)
Sol. in trichlorethylene (Radiasurf 7410, 7411)
Sol. in veg. oil (Pegosperse 100 S; Radiasurf 7410, 7411)
Disp. hot in water (Grocor 5221 SE; Pegosperse 100 S)

Ionic Nature:
Nonionic (Cithrol DGMS N/E; Clindrol SDG; DMS-33; Drewmulse DGMS; Hodag DGS; Kessco Diethylene Glycol Monostearate; Kessco Diglycol Stearate Neutral; Nikkol MYS-2; Nopalcol 1-S; Pegosperse 100 S; Radiasurf 7410, 7411; Witconol CAD)
Anionic (Cithrol DGMS S/E; Grocor 5221 SE; Kessco Diglycol Stearate S.E.)

M.W.:
470 avg. (Radiasurf 7411)
475 avg. (Radiasurf 7410)

Sp.gr.:
0.873 (98.9 C) (Radiasurf 7410)
0.878 (98.9 C) (Radiasurf 7411)
0.96 (Grocor 5221 SE; Pegosperse 100 S)
1.01 (Clindrol SDG)

Visc.:
5.90 cps (98.9 C) (Radiasurf 7411)
5.95 cps (98.9 C) (Radiasurf 7410)

M.P.:
42–45 C (Clindrol SDG)

Diethylene glycol monostearate *(cont'd.)*

42–48 C (Kessco Diglycol Stearate Neutral)
44.5–47.5 C (Kessco Diethylene Glycol Monostearate)
46–54 C (Pegosperse 100 S)
46–56 C (Grocor 5221 SE)
48 C (Schercemol DEGMS)
48–53 C (Kessco Diglycol Stearate S.E.)
51.5 C (Radiasurf 7410)
52 C (Radiasurf 7411)

Flash Pt.:
176 C (COC) (Radiasurf 7411)
191 C (COC) (Radiasurf 7410)
345 F (COC) (Kessco Diglycol Stearate S.E.)
360 F (COC) (Kessco Diglycol Stearate Neutral)
395 F (COC) (Kessco Diethylene Glycol Monostearate)

HLB:
3.5 avg. (Radiasurf 7410)
3.8 (Nopalcol 1-S)
3.8 ± 0.5 (Pegosperse 100 S)
4.0 (Nikkol MYS-2)
4.0 ± 1 (Lipo DGS-SE)
4.1 avg. (Radiasurf 7411)
4.3 (Kessco Diethylene Glycol Monostearate)
4.7 (Hodag DGS)
5.0 (DMS-33)

Acid No.:
3 max. (Radiasurf 7410, 7411)
5 max. (Kessco Diethylene Glycol Monostearate; Schercemol DEGMS)
90–110 (Lipo DGS-SE)
95–105 (Pegosperse 100 S)
103 max. (Kessco Diethylene Glycol Monostearate; Kessco Diglycol Stearate Neutral, S.E.)

Iodine No.:
0.5 max. (Kessco Diethylene Glycol Monostearate)
1 max. (Radiasurf 7410, 7411; Schercemol DEGMS)
< 2 (Pegosperse 100 S)
7 max. (Kessco Diglycol Stearate Neutral, S.E.)

Saponification No.:
160–170 (Radiasurf 7411)
160–175 (Radiasurf 7410)
160–180 (Lipo DGS-SE)
165–175 (Pegosperse 100 S; Schercemol DEGMS)

pH:
6.8–7.5 (3% aq. disp.) (Pegosperse 100 S)

Diethylene glycol monostearate *(cont'd.)*

STD. PKGS.:
40-lb net cubitainers (Clindrol SDG)
25-kg net multiply paper bags or bulk (Radiasurf 7410, 7411)
Bulk or drums (Grocor 5221 SE)

Disodium coco amide MIPA sulfosuccinate

SYNONYMS:
Butanedioic acid, sulfo-, 2-cocamido-1-methylethyl esters, disodium salts
Disodium cocamido MIPA-sulfosuccinate (CTFA)
Sulfobutanedioic acid, 2-cocamido-1-methylethyl esters, disodium salts

STRUCTURE:

$$RC-NH-CH_2CHO-CCHCH_2C-ONa$$

with carbonyl oxygens, CH_3 and SO_3Na substituents

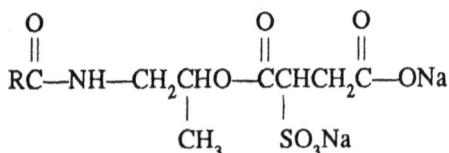

where RCO^- represents the coconut acid radical

CAS No.
68515-65-1; 68909-21-7

TRADENAME EQUIVALENTS:
Alconate CPA [Alcolac]
Cyclopol SB-CP [Cyclo]
Mackanate CP [McIntyre]
Monamate CPA-100 [Mona]
Schercopol CMIS-Na [Scher]

CATEGORY:
Mild surfactant, foaming agent

APPLICATIONS:
Bath products: (Alconate CPA); bubble bath (Cyclopol SB-CP; Schercopol CMIS-Na)
Cleansers: (Alconate CPA); body cleansers (Alconate CPA; Cyclopol SB-CP); hand cleanser (Alconate CPA)
Cosmetic industry preparations: (Monamate CPA-100); personal care products (Alconate CPA); shampoos (Alconate CPA; Cyclopol SB-CP; Schercopol CMIS-Na)
Household detergents: (Monamate CPA-100); carpet & upholstery shampoos (Alconate CPA; Schercopol CMIS-Na); dishwashing (Alconate CPA); light-duty cleaners (Schercopol CMIS-Na)

PROPERTIES:
Form:
Liquid (Cyclopol SB-CP; Mackanate CP)
Clear liquid (Alconate CPA; Schercopol CMIS-Na)

Disodium coco amide MIPA sulfosuccinate *(cont'd.)*

Powder (Monamate CPA-100)

Color:
Yellow (Cyclopol SB-CP; Schercopol CMIS-Na)
Gardner 3 (Alconate CPA)

Odor:
Mild, characteristic (Schercopol CMIS-Na)

Composition:
39–41% active (Cyclopol SB-CP)
40% active (Schercopol CMIS-Na)
40% conc. (Mackanate CP)
40% solids (Alconate CPA)
100% conc. (Monamate CPA-100)

Solubility:
Completely sol. in water (Schercopol CMIS-Na)

Ionic Nature:
Anionic (Alconate CPA; Cyclopol SB-CP)
Nonionic (Schercopol CMIS-Na)

M.W.:
491 avg. (Schercopol CMIS-Na)

Sp.gr.:
1.12 ± 0.01 (Schercopol CMIS-Na)

Density:
8.9 lb/gal (Alconate CPA)
9.4 lb/gal (Schercopol CMIS-Na)

Visc.:
400 cps max. (Schercopol CMIS-Na)

Cloud Pt.:
–5 C (Alconate CPA)
1 C max. (Schercopol CMIS-Na)

Stability:
Stable at pH 4–8; limited stability in very strong acid or alkaline systems (Alconate CPA)

pH:
5–7 (Schercopol CMIS-Na)
6.5 (5%) (Alconate CPA)

STD. PKGS.:
55-gal poly-lined drums (Schercopol CMIS-Na)

Disodium lauramido MEA-sulfosuccinate (CTFA)

SYNONYMS:

Butanedioic acid, sulfo-, 1-ester with N-(2-hydroxyethyl) dodecanamide, disodium salt

Succinic acid, sulfo-, 1-ester with N-(2-hydroxyethyl) dodecanamide, disodium salt

Sulfobutanedioic acid, 1-ester with N-(2-hydroxyethyl) dodecanamide, disodium salt

EMPIRICAL FORMULA:

$C_{18}H_{53}NO_8S \cdot 2Na$

STRUCTURE:

$$CH_3(CH_2)_{10}\overset{\overset{O}{\|}}{C}-NH-CH_2CH_2O-\overset{\overset{O}{\|}}{C}CHCH_2\overset{\overset{O}{\|}}{C}-ONa$$
$$\underset{SO_3Na}{|}$$

CAS No.

25882-44-4; 55101-78-5

TRADENAME EQUIVALENTS:

Alconate LEA [Alcolac]

Alkasurf SS-L9ME [Alkaril]

Condanol SBL203 [Dutton & Reinisch]

Cyclopol SBL203 [Cyclo]

Incrosul LMS [Croda Surfactants]

Mackanate LM-40 [McIntyre]

Merpasol L47 [Elektrochem. Fabrik Kempen]

Rewopol SBL203 [Dutton & Reinisch]

Steinapol SBL203 [Dutton & Reinisch]

CATEGORY:

Mild surfactant, detergent, foaming agent

APPLICATIONS:

Bath products: (Alconate LEA); bubble bath (Cyclopol SBL203)

Cleansers: (Alconate LEA); body cleansers (Alconate LEA; Incrosul LMS); hand cleanser (Alconate LEA)

Cosmetic industry preparations: (Condanol SBL203; Merpasol L47; Rewopol SBL203; Steinapol SBL203); personal care products (Alconate LEA); shampoos (Alconate LEA; Alkasurf SS-L9ME; Cyclopol SBL203; Incrosul LMS)

Household detergents: carpet & upholstery shampoos (Alconate LEA; Alkasurf SS-L9ME; Condanol SBL203; Cyclopol SBL203; Incrosul LMS; Merpasol L47; Rewopol SBL203; Steinapol SBL203); dishwashing (Alconate LEA); laundry detergent (Alkasurf SS-L9ME); synthetic soaps (Condanol SBL203; Rewopol SBL203; Steinapol SBL203)

Industrial cleaners: institutional cleaners (Alkasurf SS-L9ME)

PROPERTIES:

Form:

Liquid (Cyclopol SBL203; Mackanate LM-40; Merpasol L47)

Viscous liquid (Alkasurf SS-L9ME)
Opaque liquid (Alconate LEA)
Turbid liquid (Condanol SBL203; Rewopol SBL203; Steinapol SBL203)
Liquid/slurry (Incrosul LMS)

Color:
Cream white (Alkasurf SS-L9ME)
Yellow (Condanol SBL203; Cyclopol SBL203; Rewopol SBL203; Steinapol SBL203)
Gardner 3 (Alconate LEA)
Gardner 4 max. (Incrosul LMS)

Odor:
Bland (Alkasurf SS-L9ME)

Composition:
38–40% active (Cyclopol SBL203)
39% solids (Condanol SBL203; Rewopol SBL203; Steinapol SBL203)
39–41% active (Incrosul LMS)
40% active (Alkasurf SS-L9ME)
40% conc. (Mackanate LM-40; Merpasol L47)
40% solids (Alconate LEA)

Ionic Nature: Anionic

Sp.gr.:
1.00 (Alkasurf SS-L9ME)

Density:
9.0 lb/gal (Alconate LEA)

Cloud Pt.:
20 C (Alconate LEA)

Stability:
Good (Alkasurf SS-L9ME)
Stable at pH 4–8; limited stability in very strong acid or alkaline systems (Alconate LEA)

pH:
6–7 (3%) (Incrosul LMS)
6.5 (5%) (Alconate LEA)
6.5–7.5 (5% solids) (Rewopol SBL203; Steinapol SBL203)

Biodegradable: (Alkasurf SS-L9ME; Condanol SBL203; Rewopol SBL203; Steinapol SBL203)

TOXICITY/HANDLING:
Avoid prolonged contact with skin and eyes (Incrosul LMS)

STORAGE/HANDLING:
Store in a cool, dry place; warm to clarity before sampling or testing (40–50 C) (Incrosul LMS)

STD. PKGS.:
55-gal (450 lb net) polyethylene-lined Leverpak (Incrosul LMS)

Disodium lauryl ether sulfosuccinate

SYNONYMS:

Butanedioic acid, sulfo-, 4-[2-[2-[2-(dodecyloxy) ethoxy] ethoxy] ethyl] ester, disodium salt

Disodium laurethsulfosuccinate (CTFA)

Disodium lauryl alcohol polyglycol-ether sulfosuccinate

Poly (oxy-1,2-ethanediyl), α-(3-carboxy-1-oxo-3-sulfopropyl)-ω-(dodecyloxy)-,
disodium salt

Sulfobutanedioic acid, 4-[2-[2-[2-(dodecyloxy) ethoxy] ethoxy] ethyl] ester, disodium
salt

STRUCTURE:

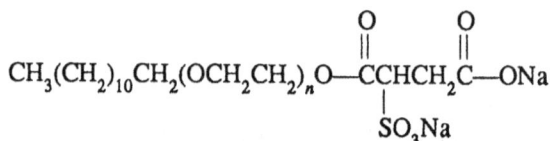

where avg. n = 1–4

CAS No.:

39354-45-5 (generic); 40754-59-4; 42016-08-0

TRADENAME EQUIVALENTS:

Alconate L-3 [Alcolac]

Alkasurf SS-LA-3 [Alkaril]

Condanol SBFA30, 40% [Dutton & Reinisch]

Cyclopol SBFA30 [Cyclo]

Emery 5320 [Emery]

Mackanate EL, L-1, L-2 [McIntyre]

Rewopol SBFA30 [Dutton & Reinisch]

Schercopol LPS [Scher]

Setacin 103 Special [Zschimmer & Schwartz]

Steinapol SBFA30, 40% [Dutton & Reinisch]

CATEGORY:

Detergent, emulsifier, wetting agent, surfactant, raw material, viscosity enhancer

APPLICATIONS:

Bath products: (Setacin 103 Special); bubble bath (Alkasurf SS-LA-3; Cyclopol
SBFA30; Emery 5320; Mackanate EL; Rewopol SBFA30; Steinapol SBFA30,
40%)

Cleansers: (Cyclopol SBFA30; Emery 5320; Mackanate L-1, L-2; Rewopol SBFA30;
Setacin 103 Special); hand cleanser (Rewopol SBFA30; Steinapol SBFA30, 40%)

Cosmetic industry preparations: (Alkasurf SS-LA-3; Setacin 103 Special); baby
preparations (Condanol SBFA30, 40%; Rewopol SBFA30); personal care products
(Alconate L-3; Rewopol SBFA30); shampoos (Alkasurf SS-LA-3; Condanol
SBFA30, 40%; Cyclopol SBFA30; Emery 5320; Mackanate EL, L-1, L-2; Rewopol
SBFA30; Setacin 103 Special; Steinapol SBFA30, 40%)

Household detergents: (Alconate L-3; Alkasurf SS-LA-3); heavy-duty cleaner

Disodium lauryl ether sulfosuccinate *(cont'd.)*

(Alconate L-3)

Industrial applications: polymers/polymerization (Alconate L-3; Alkasurf SS-LA-3)

Industrial cleaners: (Alconate L-3; Alkasurf SS-LA-3)

Pharmaceuticals: intimate hygiene products (Condanol SBFA30, 40%; Steinapol SBFA30, 40%)

PROPERTIES:

Form:

Liquid (Cyclopol SBFA30; Emery 5320; Mackanate EL, L-1, L-2)

Clear liquid (Alconate L-3; Alkasurf SS-LA-3; Condanol SBFA30, 40%; Rewopol SBFA30; Schercopol LPS; Steinapol SBFA30, 40%)

Color:

Almost colorless (Condanol SBFA30, 40%; Rewopol SBFA30; Steinapol SBFA30, 40%)

Light (Alkasurf SS-LA-3)

Yellow (Schercopol LPS)

Gardner 3 (Alconate L-3)

Odor:

Typical (Alkasurf SS-LA-3)

Composition:

30% active (Alkasurf SS-LA-3)

38–40% active (Cyclopol SBFA30)

39% solids min. (Condanol SBFA30, 40%; Rewopol SBFA30; Schercopol LPS; Steinapol SBFA30, 40%)

40% conc. (Emery 5320; Mackanate EL, L-1, L-2)

40% solids (Alconate L-3)

Solubility:

Sol. in water (Alkasurf SS-LA-3)

Ionic Nature:

Anionic (Alkasurf SS-LA-3; Condanol SBFA30, 40%; Rewopol SBFA30; Steinapol SBFA30, 40%)

Anionic/nonionic (Alconate L-3)

Sp.gr.:

1.10 (Alkasurf SS-LA-3)

Density:

9.1 lb/gal (Alconate L-3)

Visc.:

100–200 cps (Condanol SBFA30, 40%; Rewopol SBFA30; Steinapol SBFA30, 40%)

Cloud Pt.:

< –5 C (Alconate L-3)

pH:

5–7 (Schercopol LPS)

6.5 (5%) (Alconate L-3)

6.5–7.5 (5% solids) (Rewopol SBFA30; Steinapol SBFA30, 40%)

Disodium lauryl ether sulfosuccinate *(cont'd.)*

Biodegradable: (Condanol SBFA30, 40%; Rewopol SBFA30; Steinapol SBFA30, 40%)

Disodium lauryl sulfosuccinate *(CTFA)*

SYNONYMS:
 Butanedioic acid, sulfo-, 1-dodecyl ester, disodium salt
 Disodium lauryl alcohol sulfosuccinate
 Succinic acid, sulfo-, monododecyl ester, disodium salt
 Sulfobutanedioic acid, 1-dodecyl ester, disodium salt

EMPIRICAL FORMULA:
 $C_{16}H_{30}O_7S \cdot 2Na$

STRUCTURE:

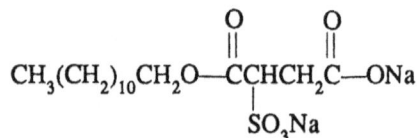

CAS No.:
 13192-12-6; 19040-44-9; 26838-05-1

TRADENAME EQUIVALENTS:
 Condanol SBF12, SBF-12 Powder [Dutton & Reinisch]
 Emcol 4400-1 [Witco/Organics]
 Mackanate LO, LO-100 [McIntyre]
 Merpasol L44 [Elektrochemische Fabrik Kempen]
 Miranate LSS [Miranol]
 Monamate LA100 [Mona]
 Steinapol SBF12, SBF12-Powder [Dutton & Reinisch]

CATEGORY:
 Surfactant, detergent, emulsifier, cleansing agent, foaming agent, foam stabilizer

APPLICATIONS:
 Bath products: bubble bath (Mackanate LO-100; Monamate LA100)
 Cleansers: body cleansers (Mackanate LO; Monamate LA100); hand cleanser
   (Mackanate LO-100; Monamate LA100)
 Cosmetic industry preparations: (Condanol SBF12; Emcol 4400-1; Merpasol L44;
   Steinapol SBF12); conditioners (Miranate LSS); personal care products (Mona-
   mate LA100); shampoos (Condanol SBF12-Powder; Mackanate LO, LO-100;
   Miranate LSS; Monamate LA100; Steinapol SBF12-Powder); shaving prepara-
   tions (Monamate LA100); toiletries (Emcol 4400-1)
 Household detergents: (Monamate LA100); carpet & upholstery shampoos (Condanol

Disodium lauryl sulfosuccinate (cont'd.)

SBF12; Merpasol L44; Monamate LA100; Steinapol SBF12); detergent/soap bars
(Condanol SBF12, SBF12-Powder; Monamate LA100; Steinapol SBF12, SBF12-
Powder); powdered detergents (Condanol SBF12-Powder; Monamate LA100;
Steinapol SBF12-Powder)

PROPERTIES:

Form:

Paste (Condanol SBF12; Emcol 4400-1; Merpasol L44; Steinapol SBF12)

Creamy solid (Miranate LSS)

Solid (Mackanate LO)

Fine powder (Monamate LA100)

Powder (Condanol SBF12-Powder; Mackanate LO-100; Steinapol SBF12-Powder)

Color:

White (Condanol SBF12, SBF12-Powder; Miranate LSS; Monamate LA100; Steinapol SBF12, SBF12-Powder)

Light yellow (Condanol SBF12-Powder; Steinapol SBF12-Powder)

Composition:

35% active (Miranate LSS)

39% solids (Condanol SBF12; Steinapol SBF12)

40% conc. (Mackanate LO; Merpasol L44)

95% solids (Condanol SBF12-Powder; Steinapol SBF12-Powder)

98% active min. (Monamate LA100)

100% conc. (Mackanate LO-100)

Solubility:

Sol. in water (Condanol SBF12-Powder; Emcol 4400-1; Steinapol SBF12-Powder)

Ionic Nature:

Anionic (Condanol SBF12, SBF12-Powder; Emcol 4400-1; Mackanate LO, LO-100;
Miranate LSS; Monamate LA100; Steinapol SBF12, SBF12-Powder)

Density:

250 g/l (Condanol SBF12-Powder; Steinapol SBF12-Powder)

Acid No.:

0–13 (Monamate LA100)

Stability:

Chemically stable under all ambient conditions for indefinite periods of time; high-
temp. storage may cause some melting (Monamate LA100)

Good in weakly acidic and weakly alkaline media (Condanol SBF12-Powder; Steinapol SBF12-Powder)

pH:

6.0–7.0 (10% sol'n.) (Monamate LA100)

6.5–7.5 (5% solids) (Steinapol SBF12, SBF12-Powder)

7.0 (10% sol'n.) (Miranate LSS)

Biodegradable: (Condanol SBF12, SBF12-Powder; Steinapol SBF12, SBF12-
Powder)

Disodium lauryl sulfosuccinate *(cont'd.)*

TOXICITY/HANDLING:
Mild primary skin irritant when tested as a 16% solid sol'n. at pH 7.0 (Miranate LSS)

Disodium N-oleyl sulfosuccinamate

TRADENAME EQUIVALENTS:
Alkasurf SS-OA [Alkaril]
Empimin MTT, MTT/A [Albright & Wilson]
CATEGORY:
Foaming agent, emulsifier, dispersant, solubilizer, foam stabilizer
APPLICATIONS:
Household detergents: soaps (Alkasurf SS-OA)
Industrial applications: carpet backing (Empimin MTT, MTT/A); polymers/ polymerization (Empimin MTT/A); rubber latex foam (Alkasurf SS-OA; Empimin MTT, MTT/A)
PROPERTIES:
Form:
Liquid (Alkasurf SS-OA; Empimin MTT)
Clear/opaque liquid (@ 20 C) (Empimin MTT/A)
Color:
Pale cream (Empimin MTT)
Amber (Alkasurf SS-OA)
Yellow/pale cream (Empimin MTT/A)
Odor:
Low (Alkasurf SS-OA)
Composition:
28% active min. in water (Empimin MTT/A)
31% active in water (Alkasurf SS-OA)
40% active (Empimin MTT)
Solubility:
Sol. in water (Alkasurf SS-OA)
Ionic Nature:
Anionic (Empimin MTT)
Density:
1.1 g/cm^3 (20 C) (Empimin MTT/A)
Visc.:
25–30 cps (70 F) (Alkasurf SS-OA)
F.P.:
< 0 C (Alkasurf SS-OA)
pH:
8.0 (5% aq.) (Empimin MTT)
8.0 ± 0.5 (5% aq. sol'n.) (Empimin MTT/A)

Disodium N-oleyl sulfosuccinamate *(cont'd.)*

STORAGE/HANDLING:
Mobile liquid at temps. down to 0 C, but sedimentation of insol. sulfosuccinamate may occur on prolonged storage; the use of apaddle stirrer or a pump recycle is recommended for bulk storage at ambient temps. (Empimin MTT/A)

Dodecylbenzene sulfonic acid (CTFA)

SYNONYMS:
Benzenesulfonic acid, dodecyl-
DDBSA
EMPIRICAL FORMULA:
$C_{18}H_{30}O_3S$
STRUCTURE:

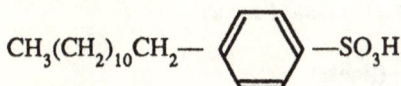

$CH_3(CH_2)_{10}CH_2$—⬡—SO_3H

CAS No.:
27176-87-0
TRADENAME EQUIVALENTS:
Alkasurf LA Acid [Alkaril] (linear)
Ardet LAS [Ardmore] (linear)
Arsul DDB [Magna/Arjay] (branched)
Arsul LAS [Magna/Arjay] (linear)
Arylan S Acid [Diamond Shamrock/Process] (branched)
Arylan SBC Acid [Lankro] (linear)
Arylan SC Acid [Lankro] (linear)
Bio Soft LAS-97 [Stepan] (linear)
Bio Soft S-100 [Stepan] (linear)
Calsoft LAS-99 [Pilot] (linear)
Carsosulf UL-100 Acid [Carson] (linear)
Conco AAS-98S [Continental] (linear)
Condasol Sulfonic Acid K [Dutton & Reinisch] (linear)
Cycloryl ABSA [Cyclo] (linear)
DDBSA 99-b [Monsanto]
DeSonate SA-H [DeSoto] (branched)
Elfan WA Sulfosäure [Akzo Chemie]
Emulsifier 99 [Pilot] (branched)
Hetsulf Acid [Heterene]
Lakeway SA [Bofors Lakeway] (linear)
Manro BA [Manro] (linear)

Dodecylbenzene sulfonic acid (cont'd.)

TRADENAME EQUIVALENTS *(cont'd.):*
Manro HA [Manro] (branched)
Manro NA [Manro] (linear)
Marlon AS₃ [Chemische Werke Huls] (linear)
Mars SA-98S [Mars] (linear)
Merpisap AS98 [Elektrochemische Fabrik Kempen] (linear)
Nansa 1042, 1042/P [Albright & Wilson/Detergents] (linear)
Nansa SBA [Albright & Wilson/Marchon]
Nansa SSA [Albright & Wilson/Marchon]
Nansa SSA/P [Albright & Wilson/Detergents] (linear)
Polyfac LAS-97 [Westvaco] (linear)
Polystep A-17 [Stepan] (branched)
Reworyl-Sulfonic Acid K [Dutton & Reinisch]
Richonic Acid B [Richardson]
Rueterg Sulfonic Acid [Finetex]
Steinaryl-Sulfonic Acid K [Dutton & Reinisch]
Sterling LA Acid C, LA Acid Reg. [Canada Packers] (linear)
Surco DDBSA [Onyx] (linear)
Tairygent CA-1 [Formosa Chem. & Fibre] (linear)
Tairygent CA-2 [Formosa Chem. & Fibre] (branched)
Tex-Wet 1197 [Intex Prod.]
Vista SA-597 [Vista] (linear)
Witco 1298 Hard Acid, 1298 Soft Acid [Witco/Organics]
Witco Acid B [Witco/Organics]

CATEGORY:
Detergent, emulsifier, penetrant, foaming agent, intermediate, surfactant, solubilizer, dispersant, wetting agent, emulsion breaker, corrosion inhibitor

APPLICATIONS:
Automobile cleaners: car shampoo (Alkasurf LA Acid; Lakeway SA)
Bath products: bubble bath (Alkasurf LA Acid)
Cleansers: (Mars SA-98S; Merpisap AS98)
Cosmetic industry preparations: shampoos (Conco AAS-98S; Cycloryl ABSA)
Degreasers: (Alkasurf LA Acid)
Farm products: (Alkasurf LA Acid; Emulsifier 99)
Household detergents: (Arsul DDB, LAS; Arylan S Acid, SBC Acid, SC Acid; Bio Soft LAS-97; Conco AAS-98S; Condasol Sulfonic Acid K; Marlon AS₃; Nansa 1042; Reworyl-Sulfonic Acid K; Richonic Acid B; Rueterg Sulfonic Acid; Steinaryl-Sulfonic Acid K; Sterling LA Acid C, LA Acid Reg.; Tex-Wet 1197; Vista SA-597; Witco 1298 Soft Acid; Witco Acid B); all-purpose cleaner (Calsoft LAS-99); built detergents (Bio Soft S-100; Cycloryl ABSA); detergent base (Hetsulf Acid; Lakeway SA; Reworyl-Sulfonic Acid K; Steinaryl-Sulfonic Acid K); dishwashing (Alkasurf LA Acid; Condasol Sulfonic Acid K; Lakeway SA; Reworyl-Sulfonic Acid K; Richonic Acid B; Steinaryl-Sulfonic Acid K; Tex-Wet

Dodecylbenzene sulfonic acid *(cont'd.)*

1197); hard surface cleaner (Cycloryl ABSA); heavy-duty cleaner (Alkasurf LA Acid; Manro BA, HA, NA); laundry detergent (Nansa 1042); light-duty cleaners (Manro BA, HA, NA); liquid detergents (Arylan SC Acid; Calsoft LAS-99; DDBSA 99-b; Merpisap AS98; Nansa 1042, SBA, SSA; Polyfac LAS-97; Sterling LA Acid C, LA Acid Reg.; Surco DDBSA); paste detergents (Alkasurf LA Acid; Merpisap AS98); powdered detergents (Alkasurf LA Acid; DDBSA 99-b; Merpisap AS98; Nansa 1042, SBA, SSA; Polyfac LAS-97; Surco DDBSA)

Industrial applications: construction (Calsoft LAS-99); dyes and pigments (Tex-Wet 1197; Witco 1298 Soft Acid); ore flotation (Calsoft LAS-99); plastics (Nansa 1042/ P, SSA/P); polymers/polymerization (Alkasurf LA Acid; Emulsifier 99; Manro BA; Nansa 1042/P, SSA/P; Polystep A-17; Witco 1298 Soft Acid; Witco Acid B); rubber (Nansa 1042/P, SSA/P; Polystep A-17)

Industrial cleaners: (Lakeway SA; Richonic Acid B; Sterling LA Acid C, LA Acid Reg.); acid cleaners (Calsoft LAS-99); all-purpose cleaners (Mars SA-98S); drycleaning compositions (Alkasurf LA Acid; Richonic Acid B); institutional cleaners (Mars SA-98S); janitorial cleaners (Mars SA-98S); metal processing surfactants (Alkasurf LA Acid; Conco AAS-98S; Witco 1298 Soft Acid; Witco Acid B); solvent cleaners (Mars SA-98S); textile cleaning (Calsoft LAS-99; Tex-Wet 1197)

PROPERTIES:

Form:

Liquid (Ardet LAS; Arsul DDB, LAS; Arylan S Acid, SBC Acid, SC Acid; Bio Soft LAS-97; Carsosulf UL-100 Acid; Conco AAS-98S; Cycloryl ABSA; DeSonate SA-H; Elfan WA Sulfosäure; Hetsulf Acid; Marlon AS $_3$; Nansa SBA, SSA; Polyfac LAS-97; Rueterg Sulfonic Acid; Sterling LA Acid C, LA Acid Reg.; Surco DDBSA; Tex-Wet 1197; Vista SA-597; Witco 1298 Hard Acid, 1298 Soft Acid; Witco Acid B)

Viscous liquid (Alkasurf LA Acid; Bio Soft S-100; Calsoft LAS-99; Condasol Sulfonic Acid K; Emulsifier 99; Lakeway SA; Manro BA, HA, NA; Mars SA-98S; Merpisap AS98; Polystep A-17; Reworyl-Sulfonic Acid K; Steinaryl-Sulfonic Acid K); (@ 20 C) (Nansa 1042, 1042/P, SSA/P)

Paste (Tairygent CA-1, CA-2)

Color:

Amber (Cycloryl ABSA)

Reddish-amber (Mars SA-98S; Polyfac LAS-97)

Dark amber (Lakeway SA)

Reddish-brown (Tex-Wet 1197)

Brown (Alkasurf LA Acid; Arylan SC Acid; Carsosulf UL-100 Acid; Conco AAS-98S; Condasol Sulfonic Acid K; Reworyl-Sulfonic Acid K; Steinaryl-Sulfonic Acid K; Tairygent CA-1, CA-2)

Dark brown (Elfan WA Sulfosäure; Manro BA, HA, NA; Merpisap AS98; Nansa SBA, SSA); (@ 20 C) (Nansa 1042, 1042/P, SSA/P)

Dark (Bio Soft S-100)

Dodecylbenzene sulfonic acid *(cont'd.)*

Klett 50 (5% neutral sol'n.) (Calsoft LAS-99)
Klett 300 (Richonic Acid B)

Odor:

SO$_3$ (Elfan WA Sulfosäure)
Characteristic (Manro BA, HA, NA)

Composition:

90% conc. (Arsul DDB)
95% active min. (Manro BA, HA; Nansa 1042, 1042/P, SSA/P)
96% active (Alkasurf LA Acid; Nansa SBA, SSA)
96% active min. (Manro NA; Surco DDBSA; Tairygent CA-1, CA-2)
96% conc. (Arylan S Acid; Vista SA-597)
96.3% active (Sterling LA Acid C, LA Acid Reg.)
97% active (Arylan SC Acid; Bio Soft S-100; Hetsulf Acid; Richonic Acid B; Tex-Wet 1197)
97% active min. (Cycloryl ABSA; Reworyl-Sulfonic Acid K; Steinaryl-Sulfonic Acid K)
97% conc. (Arylan SBC Acid; Bio Soft LAS-97; DeSonate SA-H; Rueterg Sulfonic Acid)
97% sulfonic acid (Condasol Sulfonic Acid K; Polyfac LAS-97)
97 ± 1% active (Carsosulf UL-100 Acid)
97.4% active (Polystep A-17)
97.5% sulfonic acid (Calsoft LAS-99)
98% active (DDBSA 99-b; Elfan WA Sulfosäure; Emulsifier 99; Mars SA-98S; Merpisap AS98)
98+% active (Conco AAS-98S)
98% conc. (Ardet LAS; Arsul LAS; Lakeway SA)
100% active (Marlon AS$_3$)

Solubility:

Sol. in oils (Witco 1298 Hard Acid, 1298 Soft Acid; Witco Acid B)
Sol. in water (Bio Soft S-100; Conco AAS-98S; Elfan WA Sulfosäure; Polystep A-17; Witco 1298 Hard Acid, 1298 Soft Acid; Witco Acid B); completely miscible (Mars SA-98S)

Ionic Nature:

Anionic (Ardet LAS; Arsul DDB, LAS; Arylan S Acid; Bio Soft LAS-97, S-100; Calsoft LAS-99; Conco AAS-98S; Cycloryl ABSA; Hetsulf Acid; Lakeway SA; Marlon AS$_3$; Mars SA-98S; Nansa SBA, SSA; Reworyl-Sulfonic Acid K; Rueterg Sulfonic Acid; Steinaryl-Sulfonic Acid K; Surco DDBSA; Tex-Wet 1197; Vista SA-597)

M.W.:

316 (Carsosulf UL-100 Acid)
316 avg. (DDBSA 99-b)
318 (Polyfac LAS-97)

Sp.gr.:
  1.0 (DDBSA 99-b)
  1.04 (Bio Soft S-100; Conco AAS-98S); (20 C) (Manro BA)
  1.05 (Hetsulf Acid); (25/20 C) (Surco DDBSA)
  1.06 (Carsosulf UL-100 Acid; Elfan WA Sulfosäure; Mars SA-98S); (20 C) (Manro HA, NA)

Density:
  1.05 g/cc (20 C) (Nansa 1042, 1042/P, SSA/P)
  8.7 lb/gal (Lakeway SA)
  8.8 lb/gal (Mars SA-98S; Richonic Acid B; Tex-Wet 1197)
  8.83 lb/gal (Calsoft LAS-99)
  8.84 lb/gal (Carsosulf UL-100 Acid)

Visc.:
  100 cps (Calsoft LAS-99)
  500 cps (20 C) (Elfan WA Sulfosäure)
  1440 cps (Brookfield) (Lakeway SA)
  1500 cps (20 C) (Manro BA)
  1800 cs (20 C) (Arylan SC Acid)
  1900 cs (20 C) (Nansa 1042, 1042/P, SSA/P)
  2000 cps (20 C) (Manro NA)
  2000 cs (20 C) (Nansa SSA)
  18,000 cps (20 C) (Manro HA)
  20,000 cs (20 C) (Nansa SBA)

Flash Pt.:
  150 C (PMCC) (Elfan WA Sulfosäure)
  > 200 F (Surco DDBSA)

Acid No.:
  180 ± 5 (Reworyl-Sulfonic Acid K; Steinaryl-Sulfonic Acid K)
  180–190 (Tex-Wet 1197)
  182 (Lakeway SA)
  185 (Calsoft LAS-99; Mars SA-98S)
  185 max. (Elfan WA Sulfosäure)

Neutral. Equiv.:
  300–315 (Richonic Acid B)

Stability:
  Good in hard and soft water, strong alkalis and strong acids (Alkasurf LA Acid)
  Stable to acids and alkalis (Nansa SBA, SSA)

pH:
  1.6 (1% aq. sol'n.) (Polyfac LAS-97)

Biodegradable: (Alkasurf LA Acid; Arsul LAS; Arylan SBC Acid; Bio Soft S-100; Carsosulf UL-100 Acid; Conco AAS-98S; Cycloryl ABSA; Hetsulf Acid; Manro BA; Mars SA-98S; Merpisap AS98; Polyfac LAS-97; Richonic Acid B; Tex-Wet 1197); 90% biodegradable (Condasol Sulfonic Acid K); nonbiodegradable (Emulsifier 99)

Dodecylbenzene sulfonic acid *(cont'd.)*

TOXICITY/HANDLING:

Corrosive (Conco AAS-98S; Polyfac LAS-97)

Skin and eye irritant; severely irritating (Elfan WA Sulfosäure)

Strong organic acid, hence irritating to eyes and skin (Arylan SC Acid)

Causes burns; harmful if absorbed through skin; do not get in eyes, on skin or clothing; avoid breathing mist or vapor (DDBSA 99-b)

Can cause chemical burns if spilled on skin; wear protective goggles or face mask (Polystep A-17)

STORAGE/HANDLING:

Store preferably at room temp.; etching and corrosive to metals (Elfan WA Sulfosäure)

Type 316 stainless steel, alloy, or clad; glass-lined steel; lead-lined; fiber glass-reinforced polyester; heresite-lined steel (Tex-Wet 1197)

STD. PKGS.:

210-kg net iron drums (Merpisap AS98)

45-gal drums or road tankers (Manro BA, HA, NA)

55-gal drums, tank trucks, rail cars (Polyfac LAS-97)

55-gal lined steel drums (Carsosulf UL-100 Acid)

55-gal (450 lb net) metal drums, 5000-gal tanktrucks, 8000- and 10,000-gal tankcars (DDBSA 99-b)

55-gal (450 lb net) open-head drums, tankwagons or tankcars (Conco AAS-98S)

55-gal (480 lb net) steel drums, bulk tankcar, tanktruck (Calsoft LAS-99)

475 lb net full open-head, lined steel drums (Polystep A-17)

20-ton containers (Elfan WA Sulfosäure)

Ethylene glycol distearate

SYNONYMS:
Glycol distearate (CTFA)
Octadecanoic acid, 1,2-ethanediyl ester

EMPIRICAL FORMULA:
$C_{38}H_{74}O_4$

STRUCTURE:

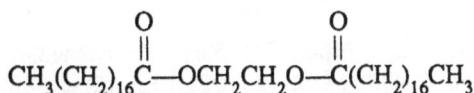

$$CH_3(CH_2)_{16}\overset{\displaystyle O}{\overset{\|}{C}}-OCH_2CH_2O-\overset{\displaystyle O}{\overset{\|}{C}}(CH_2)_{16}CH_3$$

CAS No.
627-83-8

TRADENAME EQUIVALENTS:
Cutina AGS [Henkel]
Cyclochem EGDS [Cyclo]
Drewmulse EGDS [Drew Produtos Quimicos]
Elfan L310 [Akzo Chemie]
Emerest 2355 [Emery]
Kemester EGDS [Humko Sheffield]
Kessco Ethylene Glycol Distearate [Armak]
Lexemul EGDS [Inolex]
Lipo EGDS [Lipo]
Mapeg EGDS [Mazer]
Pegosperse 50-DS [Glyco]
Radiasurf 7269 [Oleofina]
Rewopal PG280 [Rewo Chemische Werke]

CATEGORY:
Emulsifier, dispersant, thickener, opacifier, pearlescent, intermediate, solvent, stabilizer, emollient, lubricant, conditioner, softener, moisturizer, defoamer, foam modifier, solubilizer, wetting agent, plasticizer, antistat, rust inhibitor, spreading agent

APPLICATIONS:
Bath products: bubble bath (Elfan L310); bath oils (Lipo EGDS)
Cosmetic industry preparations: (Emerest 2355; Mapeg EGDS; Pegosperse 50-DS; Radiasurf 7269); conditioners (Lexemul EGDS; Pegosperse 50-DS)); creams and lotions (Cyclochem EGDS; Lipo EGDS); hair rinses (Lexemul EGDS); hair sets and sprays (Pegosperse 50-DS); makeup (Pegosperse 50-DS); shampoos (Cutina AGS; Cyclochem EGDS; Drewmulse EGDS; Elfan L310; Kessco Ethylene Glycol Distearate; Lexemul EGDS; Pegosperse 50-DS); toiletries (Pegosperse 50-DS)

Ethylene glycol distearate (cont'd.)

Farm products: insecticides/pesticides (Radiasurf 7269)

Household detergents: (Pegosperse 50-DS); liquid detergents (Emerest 2355)

Industrial applications: dyes and pigments (Drewmulse EGDS; Pegosperse 50-DS; Radiasurf 7269); lubricating/cutting oils (Kemester EGDS; Radiasurf 7269); metalworking (Kemester EGDS; Mapeg EGDS); paint mfg. (Radiasurf 7269); plastics (Radiasurf 7269); polishes and waxes (Radiasurf 7269); printing inks (Radiasurf 7269); textile/leather processing (Radiasurf 7269)

Industrial cleaners: maintenance products (Radiasurf 7269)

Pharmaceutical applications: (Mapeg EGDS; Pegosperse 50-DS; Radiasurf 7269)

PROPERTIES:

Form:

Solid (Cutina AGS; Elfan L310; Kemester EGDS; Mapeg EGDS; Radiasurf 7269)

Beads (Emerest 2355; Lipo EGDS)

Flake (Cyclochem EGDS; Drewmulse EGDS; Elfan L310; Kessco Ethylene Glycol Distearate; Lexemul EGDS; Lipo EGDS; Mapeg EGDS; Pegosperse 50-DS; Rewopal PG280)

Color:

White (Cyclochem EGDS; Pegosperse 50-DS; Radiasurf 7269)

White to off-white (Lipo EGDS)

White to cream (Mapeg EGDS)

Light yellow (Elfan L310)

Gardner 1 (Emerest 2355)

Gardner 2 max. (Kemester EGDS)

Composition:

48% active (Drewmulse EGDS)

100% active (Emerest 2355)

100% conc. (Kemester EGDS; Lexemul EGDS; Mapeg EGDS; Rewopal PG280)

Solubility:

Sol. cloudy in benzene (@ 10%, 75 C) (Radiasurf 7269)

Sol. hot in ethanol (Pegosperse 50-DS)

Sol. cloudy in hexane (@ 10%, 75 C) (Radiasurf 7269)

Sol. in isopropanol (Emerest 2355; Mapeg EGDS); sol. cloudy in isopropanol (@ 10%, 75 C) (Radiasurf 7269)

Sol. in min. oil (Emerest 2355; Mapeg EGDS); sol. hot (Pegosperse 50-DS); sol. cloudy (@ 10%, 75 C) (Radiasurf 7269)

Sol. in soybean oil (Mapeg EGDS)

Sol. in toluol (Emerest 2355; Mapeg EGDS)

Sol. in trichlorethylene (@ 10%, 75 C) (Radiasurf 7269)

Sol. hot in veg. oils (Pegosperse 50-DS); sol. cloudy (@ 10%, 75 C) (Radiasurf 7269)

Disp. in water (Elfan L310); insol. (Pegosperse 50-DS)

Ionic Nature:

Nonionic (Emerest 2355; Kessco Ethylene Glycol Distearate; Lipo EGDS; Mapeg EGDS; Pegosperse 50-DS; Rewopal PG280)

Anionic (Elfan L310)
Sp.gr.:
0.96 (Elfan L310)
F.P.:
50 C (Emerest 2355)
M.P.:
58–63 C (Pegosperse 50-DS)
60–63 C (Kessco Ethylene Glycol Distearate)
61 C (Cyclochem EGDS)
63 C (Mapeg EGDS)
64–66 C (Elfan L310)
≈ 65 C (Radiasurf 7269)
Flash Pt.:
390 F (COC) (Kessco Ethylene Glycol Distearate)
HLB:
1.0 ± 1 (Lipo EGDS; Pegosperse 50-DS)
1.4 (Mapeg EGDS)
1.5 (Lexemul EGDS; Radiasurf 7269)
2.1 (Emerest 2355)
Acid No.:
3 max. (Pegosperse 50-DS; Radiasurf 7269)
6 max. (Elfan L310; Kemester EGDS; Mapeg EGDS)
7 max. (Lipo EGDS)
8 (Emerest 2355)
15 max. (Kessco Ethylene Glycol Distearate)
20 max. (Cyclochem EGDS)
Iodine No.:
0.5 max. (Kessco Ethylene Glycol Distearate)
1 max. (Mapeg EGDS; Radiasurf 7269)
Saponification No.:
185 (Emerest 2355)
185–200 (Pegosperse 50-DS)
190–199 (Mapeg EGDS)
190–200 (Kemester EGDS; Radiasurf 7269)
190–205 (Lipo EGDS)
195 (Cyclochem EGDS)
200 max. (Elfan L310)
Hydroxyl No.:
ca. 30 (Elfan L310)
33–43 (Kemester EGDS)
STD. PKGS.:
25 kg paper bags (Elfan L310)
25 kg net multiply paper bags or bulk (Radiasurf 7269)

Glyceryl monolaurate

SYNONYMS:
 Dodecanoic acid, 2,3-dihydroxypropyl ester
 Dodecanoic acid, monoester with 1,2,3-propanetriol
 Glycerol monolaurate
 Glyceryl laurate (CTFA)
EMPIRICAL FORMULA:
 $C_{15}H_{30}O_4$
STRUCTURE:

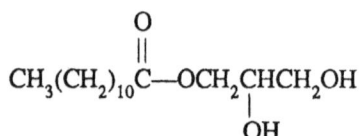

CAS No.:
 142-18-7; 27215-38-9
TRADENAME EQUIVALENTS:
 Aldo MLD [Glyco]
 Alkamuls GML, GML-45 [Alkaril]
 Grindtek ML90 [Grinsted]
 Hodag GML [Hodag]
 Kessco Glycerol Monolaurate [Armak]
 Lamesoft LMG [Chem. Fabrik Grunau]
 Monomuls 90-L12 [Chem. Fabrik Grunau]
 Tegin 4480 [Th. Goldschmidt AG]

CATEGORY:
 Emulsifier, opacifier, stabilizer, thickener, dispersant, antifoam agent, emollient, lubricant, antistat, antifogging agent

APPLICATIONS:
 Bath products: bubble bath (Lamesoft LMG; Monomuls 90-L12)
 Cosmetic industry preparations: (Hodag GML; Kessco Glycerol Monolaurate; Tegin 4480); creams and lotions (Monomuls 90-L12); shampoos (Lamesoft LMG; Monomuls 90-L12)
 Food applications: (Hodag GML); food emulsifying (Alkamuls GML)
 Industrial applications: mold release (Alkamuls GML-45); textile/leather processing (Alkamuls GML)
 Pharmaceutical applications: (Hodag GML; Tegin 4480)

Glyceryl monolaurate *(cont'd.)*

PROPERTIES:
Form:
Liquid (Lamesoft LMG)
Liquid to paste (Alkamuls GML-45)
Paste (Alkamuls GML; Hodag GML)
Solid (Kessco Glycerol Monolaurate; Monomuls 90-L12)
Semisolid (Tegin 4480)
Soft solid (Aldo MLD)
Block (Grindtek ML90)
Color:
White (Kessco Glycerol Monolaurate)
Whitish (Grindtek ML90)
Cream (Aldo MLD)
Composition:
25% conc. (Lamesoft LMG)
40% monoglyceride min. (Alkamuls GML-45)
100% active (Alkamuls GML)
100% conc. (Aldo MLD; Hodag GML; Monomuls 90-L12; Tegin 4480)
Solubility:
Disp. in aromatic solvent @ 10% (Alkamuls GML-45)
Sol. in ethanol (Aldo MLD; Grindtek ML90)
Sol. in ethyl acetate (Aldo MLD)
Sol. in min. oils (Aldo MLD); disp. @ 10% (Alkamuls GML-45)
Disp. in min. spirits @ 10% (Alkamuls GML-45)
Sol. in naphtha (Aldo MLD)
Disp. in perchloroethylene @ 10% (Alkamuls GML-45)
Sol. warm in propylene glycol (Grindtek ML90)
Sol. warm in toluene (Grindtek ML90)
Sol. in toluol (Aldo MLD)
Sol. in veg. oils (Aldo MLD)
Disp. in water (Aldo MLD); insol. @ 10% (Alkamuls GML-45)
Partly sol. in white spirit (Grindtek ML90)

Ionic Nature: Nonionic

Sp.gr.:
0.97 (Aldo MLD)

Density:
0.97 g/ml (Alkamuls GML-45)

M.P.:
21–26 C (Aldo MLD)
53.9 C (Kessco Glycerol Monolaurate)
Flash Pt.:
425 F (COC) (Kessco Glycerol Monolaurate)

Glyceryl monolaurate (cont'd.)

HLB:
3.0 (Alkamuls GML, GML-45; Hodag GML)
4.0 (Tegin 4480)
5.3 (Grindtek ML90)
6.8 (Aldo MLD)
Acid No.:
< 5 (Aldo MLD)
5 max. (Kessco Glycerol Monolaurate)
Iodine No.:
1 max. (Kessco Glycerol Monolaurate)
< 9 (Aldo MLD)
Saponification No.:
185–195 (Aldo MLD)
pH:
7.5–8.5 (5% aq.) (Aldo MLD)

Glyceryl monooleate

SYNONYMS:
Glycerol monooleate
Glyceryl oleate (CTFA)
Monoolein
9-Octadecenoic acid, 2,3-dihydroxypropyl ester
9-Octadecenoic acid, monoester with 1,2,3-propanetriol
EMPIRICAL FORMULA:
$C_{21}H_{40}O_4$
STRUCTURE:

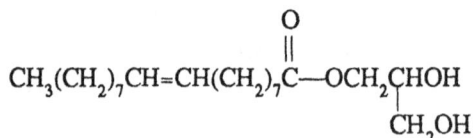

CAS No.:
111-03-5; 25496-72-4
TRADENAME EQUIVALENTS:
Aldo HMO, MO, MO Tech. [Glyco]
Aldo MOD [Glyco] (SE grade)
Alkamuls GMO [Alkaril]
Alkamuls GMO-45, GMO-45LG [Alkaril]
Alkamuls GMO-55LG [Alkaril]
Atmos 300 [ICI]

TRADENAME EQUIVALENTS *(cont'd.)*
Capmul GMO [Stokely-Van Camp]
CPH-31-N, -205-NX, -362-N [C.P. Hall]
Cyclochem GMO [Cyclo]
Drewmulse 85, GMO [PVO]
Dur-Em 114, 204, GMO [Durkee]
Emerest 2421 [Emery]
Grocor 2000 [A. Gross]
Hodag GMO, GMO-D [Hodag]
Kessco Glycerol Monooleate [Armak]
Mazol 300, GMO [Mazer]
Monomuls 90-018 [Grunau]
Radiamuls 152 [Oleofina]
Radiamuls MG 2152 [Oleofina]
Tegin O Special [Goldschmidt]
Witco 942 [Witco/Organics]

CATEGORY:
Emulsifier, wetting agent, mold release agent, rust preventive, lubricant, processing aid, antistat, antifogging agent, emollient, opacifier, thickener, dispersant, stabilizer, penetrant, antifoam agent, antiblocking agent

APPLICATIONS:
Cosmetic industry preparations: (Aldo MO, MOD; Cyclochem GMO; Drewmulse GMO; Dur-Em 114, GMO; Hodag GMO, GMO-D; Kessco Glycerol Monooleate; Monomuls 90-018); conditioners (Drewmulse GMO); cosmetic base (Drewmulse GMO); creams and lotions (Alkamuls GMO; Drewmulse GMO; Dur-Em 114, GMO); hair rinses (Drewmulse GMO; Dur-Em 114); makeup (Dur-Em 114); shampoos (Alkamuls GMO); shaving preparations (Cyclochem GMO; Dur-Em 114, GMO); sunscreen products (Dur-Em 114)
Farm products: agricultural oils (Alkamuls GMO); animal feed (Radiamuls MG2152); insecticides/pesticides (Emerest 2421; Dur-Em GMO)
Food applications: (Capmul GMO; Hodag GMO, GMO-D; Mazol 300, GMO; Monomuls 90-018); food emulsifying (Aldo MO, MOD; Atmos 300; Drewmulse 85; Dur-Em 204; Radiamuls 152, MG2152)
Industrial applications: industrial processing (Aldo MO, MOD; Alkamuls GMO-55LG); lubricating/cutting oils (Alkamuls GMO-45, GMO-45LG; CPH-205-NX; CPH-362-N; Emerest 2421; Grocor 2000); metalworking (CPH-205-NX; CPH-362-N); paint mfg. (Dur-Em GMO; Grocor 2000; Witco 942); petroleum industry (Emerest 2421); plastics (Alkamuls GMO-55LG; Grocor 2000); textile/leather processing (Alkamuls GMO, GMO-45, GMO-45LG; Emerest 2421; Grocor 2000)
Industrial cleaners: drycleaning compositions (Dur-Em GMO)
Pharmaceutical applications: (Drewmulse GMO; Hodag GMO, GMO-D; Monomuls 90-018)

Glyceryl monooleate *(cont'd.)*

PROPERTIES:
Form:

Liquid (Aldo MOD; Alkamuls GMO; Capmul GMO; CPH-31-N; CPH-205-NX; CPH-362-N; Cyclochem GMO; Drewmulse GMO; Emerest 2421; Grocor 2000; Hodag GMO, GMO-D; Kessco Glycerol Monooleate; Mazol 300, GMO; Radiamuls 152, MG2152)

Clear liquid (Atmos 300)

Semiliquid (Aldo HMO; Drewmulse 85)

Liquid to paste (Alkamuls GMO-45, GMO-45LG, GMO-55LG)

Paste (Radiamuls MG2152; Tegin O Special)

Soft solid (Aldo MO, MO Tech.)

Solid (Kessco Glycerol Monooleate; Monomuls 90-018)

Color:

Amber (Aldo MO Tech.; Alkamuls GMO-45, GMO-45LG, GMO-55LG; Cyclochem GMO)

Light/pale yellow (Alkamuls GMO; Drewmulse 85; Grocor 2000)

Yellow (Aldo HMO, MO, MOD; Atmos 300; Kessco Glycerol Monooleate; Mazol 300, GMO)

Gardner 6 (Drewmulse GMO)

Gardner 11 (Emerest 2421)

Odor:

Mild (Grocor 2000)

Composition:

40% alpha monoglyceride content (Drewmulse 85; Dur-Em 114)

40% monoglyceride (Alkamuls GMO-45, GMO-45LG; Drewmulse GMO; Radiamuls 152, MG2152)

42% alpha monoglyceride min. (Dur-Em GMO)

50% monoglyceride min. (Alkamuls GMO-55LG)

97.5% active (Grocor 2000)

100% active (Alkamuls GMO; Atmos 300; Emerest 2421)

100% conc. (Aldo MO, MOD, MO Tech.; Capmul GMO; CPH-31-N; CPH-205-NX; CPH-362-N; Hodag GMO, GMO-D; Mazol 300, GMO; Monomuls 90-018; Tegin O Special)

Solubility:

Miscible in certain proportions with acetone (Aldo MO, MO Tech.)

Sol. in aliphatic solvent (@ 10%) (Alkamuls GMO-45, GMO-45LG, GMO-55LG)

Sol. in cottonseed oil (Atmos 300)

Disp. in ethanol (Mazol 300, GMO); miscible in certain proportions (Aldo MO, MO Tech.)

Sol. in ethyl acetate (Aldo MO, MO Tech.)

Sol. in hydrocarbons (Grocor 2000)

Sol. in isopropanol (Atmos 300)

Miscible in certain proportions with methanol (Aldo MO, MO Tech.)

Glyceryl monooleate (cont'd.)

Sol. in min. oils (Aldo MOD; Atmos 300; Emerest 2421); miscible in certain proportions (Aldo MO, MO Tech.); disp. (Alkamuls GMO-45, GMO-45LG, GMO-55LG)

Disp. in min. spirits (Alkamuls GMO-45, GMO-45LG, GMO-55LG)

Sol. in naphtha (Aldo MOD); misc. in certain proportions (Aldo MO, MO Tech.)

Sol. in oils (Grocor 2000)

Sol. in perchloroethylene (@ 10%) (Alkamuls GMO-45, GMO-45LG, GMO-55LG)

Sol. in propylene glycol (Mazol 300); disp. (Mazol GMO)

Sol. in soybean oil (Mazol 300, GMO)

Sol. in toluol (Aldo MOD; Emerest 2421); miscible in certain proportions (Aldo MO, MO Tech.)

Sol. in veg. oils (Aldo MOD); miscible in certain proportions (Aldo MO, MO Tech.)

Disp. in water (Aldo MOD); insol. (Alkamuls GMO-45, GMO-45LG, GMO-55LG)

Ionic Nature:

Nonionic (Aldo MO, MOD, MO Tech.; Alkamuls GMO; Atmos 300; Emerest 2421; Dur-Em 114, GMO; Grocor 2000; Kessco Gycerol Monooleate; Mazol 300, GMO; Radiamuls 152)

Anionic (Tegin O Special)

Sp.gr.:

0.945–0.953 (Kessco Glycerol Monooleate)

0.949 (Grocor 2000)

0.95 (Aldo MO, MOD, MO Tech.)

0.96 (Atmos 300)

Density:

0.93 g/ml (Alkamuls GMO-45LG)

0.96 g/ml (Alkamuls GMO-55LG)

0.98 g/ml (Alkamuls GMO-45)

7.9 lb/gal (Emerest 2421; Kessco Glycerol Monooleate)

Visc.:

91 cs (38 C) (Emerest 2421)

130 cps (Atmos 300)

204 cps (Kessco Glycerol Monooleate)

375 SUS (100 F) (Grocor 2000)

F.P.:

6 C (Emerest 2421)

M.P.:

< 0 C (Aldo MOD)

9 C (Cyclochem GMO)

10 C max. (Radiamuls MG2152)

< 20 C (Aldo HMO)

< 25 C (Aldo MO, MO Tech.; Alkamuls GMO)

43–49 C (Dur-Em 114)

114–121 F (Dur-Em 204)

Glyceryl monooleate (cont'd.)

Flash Pt.:
   235 C (Emerest 2421)
   > 300 F (Atmos 300)
   435 F (COC) (Kessco Glycerol Monooleate)
Fire Pt.:
   > 300 F (Atmos 300)
Cloud Pt.:
   10 C max. (Radiamuls 152)
   5 F (Grocor 2000)
HLB:
   2.4 (Mazol GMO)
   2.7 (Hodag GMO, GMO-D)
   2.8 (Atmos 300; Dur-Em 114, GMO; Radiamuls 152)
   3.0 (Aldo HMO; Alkamuls GMO-45, GMO-45LG, GMO-55LG; CPH-205-NX;
      Mazol 300)
   3.4 (Aldo MO, MO Tech.; Alkamuls GMO; Capmul GMO; Drewmulse 85, GMO)
   3.7 (CPH-31-N)
   3.8 (Kessco Glycerol Monooleate)
   4.7 (CPH-362-N)
   5.0 (Aldo MOD)
Acid No.:
   4 max. (Dur-Em GMO)
   5 max. (Aldo MO, MO Tech.; Kessco Glycerol Monooleate; Mazol 300, GMO)
   6 (Emerest 2421)
   < 8 (Aldo MOD)
   9 max. (Cyclochem GMO)
Iodine No.:
   65–69 (Aldo MOD)
   68–73 (Aldo MO, MO Tech.)
   70 (Dur-Em GMO)
   72–77 (Radiamuls 152)
   73 (Emerest 2421)
   75 max. (Dur-Em 114)
   77 max. (Kessco Glycerol Monooleate)
   90 (Mazol 300, GMO)
Saponification No.:
   141–147 (Aldo MOD)
   145–155 (Aldo HMO)
   150–160 (Mazol 300)
   160–180 (Drewmulse 85)
   165–178 (Radiamuls 152)
   167–177 (Mazol GMO)
   170 (Cyclochem GMO; Emerest 2421)

170–180 (Aldo MO, MO Tech.)
Ref. Index:
1.4689 (Grocor 2000)
pH:
4.5–6.5 (Aldo MO, MO Tech.)
8.5–9.5 (5% aq.) (Aldo MOD)
STD. PKGS.:
SS drums (Grocor 2000)
190 kg net bung drums or bulk (Radiamuls 152, MG2152)

Glyceryl monostearate

SYNONYMS:
2,3-Dihydroxypropyl octadecanoate
Glycerol monostearate
Glyceryl stearate (CTFA)
GMS
Monostearin
Octadecanoic acid, 2,3-dihydroxypropyl ester
Octadecanoic acid, monoester with 1,2,3-propanetriol
EMPIRICAL FORMULA:
$C_{21}H_{42}O_4$
STRUCTURE:

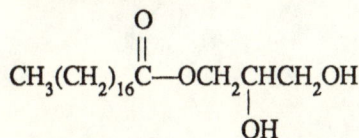

CAS No.:
123-94-4; 11099-07-3; 31566-31-1
TRADENAME EQUIVALENTS:
Aldo HMS, MS, MSLG [Glyco]
Alkamuls GMS [Alkaril]
Atmos 150 [ICI]
Atmul 84, 84K, 124 [ICI]
Cerasynt 945, SD, WM [Van Dyk]
Cithrol GMS A/S ES 0743, [Croda]
Cithrol GMS N/E [Croda]
CPH-53-N [C.P. Hall]
Cutina GMS, MD, MD-A [Henkel/Canada]
Cyclochem GMS [Cyclo]

Glyceryl monostearate *(cont'd.)*

TRADENAME EQUIVALENTS *(cont'd.)*:

Drewmulse TP, V [PVO]
Dur-Em 117 [Durkee]
Emerest 2400, 2401 [Emery]
Empilan GMS NSE 32, GMS NSE 40 [Albright & Wilson/Marchon]
Geleol [Gattefosse]
Graden Glycerol Mono Stearate [Graden]
Grindtek MSP40, MSP40F, MSP52, MSP90 [Grinsted]
Grocor 5500, 6000, 6000E [A. Gross]
Hodag GMS [Hodag]
Imwitor 191, 900K [Dynamit Nobel]
Kessco Glycerol Monostearate 860, DH-1, Pure [Armak]
Lexemul 55G, 503, 515 [Inolex]
Lipo GMS 450 [Lipo]
Mazol GMS [Mazer]
Radiamuls 142, 600, 601, 900 [Oleofina]
Radiamuls MG 2141, MG2142, MG2600, MG2900 [Oleofina]
Rewomul MG [Rewo Chem. Werke]
Schercemol GMS [Scher]
Tegin 90, 90NSE, 515, GRB, M, MAV, NSE [Goldschmidt]
Witconol MST [Witco/Organics]

Self-emulsifying grades:
Ahcovel Base N-15 [ICI United States]
Aldo MSD [Glyco]
Cerasynt Q [Van Dyk]
Cithrol GMS Acid Stable, GMS S/E [Croda]
CPH-250-SE [C.P. Hall]
Cyclochem GMS21, GMS165 [Cyclo]
Drewmulse V-SE [PVO]
Dur-Em 207-E [Durkee]
Emerest 2407 [Emery]
Empilan GMS LSE 32, GMS SE 32, GMS SE 40 [Albright & Wilson/Marchon]
Grindtek MSP 32-6 [Grinsted]
Grocor 6000 SE [A. Gross]
Imwitor 960 [Dynamit Nobel]
Kessco Glycerol Monostearate SE [Armak]
Lexemul 530, 561, AR, AS, T [Inolex]
Lipo GMS 470 [Lipo]
Mazol GMS-D [Mazer]
Radiamuls 141, 341 [Oleofina]
Tegin, Special SE [Goldschmidt]
Witconol CA, RHT [Witco/Organics]

Glyceryl monostearate *(cont'd.)*

CATEGORY:
   Emulsifier, coemulsifier, wetting agent, opacifier, stabilizer, viscosity builder, thickener, lubricant, emollient, plasticizer, antitack agent, suspending agent, dispersant, antistat, antifogging agent, binding agent, flow control agent, bodying agent

APPLICATIONS:
   Cosmetic industry preparations: (Aldo MS; Cerasynt SD; Cithrol GMS A/S ES 0743, GMS N/E, GMS S/E, GMS Acid Stable; Cutina GMS, MD, MD-A; ; Drewmulse TP, V, V-SE; Dur-Em 117, 207-E; Emerest 2400, 2401; Grocor 5500, 6000, 6000SE; Hodag GMS; Imwitor 191, 900K; Kessco Glycerol Monostearate 860, DH-1, Pure, SE; Lexemul 55G, 530, 561, AR; Mazol GMS, GMS-D; Schercemol GMS; Tegin 90NSE, 515, MAV, NSE, Special SE; Witconol CA, MST); bleaches (Cithrol GMS A/S ES 0743; Lexemul 561); conditioners (Drewmulse TP, V, V-SE; Schercemol GMS); cosmetic base (Drewmulse TP, V, V-SE; Imwitor 960); creams and lotions (Alkamuls GMS; Cithrol GMS A/S ES 0743; Cyclochem GMS21, GMS165; Drewmulse TP, V, V-SE; Dur-Em 117; Emerest 2400, 2401, 2407; Lexemul 55G, 530, 561; Lipo GMS 450, GMS 470; Rewomul MG; Schercemol GMS); hair rinses (Cithrol GMS A/S ES 0743; Drewmulse TP, V, V-SE; Lexemul 530; Schercemol GMS); makeup (Imwitor 191, 900K); personal care products (Cerasynt Q; Schercemol GMS); shampoos (Alkamuls GMS; Schercemol GMS); shaving preparations (Lexemul 530); toiletries (Cyclochem GMS21; Lexemul AR; Mazol GMS, GMS-D)
   Farm products: agricultural oils/sprays (Alkamuls GMS)
   Food applications: (Hodag GMS); food emulsifying (Aldo MS; Atmos 150; Atmul 84, 124; Dur-Em 207-E; Empilan GMS LSE 32, GMS NSE 32, GMS NSE 40; Grocor 6000E; Mazol GMS, GMS-D; Radiamuls 141, 142, 341, 600, 601, 900; Radiamuls MG2142, MG2600, MG2900); kosher grade (Atmul 84K)
   Industrial applications: (Cithrol GMS N/E, GMS S/E, GMS Acid Stable; Witconol CA, MST, RHT); aerosols (Witconol CA, MST); dyes and pigments (Dur-Em 117); industrial processing (Aldo MS; Grindtek MSP32-6, MSP40, MSP40F, MSP52, MSP90); lubricating/cutting oils (Emerest 2400, 2401; Mazol GMS, GMS-D); paper mfg. (Grocor 6000); plastics (Dur-Em 117; Grindtek MSP32-6, MSP40, MSP40F, MSP52, MSP90; Grocor 6000SE; Mazol GMS, GMS-D); polymers/ polymerization (Witconol MST); rubber (Grocor 6000, 6000SE); textile/leather processing (Ahcovel Base N-15; Alkamuls GMS; Dur-Em 117; Emerest 2400, 2401; Grocor 5500, 6000, 6000SE; Mazol GMS, GMS-D)
   Pharmaceutical applications: (Cerasynt SD; Cithrol GMS A/S ES 0743, GMS N/E, GMS S/E, GMS Acid Stable; Cutina MD-A; Drewmulse TP, V, V-SE; Dur-Em 207-E; Hodag GMS; Mazol GMS, GMS-D; Tegin 90 NSE, MAV, NSE, Special SE; Witconol MST); antiperspirants/deodorants (Lexemul 561, AR); ointments (Cithrol GMS A/S ES 0743); tablet mfg. (Imwitor 191; Kessco Glycerol Monostearate 860, DH-1, Pure, SE); topical pharmaceuticals (Lexemul 55G, AR)

Glyceryl monostearate *(cont'd.)*

PROPERTIES:
Form:
  Liquid (Radiamuls MG2142, MG2600, MG2900)
  Paste (Witconol RHT)
  Beads (Aldo HMS, MS, MSLG; Atmos 150; Atmul 84; Cutina GMS; Drewmulse V, V-SE; Emerest 2400, 2401, 2407; Grocor 5500, 6000, 6000E, 6000SE; Lipo GMS 450, GMS 470)
  Flake (Atmul 84, 84K, 124; Cerasynt 945, Q, SD, WM; CPH-53-N, CPH-250-SE; Cyclochem GMS, GMS21, GMS165; Drewmulse TP; Graden Glycerol Mono Stearate; Grocor 5500; Imwitor 900K; Kessco Glycerol Monostearate 860, DH-1, Pure, SE; Lexemul 55G, 503, 505, 530, 561, AR, AS, T; Lipo GMS 450, GMS 470; Mazol GMS, GMS-D; Radiamuls 141, 142, 341, 600, 601, 900; Radiamuls MG2142, MG2600, MG2900; Rewomul MG; Schercemol GMS; Tegin 90)
  Powder (Dur-Em 207-E; Grindtek MSP32-6, MSP40, MSP40F, MSP52, MSP90; Grocor 5500; Imwitor 191, 900K, 960; Radiamuls 141, 142, 341, 600, 601, 900; Radiamuls MG2142, MG2600, MG2900; Tegin 90 NSE, 515, GRB, MAV, NSE, Special SE)
  Wax-like powder (Empilan GMS LSE 32, GMS NSE 32, GMS NSE 40, GMS SE 32, GMS SE 40)
  Granular (Cutina MD-A)
  Solid (Ahcovel Base N-15; Alkamuls GMS; Cithrol GMS N/E, GMS S/E, GMS Acid Stable; Cutina MD; Geleol; Grocor 5500, 6000, 6000E, 6000SE; Hodag GMS; Mazol GMS, GMS-D; Tegin M; Witconol MST)
  Waxy solid (Imwitor 900K)
  Wax (Witconol CA)
Color:
  White (Aldo HMS, MS, MSLG; Cyclochem GMS; Dur-Em 207-E; Empilan GMS LSE 32, GMS NSE 32, GMS NSE 40, GMS SE 32, GMS SE 40; Grocor 5500, 6000, 6000E, 6000SE; Kessco Glycerol Monostearate 860, DH-1, Pure; Lexemul 55G; Lipo GMS 450, GMS 470; Mazol GMS, GMS-D)
  Whitish (Grindtek MSP32-6, MSP40, MSP40F, MSP52, MSP90)
  Ivory white (Atmos 150; Atmul 84, 84K, 124)
  Off-white (Cyclochem GMS21, GMS165)
  White to cream (Imwitor 900K; Kessco Glycerol Monostearate SE; Lexemul 530, 561, AR, AS; Schercemol GMS)
  Gardner 1 (Emerest 2400)
  Gardner 2 (Emerest 2401)
  Gardner 3 (Cithrol GMS A/S ES 0743; Emerest 2407)
  Gardner 4 (Drewmulse TP); 4 max. (Imwitor 191, 960)
  Gardner 6 (Drewmulse V, V-SE)
  Lovibond 40Y/4R max. (Graden Glycerol Mono Stearate)
Odor:
  Low (Lexemul 530)

Bland (Atmos 150; Atmul 84, 84K, 124; Grocor 6000SE)
Mild (Grocor 5500, 6000, 6000E)
Mild fatty (Lexemul 55G, AR, AS)
Mild ethoxylate (Lexemul 561)
Slightly fatty (Imwitor 900K)

Taste:
Bland (Atmos 150; Atmul 84, 84K, 124)

Composition:
30.0% mono (Drewmulse V-SE)
30–40% 1-monoglycerides (Imwitor 960)
32.5% monoglyceride min. (Empilan GMS LSE 32, GMS NSE 32, GMS SE 32)
38% monoglyceride min. (Empilan GMS NSE 40, GMS SE 40)
40% monoglyceride (Drewmulse TP, V; Radiamuls 141, 142, 341; Radiamuls MG2142)
40% monoglyceride content min. (Dur-Em 117; Schercemol GMS)
42–48% monoglycerides (Imwitor 900K)
60% monoglyceride min. (Radiamuls 600, 601; Radiamuls MG2600)
90% 1-monoglycerides (Imwitor 191)
90% monoglyceride min. (Radiamuls 900; Radiamuls MG2900)
100% active (Dur-Em 207-E; Emerest 2400, 2401, 2407; Grocor 5500; Lipo GMS 450, GMS 470)
100 % conc. (Ahcovel Base N-15; Aldo MS; Cerasynt 945, Q, SD, WM; Cithrol GMS N/E, GMS S/E, GMS Acid Stable; CPH-53-N, CPH-250-SE; Cutina GMS, MD-A; Geleol; Grocor 6000E, 6000SE; Hodag GMS; Lexemul 55G, 503, 515, 530, 561, AR, AS, T; Rewomul MG; Tegin 90, 90NSE, 515, GRB, M, MAV, NSE, Special SE; Witconol CA, MST, RHT)

Solubility:
Sol. in alcohols (Schercemol GMS); sol. above its melting pt. (Atmul 84, 84K)
Sol. in aromatic hydrocarbons (Schercemol GMS)
Sol. in benzene (@ 10%, 75 C) (Radiamuls 142, 600, 601)
Sol. in carbon tetrachloride (@ 60 C) (Lexemul 530)
Sol. in chlorinated hydrocarbons (Schercemol GMS)
Sol. in chloroform (Imwitor 900K)
Sol. in esters (Schercemol GMS)
Sol. hot in ethanol (Aldo MS); sol. warm (Grindtek MSP90); (@ 60 C) (Imwitor 900K)
Sol. in ether (Imwitor 900K)
Sol. in ethyl acetate (@ 60 C) (Lexemul 530)
Sol. in fats (Imwitor 900K)
Disp. in glycerol (Schercemol GMS)
Disp. in glycols (Schercemol GMS)
Sol. in hexane (@ 10%, 75 C) (Radiamuls 600, 601); sol. cloudy (@ 10%, 75 C) (Radiamuls 141)
Sol. in isopropanol (Emerest 2400, 2401, 2407; Grocor 5500, 6000, 6000E, 6000SE);

Glyceryl monostearate *(cont'd.)*

sol. above its melting pt. (Atmos 150; Atmul 124); (@ 60 C) (Lexemul 530); (@ 10%, 75 C) (Radiamuls 141, 142, 600, 601)

Sol. hot in methanol (Aldo MS)

Sol. in min. oil (Schercemol GMS); sol. hot (Aldo MS; Emerest 2400, 2401, 2407); sol. above its melting pt. (Atmos 150; Atmul 84, 84K, 124); (@ 60 C) (Lexemul 530); (@ 10%, 75 C) (Radiamuls 141, 142, 600, 601)

Sol. hot in naphtha (Aldo MS)

Sol. in oils (Imwitor 900K)

Sol. in most organic solvents (Schercemol GMS)

Disp. in polyols (Schercemol GMS)

Sol. warm in toluene (Grindtek MSP40, MSP40F, MSP52, MSP90); disp. warm (Grindtek MSP32-6)

Sol. hot in toluol (Aldo MS; Emerest 2400, 2401, 2407)

Sol. in trichlorethylene (@ 10%, 75 C) (Radiamuls 141, 142, 600, 601)

Sol. hot in veg. oil (Aldo MS); sol. above its melting pt. (Atmos 150; Atmul 84, 84K, 124); (@ 10%, 75 C) (Radiamuls 141, 142, 600, 601)

Disp. in water (Grindtek MSP32-6; Grocor 6000SE); disp. hot (Aldo MS); disp. @ 60 C (Lexemul 530, 561, AR, AS); insol. (Graden Glycerol Mono Stearate; Schercemol GMS)

Sol. in waxes (Imwitor 900K)

Disp. warm in white spirit (Grindtek MSP32-6, MSP40, MSP40F, MSP52)

Ionic Nature:

Nonionic (Aldo MS, MSD; Alkamuls GMS; Dur-Em 117; Emerest 2400, 2401; Empilan GMS LSE 32, GMS NSE 32, GMS NSE 40, GMS SE 32, GMS SE 40; Grocor 5500, 6000, 6000E; Kessco Glycerol Monostearate 860, DH-1, Pure, SE; Lexemul 55G, 561, Lipo GMS450, GMS470; Radiamuls 141, 142, 341, 600, 601, 900; Rewomul MG; Schercemol GMS; Witconol CA, MST, RHT)

Anionic (Cyclochem GMS21; Grocor 6000SE; Lexemul' AS); modified anionic (Emerest 2407; Lexemul 530)

Cationic (Lexemul AR)

M.W.:

342 (theoret.) (Schercemol GMS)

358 (of monoglyceride) (Emerest 2407)

460 avg. (Radiamuls 600, 601)

509 avg. (Radiamuls 141)

515 avg. (Radiamuls 142)

Sp.gr.:

0.885 (98.9 C) (Radiamuls 142)

0.893 (98.9 C) (Radiamuls 600)

0.900 (98.9 C) (Radiamuls 601)

0.902 (98.9 C) (Radiamuls 141)

0.91 (Grocor 6000, 6000E)

0.92 (Grocor 6000SE)

0.97 (Aldo MS; Grocor 5500)
Visc.:
   10.7 cps (98.9 C) (Radiamuls 601)
   12.0 cps (98.9 C) (Radiamuls 142)
   13.4 cps (98.9 C) (Radiamuls 600)
   20.1 cps (98.9 C) (Radiamuls 141)
M.P.:
   52 C (Cithrol GMS A/S ES 0743; Cyclochem GMS21; Radiamuls 601)
   53–57 C (Graden Glycerol Mono Stearate)
   53–58 C (Empilan GMS NSE 40)
   54–58 C (Kessco Glycerol Monostearate SE)
   54–59 C (Empilan GMS NSE 32)
   55 C (Cyclochem GMS165)
   55–58 C (Kessco Glycerol Monostearate DH-1)
   55–59 C (Lexemul 55G)
   55–60 C (Empilan GMS LSE 32, GMS SE 32; Schercemol GMS)
   56 C (Cyclochem GMS; Lexemul 530)
   56–59 C (Grocor 5500)
   56–60 C (Imwitor 900K; Radiamuls 341; Radiamuls MG2142)
   56–61 C (Imwitor 191, 960; Empilan GMS SE 40)
   56.5–58.5 (Kessco Glycerol Monostearate Pure)
   57–61 C (Aldo MS)
   57.5 C (Emerest 2400)
   58 C (Emerest 2401, 2407; Grocor 6000, 6000E, 6000SE; Radiamuls 141, 142)
   58–60 C (Radiamuls MG2600)
   58.5–61.5 (Kessco Glycerol Monostearate 860)
   58–62 C (Aldo MSLG)
   59 C (Radiamuls 600)
   60 C (Lexemul AR, AS)
   60–70 C (Radiamuls MG2900)
   61–68 C (Aldo HMS)
   62–65 C (Dur-Em 117)
   70 C (Radiamuls 900)
   140 F (Atmos 150; Atmul 84, 124)
   140–146 F (Dur-Em 207-E)
   144 F (Atmul 84K)
Solidification Pt.:
   56–61 C (Imwitor 960)
   63–68 C (Imwitor 191)
Flash Pt.:
   198 C (COC) (Radiamuls 141)
   217 C (COC) (Radiamuls 601)
   223 C (COC) (Radiamuls 600)

Glyceryl monostearate *(cont'd.)*

233 C (COC) (Radiamuls 142)
> 300 F (Atmos 150; Atmul 84, 84K, 124)
410 F (COC) (Kessco Glycerol Monostearate DH-1, Pure)
450 F (COC) (Kessco Glycerol Monostearate 860)
Fire Pt.:
> 300 F (Atmos 150; Atmul 84, 84K, 124)
HLB:
2.7 (Hodag GMS)
2.8 (Aldo HMS; Atmul 84, 84K; Dur-Em 117; Grindtek MSP40, MSP40F; Mazol
   GMS)
3.0 (Radiamuls 142)
3.2 (Atmos 150; Lexemul 503)
3.3 (Aldo MSLG; Radiamuls 600, 601)
3.4 (Alkamuls GMS)
3.5 (Atmul 124; Radiamuls 141)
3.6 (Aldo MS)
3.6 ± 1 (Lipo GMS 450)
3.7 (CPH-53-N; Grindtek MSP32-6)
3.8 (Drewmulse TP, V; Grindtek MSP52; Imwitor 900K; Kessco Glycerol Mono–
   stearate 860, DH-1, Pure; Lexemul 515; Tegin 515, M, MAV NSE)
4.3 (Grindtek MSP90)
4.5 (Tegin 90, 90NSE)
4.6 (Radiamuls 900)
5.0 (Radiamuls 341)
5.5 (Lexemul T)
5.8 (Aldo MSD)
5.8 ± 1 (Lipo GMS 470)
6.0 (Mazol GMS-D)
8.4 (Drewmulse V-SE)
Acid No.:
2.0 max. (Cyclochem GMS165; Lexemul 55G, 561; Mazol GMS, GMS-D)
2.5 max. (Imwitor 191)
< 3 (Aldo MS)
3.0 (Emerest 2400, 2401)
3.0 max. (Imwitor 900K; Kessco Glycerol Monostearate 860, Pure, SE; Schercemol
   GMS)
< 3.5 (Aldo MSD)
5.0 max. (Cyclochem GMS; Kessco Glycerol Monostearate DH-1; Lipo GMS 450,
   GMS 470)
6.0 max. (Imwitor 960)
14–18 (Lexemul AS)
18.0 max. (Cyclochem GMS21)
20 (Emerest 2407); 20 max. (Lexemul 530)

25–31 (Lexemul AR)
Iodine No.:
0.5 max. (Emerest 2400; Kessco Glycerol Monostearate Pure)
1.0 (Emerest 2401, 2407)
1.0 max. (Graden Glycerol Mono Stearate; Lexemul 55G; Radiamuls 141, 142, 600, 900)
1.5 max. (Lexemul 561)
2.0 max. (Kessco Glycerol Monostearate 860; Mazol GMS, GMS-D; Radiamuls 341; Schercemol GMS)
3.0 max. (Imwitor 191, 900K, 960; Lexemul 530, AR, AS)
< 5.0 (Aldo MS, MSD; Dur-Em 117)
6.0 max. (Kessco Glycerol Monostearate DH-1)
50–55 (Radiamuls 601)
Saponification No.:
90–100 (Lexemul 561)
93 (Cyclochem GMS165)
≈ 96 (Cithrol GMS A/S ES 0743)
138–152 (Lipo GMS 470)
140–150 (Aldo MSD)
142 (Mazol GMS-D)
145–160 (Radiamuls 141)
146–154 (Lexemul 530)
153 (Emerest 2401, 2407)
153–162 (Lexemul AS)
155–170 (Imwitor 191; Radiamuls 900)
155–175 (Imwitor 960)
157 (Cyclochem GMS21)
158–165 (Aldo MS)
160–170 (Lexemul 55G)
160–176 (Imwitor 900K; Schercemol GMS)
164–174 (Radiamuls 341, 601)
165–175 (Aldo HMS, MSLG; Radiamuls 600)
166–174 (Lexemul AR)
167–177 (Graden Glycerol Mono Stearate)
169–176 (Radiamuls 142)
169–182 (Lipo GMS 450)
170 (Cyclochem GMS)
172 (Mazol GMS)
173 (Emerest 2400)
Stability:
Acid-stable (Cyclochem GMS 21, GMS 165; Kessco Glycerol Monostearate SE; Lexemul AR)
Stable in acids, solutions of electrolytes (Cithrol GMS A/S ES 0743)

Glyceryl monostearate *(cont'd.)*

Stable over wide pH range (Schercemol GMS)

pH:

4.3 (3% aq. disp.) (Lexemul 561)

5.5 (3% aq. suspension) (Lexemul 55G)

6.0–8.0 (10% aq.) (Empilan GMS NSE 32, GMS NSE 40)

7.0–9.0 (10% aq.) (Empilan GMS LSE 32, GMS SE 32, GMS SE 40)

7.6–8.6 (3% aq.) (Aldo MS)

9.2–10.2 (Aldo MSD)

STORAGE/HANDLING:

Store in well-closed bags, dry and below 30 C (Imwitor 191, 960)

Avoid storage above 50 C (Lexemul 561)

STD. PKGS.:

Bulk (Radiamuls MG2142, MG2600, MG2900—liquid)

Drums, bulk (Grocor 5500)

Bags or drums (Grocor 6000SE)

Paper bags, drums, or bulk (Grocor 6000, 6000E)

25-kg net bags (Imwitor 191, 960; Radiamuls MG2142, MG2600, MG2900—powder
or flake)

25-kg net cartons (Imwitor 900K)

25-kg net bags or bulk (Radiamuls 141, 142, 341, 600, 601, 900)

47 gal fiber Leverpak (Atmos 150)

50 lb bags (Atmul 124)

50 lb multiwall bags (Graden Glycerol Mono Stearate)

50 lb carton (Dur-Em 207-E)

100 lb net fiber drums (Lexemul 55G, 530, 561, AR, AS)

Isopropylamine dodecylbenzenesulfonate (CTFA)

SYNONYMS:
Benzenesulfonic acid, dodecyl-, compd. with 2-propanamine (1:1)
Dodecylbenzene isopropylamine sulfonate
Dodecylbenzenesulfonic acid, comp. with 2-propanamine (1:1)
Dodecylbenzenesulfonic acid, isopropylamine salt

EMPIRICAL FORMULA:
$C_{18}H_{30}O_3S \cdot C_3H_9N$

STRUCTURE:

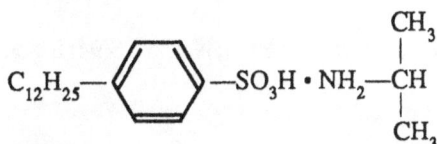

CAS No.:
26264-05-1

TRADENAME EQUIVALENTS:
Alkasurf IPAM [Alkaril]
Calimulse PRS [Pilot]
Cindet GE [Cindet]
Conco AAS Special 3, 3H [Continental]
Manro HCS [Manro]
Nansa YS 94 [Albright & Wilson/Marchon]
Richonate YLA [Richardson]
Siponate 330 [Alcolac]

CATEGORY:
Emulsifier, surfactant, solubilizer, coupling agent, detergency aid, dispersing aid, solvent, penetrant, lubricant

APPLICATIONS:
Automobile cleaners: (Siponate 330)
Cleansers: hand cleanser (Calimulse PRS; Manro HCS; Nansa YS 94)
Degreasers: (Alkasurf IPAM; Calimulse PRS; Nansa YS 94; Siponate 330)
Farm products: agricultural sprays (Calimulse PRS; Siponate 330); herbicides/fungicides (Alkasurf IPAM; Siponate 330); insecticides/pesticides (Alkasurf IPAM; Siponate 330)
Household detergents: (Nansa YS 94; Richonate YLA); dishwashing (Richonate YLA)
Industrial applications: dyes and pigments (Calimulse PRS; Cindet GE; Siponate 330);

Isopropylamine dodecylbenzenesulfonate *(cont'd.)*

lubricating/cutting oils (Calimulse PRS; Siponate 330); paint mfg. (Calimulse PRS; Siponate 330); petroleum industry (Alkasurf IPAM; Calimulse PRS; Conco AAS Special 3; Siponate 330); textile/leather processing (Alkasurf IPAM; Cindet GE)

Industrial cleaners: (Richonate YLA); drycleaning compositions (Alkasurf IPAM; Calimulse PRS; Conco AAS Special 3; Richonate YLA; Siponate 330); institutional cleaners (Conco AAS Special 3); metal processing surfactants (Calimulse PRS; Siponate 330); solvent cleaners (Calimulse PRS; Conco AAS Special 3; Nansa YS 94; Cindet GE)

PROPERTIES:

Form:

Liquid (Siponate 330)

Clear viscous liquid (Alkasurf IPAM; Calimulse PRS; Manro HCS)

Viscous liquid (Cindet GE; Nansa YS 94)

Color:

Pale amber (Calimulse PRS)

Amber (Cindet GE; Conco AAS Special 3, 3H; Manro HCS; Nansa YS 94; Siponate 330)

Pale yellow (Alkasurf IPAM)

Gardner 5 (Richonate YLA)

Odor:

Mild (Manro HCS)

Typical (Conco AAS Special 3, 3H)

Composition:

89.5% active min. (Manro HCS)

90% active (Siponate 330)

90% amine sulfonate (Calimulse PRS)

94% active (Cindet GE; Nansa YS 94)

95% active (Alkasurf IPAM; Richonate YLA)

97% active min. (Conco AAS Special 3, 3H)

Solubility:

Sol. in alcohol (Conco AAS Special 3)

Sol. in aromatic solvents (Conco AAS Special 3)

Sol. in chlorinated solvents (Conco AAS Special 3)

Sol. in kerosene (Conco AAS Special 3)

Sol. in min. spirits (Conco AAS Special 3)

Sol. in oils (Alkasurf IPAM)

Sol. in pine oil (Conco AAS Special 3)

Sol. in mamy solvents (Alkasurf IPAM; Cindet GE)

Sol. in low concs. in water (Cindet GE); insol. in water (Alkasurf IPAM)

Ionic Nature:

Anionic

Sp.gr.:

1.03 (Siponate 330); @ 20 C (Manro HCS)

Isopropylamine dodecylbenzenesulfonate (cont'd.)

1.06 (Alkasurf IPAM)

Density:
8.4 lb/gal (Conco AAS Special 3)
8.5 lb/gal (Calimulse PRS)
8.6 lb/gal (Richonate YLA)

Visc.:
3000 cps (20 C) (Manro HCS)
6500 cps (Siponate 330)
≈ 10,000 cs (20 C) (Nansa YS 94)

Cloud Pt.:
< 0 C (quick cool) (Nansa YS 94)

HLB:
11.7 (Siponate 330)

pH:
3–6 (5% sol'n.) (Siponate 330)
4.8 (Calimulse PRS)
5–6 (2% aq.) (Manro HCS)
5.0–7.0 (Richonate YLA)
5.0–8.0 (2% aq.) (Nansa YS 94)
6.1–7.5 (10% sol'n.) (Conco AAS Special 3)
7.2–7.8 (1% sol'n.) (Cindet GE)

Biodegradable: (Manro HCS)

STD. PKGS.:
45-gal drums or road tankers (Manro HCS)
55-gal (450 lb net) lined steel drums (Calimulse PRS)

Lauryl betaine (CTFA)

SYNONYMS:

1-Dodecanaminium N-(carboxymethyl)-N,N-dimethyl-, hydroxide, inner salt

Lauryl dimethyl glycine

EMPIRICAL FORMULA:

$C_{16}H_{33}NO_2$

STRUCTURE:

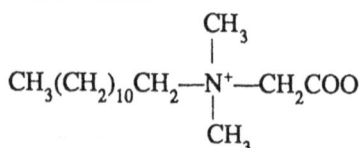

CAS No.:

683-10-3

TRADENAME EQUIVALENTS:

Empigen BB [Albright & Wilson/Marchon]

Lorapon AM-DML [Dutton & Reinisch]

Mirataine LDMB [Miranol]

Rewoteric AM-DML [Rewo Chemische Werke]

Swanol AM-301 [Nikko]

CATEGORY:

Detergent, foam booster, foam stabilizer, wetting agent, thickener, antistat, conditioner, dispersant

APPLICATIONS:

Bath products: bubble bath (Rewoteric AM-DML)

Cosmetic industry preparations: personal care products (Mirataine LDMB); shampoos (Empigen BB; Lorapon AM-DML; Mirataine LDMB; Rewoteric AM-DML); toiletries (Empigen BB)

Household detergents: hard surface cleaner (Lorapon AM-DML; Rewoteric AM-DML)

Industrial cleaners: (Lorapon AM-DML; Rewoteric AM-DML)

PROPERTIES:

Form:

Liquid (Empigen BB; Lorapon AM-DML; Mirataine LDMB; Rewoteric AM-DML; Swanol AM-301)

Color:

Almost water-white (Empigen BB)

Composition:
  30% conc. (Empigen BB; Mirataine LDMB)
  35% conc. (Swanol AM-301)
  40% active (Lorapon AM-DML)
  40% conc. (Rewoteric AM-DML)
Ionic Nature:
  Amphoteric (Empigen BB; Mirataine LDMB; Swanol AM-301)
Stability:
  Excellent acid and alkali stability (Empigen BB)

Lauryl dimethyl amine

SYNONYMS:
  N,N-Dimethyl-1-dodecanamine
  Dimethyl lauramine (CTFA)
  Dimethyl lauryl amine
  1-Dodecanamine, N,N-dimethyl-
  Dodecyl dimethyl amine
EMPIRICAL FORMULA:
  $C_{14}H_{31}N$
STRUCTURE:

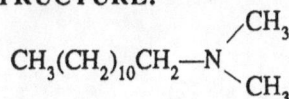

CAS No.:
  112-18-5
TRADENAME EQUIVALENTS:
  Adma 2 [Ethyl Corp.]
  Armeen DM12D [Akzo Chemie]
  Empigen AB [Albright & Wilson/Marchon]
  Nissan Tertiary Amine BB [Nippon Oil & Fats]
  Onamine 12 [Onyx]
Distilled:
  Lilamin 312D [Lilachim S.A.]
CATEGORY:
  Detergent, intermediate, corrosion inhibitor, emulsifier, catalyst, curing agent, germi-
  cide, dispersant, extraction reagent
APPLICATIONS:
  Household detergents: surfactant intermediate (Armeen DM12D; Nissan Tertiary
    Amine BB)
  Industrial applications: dyes and pigments (Lilamin 312D); industrial processing

Lauryl dimethyl amine *(cont'd.)*

(Empigen AB); lubricating/cutting oils (Lilamin 312D; Nissan Tertiary Amine BB; Onamine 12); ore flotation (Empigen AB); petroleum industry (Lilamin 312D); plastics (Empigen AB; Nissan Tertiary Amine BB); polymers/polymerization (Lilamin 312D); printing inks (Lilamin 312D); rubber (Lilamin 312D)

PROPERTIES:

Form:

Liquid (Adma 2; Armeen DM12D; Lilamin 312D; Nissan Tertiary Amine BB; Onamine 12); (@ 20 C) (Empigen AB)

Color:

Pale straw (@ 20 C) (Empigen AB)

Light yellow (Nissan Tertiary Amine BB)

Yellow (Armeen DM12D)

Gardner 2 max. (Lilamin 312D)

Odor:

Amine (Armeen DM12D)

Characteristic (Empigen AB)

Composition:

91% tertiary amine min. (Empigen AB)

95% active (Lilamin 312D)

95% tertiary amine min. (Nissan Tertiary Amine BB)

98% active (Armeen DM12D)

100% conc. (Adma 2; Onamine 12)

Solubility:

Insol. in water (Armeen DM12D)

Ionic Nature:

Cationic (Adma 2; Armeen DM12D; Empigen AB; Onamine 12)

M.W.:

$\approx$ 220 (Lilamin 312D)

Sp.gr.:

0.78 (Armeen DM12D); (25/20 C) (Onamine 12)

Density:

0.79 g/cc (20 C) (Empigen AB)

Visc.:

3.30 cs (Armeen DM12D)

F.P.:

–15 C (Armeen DM12D)

B.P.:

80–115 C (3 mm Hg) (Armeen DM12D)

Flash Pt.:

122 C (COC) (Armeen DM12D)

> 200 F (Onamine 12)

Iodine No.:

1 max. (Lilamin 312D; Nissan Tertiary Amine BB)

3 max. (Armeen DM12D)

TOXICITY/HANDLING:

Skin irritant, severe eye irritant (Armeen DM12D)

Caustic—avoid contact with skin (Empigen AB)

STORAGE/HANDLING:

Avoid contact with strong oxidizing agents (Armeen DM12D)

STD. PKGS.:

200 L. bung-type steel drums (Armeen DM12D)

130-kg can, 160-kg drum (Nissan Tertiary Amine BB)

Lauryl pyridinium chloride (CTFA)

SYNONYMS:

1-Dodecylpyridinium chloride

Pyridinium, 1-dodecyl-, chloride

EMPIRICAL FORMULA:

$C_{17}H_{30}N \cdot Cl$

STRUCTURE:

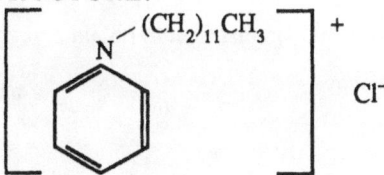

CAS No.:

104-74-5

TRADENAME EQUIVALENTS:

Charlab LPC [Catawba-Charlab]

Dehyquart C, C Crystals [Henkel]

CATEGORY:

Surfactant, wetting agent, antimicrobial, sequestrant, corrosion inhibitor, emulsifier

APPLICATIONS:

Cosmetic industry preparations: conditioners (Dehyquart C Crystals); hair care products (Dehyquart C, C Crystals)

Industrial applications: industrial processing (Dehyquart C)

Industrial cleaners: sanitizers/germicides (Dehyquart C, C Crystals)

Pharmaceutical applications: deodorant (Dehyquart C,)

PROPERTIES:

Form:

Paste (Dehyquart C)

Lauryl pyridinium chloride (cont'd.)

Solid (Charlab LPC)
Powder (Dehyquart C Crystals)
Color:
White (Dehyquart C,)
Light brown (Charlab LPC)
Odor:
Pyridine (Charlab LPC)
Composition:
80–82% active (Dehyquart C)
90–94% active (Dehyquart C Crystals)
97–100% active (Charlab LPC)
Solubility:
Sol. in water (Charlab LPC)
Ionic Nature: Cationic
F.P.:
45 C (Charlab LPC)
TOXICITY/HANDLING:
Skin irritant; solid material or powder may cause a rash with certain individuals (Charlab LPC)
Irritating to skin and eyes in conc. form (Dehyquart C, C Crystals)
STORAGE/HANDLING:
Store in sealed containers below 30 C and protected against frost to obtain 3 yr shelf life (Dehyquart C, C Crystals)
STD. PKGS.:
55 lb net fiber drums (Dehyquart C, C Crystals)

Lauryl trimethyl ammonium chloride

SYNONYMS:
1-Dodecanaminium, N,N,N-trimethyl-, chloride
Dodecyl trimethyl ammonium chloride
Laurtrimonium chloride (CTFA)
N,N,N-trimethyl-1-dodecanaminium chloride
Trimethyl dodecyl ammonium chloride

EMPIRICAL FORMULA:
$C_{15}H_{34}N \cdot Cl$

Lauryl trimethyl ammonium chloride (cont'd.)

STRUCTURE:

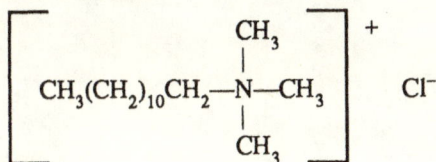

CAS No.:
112-00-5

TRADENAME EQUIVALENTS:
Arquad 12-33, 12-50 [Armak]
Chemquat 12-33, 12-50 [Chemax]
Dehyquart LT [Henkel KGaA]
Nissan Cation BB [Nippon Oil & Fats]

CATEGORY:
Emulsifier, foaming agent, wetting agent, dispersant, corrosion inhibitor, antistat, germicide, coagulant, softener

APPLICATIONS:
Cosmetic industry preparations: conditioners (Arquad 12-33, 12-50); hair rinses (Arquad 12-33, 12-50); shampoos (Dehyquart LT)
Farm products: herbicides (Nissan Cation BB)
Food applications: (Nissan Cation BB)
Industrial applications: dyes and pigments (Nissan Cation BB); latex foam (Arquad 12-33, 12-50; Chemquat 12-33, 12-50); ore flotation (Arquad 12-33, 12-50); paper mfg. (Nissan Cation BB); petroleum industry (Nissan Cation BB); plastics (Nissan Cation BB); polishes and waxes (Arquad 12-33, 12-50); textile/leather processing (Arquad 12-33, 12-50; Nissan Cation BB); water/sewage treatment (Arquad 12-33, 12-50; Nissan Cation BB)
Industrial cleaners: metal processing surfactants (Arquad 12-33, 12-50)

PROPERTIES:
Form:
Liquid (Arquad 12-33, 12-50; Nissan Cation BB)
Color:
Light yellow (Nissan Cation BB)
Gardner 1 (Arquad 12-50)
Gardner 7 (Arquad 12-33, 12-50)
Composition:
30% active min. (Nissan Cation BB)
33% active (Chemquat 12-33)
33% active in aq. isopropanol (Arquad 12-33)
50% active (Chemquat 12-50)
50% active in aq. isopropanol (Arquad 12-50)
Ionic Nature: Cationic

Lauryl trimethyl ammonium chloride (cont'd.)

M.W.:
263 (of active) (Arquad 12-33, 12-50)

Sp.gr.:
0.89 (Arquad 12-50)
0.98 (Arquad 12-33)

F.P.:
5 F (Arquad 12-33)
13 F (Arquad 12-50)

Flash Pt.:
< 80 F (Arquad 12-50)
140 F (Arquad 12-33)

HLB:
17.1 (Arquad 12-33, 12-50)

Stability:
Good (Arquad 12-33, 12-50)

Surface Tension:
33 dynes/cm (0.1% conc.) (Arquad 12-50)

Biodegradable: (Arquad 12-33, 12-50)

TOXICITY/HANDLING:
Skin irritant, severe eye irritant; protective clothing, goggles, gloves should be worn (Arquad 12-33, 12-50)

STORAGE/HANDLING:
Flammable (Arquad 12-33, 12-50)

STD. PKGS.:
18-kg Atron can (Nissan Cation BB)
55-gal nonreturnable epoxy-phenolic lined drums; bulk SS tank trucks (Arquad 12-33, 12-50)

Magnesium lauryl ether sulfate

SYNONYMS:
Magnesium laureth sulfate (CTFA)

STRUCTURE:
$[CH_3(CH_2)_{10}CH_2(OCH_2CH_2)_nOSO_3]_2Mg$
where $n = 1-4$

CAS No.:
RD No. 977064-40-6

TRADENAME EQUIVALENTS:
Drewpon ESG [Drew Produtos Quimicos]
Empicol EGB [Albright & Wilson/Detergents]
Sactipon 2 OMG [Lever Industriel]
Sactol 2 OMG [Lever Industriel]
Texapon MG [Henkel KGaA]
Tylorol MG [Thomas Triantaphyllou]

CATEGORY:
Detergent

APPLICATIONS:
Bath products: (Texapon MG)
Cosmetic industry preparations: (Tylorol MG); shampoos (Drewpon ESG; Texapon MG); toiletries (Empicol EGB)
Household detergents: (Empicol EGB)

PROPERTIES:
Form:
Liquid (Drewpon ESG; Empicol EGB; Sactol 2 OMG; Tylorol MG))

Composition:
25% conc. (Empicol EGB)
26–28% active (Drewpon ESG)
28% conc. (Sactipon 2 OMG; Sactol 2 OMG)
29–31% conc. (Texapon MG)
30% conc. (Tylorol MG)

Ionic Nature:
Anionic (Drewpon ESG; Empicol EGB; Sactipon 2 OMG; Sactol 2 OMG; Texapon MG)

Magnesium lauryl sulfate (CTFA)

SYNONYMS:
Magnesium monododecyl sulfate
Sulfuric acid, monododecyl ester, magnesium salt

EMPIRICAL FORMULA:
$C_{12}H_{26}O_4S \cdot \frac{1}{2}Mg$

STRUCTURE:
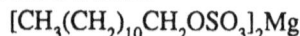
$[CH_3(CH_2)_{10}CH_2OSO_3]_2Mg$

CAS No.:
3097-08-3

TRADENAME EQUIVALENTS:
Carsonol MLS [Carson, Lonza]
Conco Sulfate M [Continental]
Cycloryl MG [Cyclo]
Drewpon MG [Drew Produtos]
Empicol ML 26 [Albright & Wilson/Detergent]
Maprofix MG [Onyx]
Norfox MLS [Norman, Fox & Co.]
Polystep B-9 [Stepan]
Richonol Mg [Richardson]
Sipon LM, MLS [Alcolac]
Standapol MG [Henkel, Henkel/Canada]
Stepanol MG [Stepan]
Surco MG-LS [Onyx]

CATEGORY:
Detergent, base, emulsifier, foaming agent, lubricant, solubilizer, wetting agent

APPLICATIONS:
Automobile cleaners: car shampoo (Carsonol MLS; Stepanol MG)
Bath products: bubble bath (Carsonol MLS; Sipon LM; Standapol MG; Stepanol MG)
Cleansers: (Standapol MG); cleansing creams (Carsonol MLS; Stepanol MG)
Cosmetic industry preparations: (Conco Sulfate M); personal care products (Carsonol MLS; Stepanol MG); shampoos (Carsonol MLS; Cycloryl MG; Drewpon MG; Empicol ML26; Richonol Mg; Sipon LM; Standapol MG; Stepanol MG); shaving preparations (Carsonol MLS; Stepanol MG)
Farm products: insecticides (Stepanol MG)
Food applications: dairy cleaners (Stepanol MG); fruit washing (Stepanol MG); vegetable scrubber (Carsonol MLS)
Household detergents: (Conco Sulfate M); carpet & upholstery shampoos (Carsonol MLS; Cycloryl MG; Maprofix MG; Norfox MLS; Sipon LM; Stepanol MG; Surco MG-LS); liquid detergents (Carsonol MLS; Stepanol MG)
Industrial applications: industrial processing (Polystep B-9); textile/leather processing (Stepanol MG)
Industrial cleaners: (Carsonol MLS; Stepanol MG); metal processing surfactants (Stepanol MG); railway cleaners (Stepanol MG)

Pet shampoos: (Carsonol MLS)

Pharmaceutical applications: (Stepanol MG); tablet mfg. (Sipon MLS); toothpaste (Empicol ML26)

PROPERTIES:

Form:

Liquid (Conco Sulfate M; Cycloryl MG; Drewpon MG; Empicol ML26; Maprofix MG; Norfox MLS; Richonol Mg; Standapol MG; Surco MG-LS)

Clear liquid (Polystep B-9; Sipon LM)

Clear, viscous liquid (Stepanol MG)

Viscous liquid (Carsonol MLS)

Spray-dried (Sipon MLS)

Color:

Pale straw (Empicol ML 26)

Pale yellow (Cycloryl MG; Stepanol MG)

Pale/low (Carsonol MLS; Polystep B-9)

Straw yellow (Standapol MG)

Gardner 3 (Richonol Mg)

Odor:

Low (Carsonol MLS)

Composition:

27% active (Sipon LM)

27–29% active (Maprofix MG)

28% active (Polystep B-9)

28–30% active (Cycloryl MG; Drewpon MG; Standapol MG; Stepanol MG)

30% active (Carsonol MLS)

30% conc. (Conco Sulfate M; Norfox MLS)

88% active (Sipon MLS)

Solubility:

Sol. in water (Carsonol MLS; Stepanol MG)

Miscible with water; may form slightly hazy solution due to residual alkalinity; add small amount of citric acid to form clear solution (Empicol ML 26)

Ionic Nature:

Anionic (Conco Sulfate M; Cycloryl MG; Drewpon MG; Empicol ML26; Maprofix MG; Norfox MLS; Sipon LM; Standapol MG; Stepanol MG)

Density:

8.57 lb/gal (Carsonol MLS)

8.7 lb/gal (Richonol Mg)

1.0 g/cc (Empicol ML 26)

Sp.gr.:

1.02 (Surco MG-LS)

1.02 @ 25/20 C (Maprofix MG)

Visc.:

50 cps (Sipon LM)

Magnesium lauryl sulfate *(cont'd.)*

150 cs (Empicol ML 26)
pH:
6.0–7.0 (Richonol Mg)
6.0–7.0 (10% aq.) (Standapol MG)
6.5 (3% sol'n.) (Sipon MLS)
6.5 (10% sol'n.) (Sipon LM)
6.5–7.5 (10% sol'n.) (Carsonol MLS; Stepanol MG)
Flash Pt.:
> 200 F (Maprofix MG, Surco MG-LS)
Cloud Pt.:
2 C (Sipon LM)
3 C max. (Standapol MG)
50 F (Richonol Mg)
Biodegradable: (Carsonol MLS)

Monoethanolamine lauryl ether sulfate

SYNONYMS:
MEA-laureth sulfate (CTFA)
STRUCTURE:

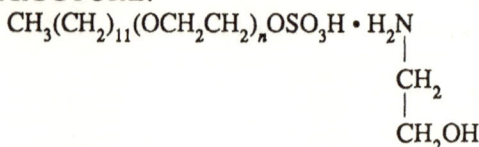

$$CH_3(CH_2)_{11}(OCH_2CH_2)_nOSO_3H \cdot H_2N$$
$$CH_2$$
$$CH_2OH$$

where $n = 1-4$
CAS No.:
RD No.: 977067-77-8
TRADENAME EQUIVALENTS:
Akyposal MLES35 [Chem-Y GmbH]
Drewpon ECM [Drew Produtos Quimicos]
Sactipon 2OM [Lever Industriel]
Sactol 2OM [Lever Industriel]
Tylorol LM [Thomas Triantaphyllou]
CATEGORY:
Detergent, base
APPLICATIONS:
Bath products: bubble bath (Akyposal MLES35)
Cosmetic industry preparations: shampoos (Akyposal MLES35; Drewpon ECM; Tylorol LM)

Monoethanolamine lauryl ether sulfate *(cont'd.)*

PROPERTIES:
Form:
Liquid (Akyposal MLES35; Drewpon ECM; Sactipon 2OM; Sactol 2OM; Tylorol LM)
Composition:
27% conc. (Sactipon 2OM; Sactol 2OM)
30% conc. (Tylorol LM)
30–32% active (Drewpon ECM)
35% conc. (Akyposal MLES35)
Ionic Nature:
Anionic (Akyposal MLES35; Drewpon ECM; Sactipon 2M; Sactol 2OM; Tylorol LM)

Monoethanolamine lauryl sulfate

SYNONYMS:
Lauryl sulfate monoethanolamine salt
MEA-lauryl sulfate (CTFA)
Sulfuric acid, monododecyl ester, compd. with 2-aminoethanol (1:1)
EMPIRICAL FORMULA:
$C_{12}H_{26}O_4S \cdot C_2H_7NO$
STRUCTURE:
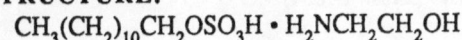
$CH_3(CH_2)_{10}CH_2OSO_3H \cdot H_2NCH_2CH_2OH$
CAS No.:
4722-98-9
TRADENAME EQUIVALENTS:
Alkasurf MLS [Alkaril]
Condanol MLS-35 [Dutton & Reinisch]
Cycloryl SA [Cyclo]
Drewpon MEA [Drew Produtos Quimicos]
Elfan 240M [Akzo Chemie]
Empicol EL [Albright & Wilson/Marchon]
Empicol LQ33 [Albright & Wilson/Australia]
Empicol LQ70 [Albright & Wilson/Detergents]
Empicol XT45 [Albright & Wilson/Marchon]
Manro ML 33S [Manro]
Rewopol MLS30, MLS35 [Rewo Chemische Werke]
Sactipon 2M [Lever Industriel]
Sactol 2M [Lever Industriel]
Standapol MLS [Henkel]
Sulfatol 33MO [Aarhus Oliefabrik]
Sulphonated Lorol Liquid MA, Liquid MR [Ronsheim & Moore]

Monoethanolamine lauryl sulfate *(cont'd.)*

TRADENAME EQUIVALENTS *(cont'd.):*
Texapon MLS [Henkel KGaA]
Zoharpon LAM [Zohar Deterg. Factory]

CATEGORY:
Detergent, surfactant, wetting agent, foaming agent

APPLICATIONS:
Bath products: bubble bath (Elfan 240M; Empicol LQ70; Manro ML 33S; Rewopol MLS30, MLS35; Standapol MLS; Sulfatol 33MO; Zoharpon LAM); shower gels (Empicol LQ33)

Cleansers: (Standapol MLS); body cleansers (Empicol LQ33); hand cleanser (Empicol LQ33)

Cosmetic industry preparations: (Empicol LQ70; Manro ML 33S); shampoo base (Condanol MLS35; Empicol XT45); shampoos (Alkasurf MLS; Cycloryl SA; Drewpon MEA; Elfan 240M; Empicol EL, LQ33, LQ70; Manro ML 33S; Rewopol MLS30, MLS35; Standapol MLS; Sulfatol 33MO; Sulphonated Lorol Liquid MA, Liquid MR; Texapon MLS; Zoharpon LAM); toiletries (Empicol LQ33, LQ70; Manro ML 33S; Sulfatol 33MO)

Household detergents: carpet & upholstery shampoos (Sulfatol 33MO); dishwashing (Sulfatol 33MO); light-duty cleaners (Elfan 240M; Sulfatol 33MO); liquid detergents (Empicol LQ33; Rewopol MLS35)

Industrial applications: (Sulfatol 33MO)

Industrial cleaners: wax strippers (Alkasurf MLS)

Pharmaceutical applications: toothpaste (Elfan 240M)

PROPERTIES:

Form:
Liquid (Alkasurf MLS; Cycloryl SA; Drewpon MEA; Elfan 240M; Empicol LQ70, XT45; Rewopol MLS35; Sactipon 2M; Sactol 2M; Sulfatol 33MO; Sulphonated Lorol Liquid MA, Liquid MR; Texapon MLS; Zoharpon LAM)

Clear/opalescent viscous liquid (Empicol LQ33; Manro ML 33S)

Clear, viscous liquid (Standapol MLS)

Viscous liquid (Condanol MLS35; Empicol EL; Rewopol MLS30)

Color:
Light amber (Empicol LQ33)

Amber (Empicol EL)

Pale/light yellow (Alkasurf MLS; Condanol MLS35; Empicol XT45; Manro ML 33S; Standapol MLS)

Odor:
Low (Alkasurf MLS)

Slight, typical (Condanol MLS35)

Negligible (Manro ML 33S)

Composition:
26–27% active (Empicol XT45)

28% conc. (Cycloryl SA; Texapon MLS)

Monoethanolamine lauryl sulfate (cont'd.)

29% conc. (Elfan 240M; Sactipon 2M; Sactol 2M; Zoharpon LAM)
30% active (Sulfatol 33MO)
30% conc. (Rewopol MLS30)
32% active (Alkasurf MLS)
32% active min. (Manro ML 33S)
32% conc. (Sulphonated Lorol Liquid MA; Liquid MR)
32–34% active (Drewpon MEA; Standapol MLS)
33.5 ± 1.0% active (Empicol LQ33)
35% active in water (Condanol MLS35)
35% conc. (Rewopol MLS35)
50% active (Empicol EL)
70% conc. (Empicol LQ70)
Ionic Nature:
Anionic (Cycloryl SA; Drewpon MEA; Elfan 240M; Empicol EL, LQ70; Rewopol
MLS30, MLS35; Zoharpon LAM)
Sp.gr.:
1.0 (Condanol MLS35)
1.025 (20 C) (Empicol EL)
1.03 (20 C) (Manro ML 33S)
1.05 (20 C) (Empicol LQ33, XT45)
Visc.:
17 cSt max. (50% aq. sol'n., #1A U-tube) (Empicol LQ33)
100 cs max. (20 C) (Empicol XT45)
1143 cs (50% sol'n., 20 C) (Empicol EL)
3000–5000 cps (Standapol MLS)
7500 cps (20 C) (Manro ML 33S)
Cloud Pt.:
–2 C (Empicol EL)
< 0 C (Empicol XT45)
≈ 0 C (Standapol MLS)
2 C max. (Quick Cool, 50% aq. sol'n.) (Empicol LQ33)
pH:
6.3–7.3 (10% aq.) (Manro ML 33S)
7.0 ± 0.5 (5% aq. sol'n.) (Empicol LQ33)
7.0–8.0 (10% aq.) (Standapol MLS)
9.0–9.5 (5% aq.) (Empicol XT45)
Biodegradable: (Sulphonated Lorol Liquid MA, Liquid MR)
STD. PKGS.:
Lacquered drums (Condanol MLS35)
200-kg net lined, open-head mild-steel drums (Empicol LQ33)
45-gal drums or road tankers (Manro ML 33S)

Monotriethanolamine N-cocoyl-L-glutamate

TRADENAME EQUIVALENTS:
Acylglutamate CT-12 [Ajinomoto]
Amisoft C-T-12 [Ajinomoto]
CATEGORY:
Detergent, emulsifier, emollient, bactericide
APPLICATIONS:
Cosmetic industry preparations: (Amisoft C-T-12); personal care products (Acylglutamate CT-12)
PROPERTIES:
Form:
Liquid (Acylglutamate CT-12; Amisoft C-T-12)
Composition:
20% active (Acylglutamate CT-12; Amisoft C-T-12)
Stability:
Good in hard water (Acylglutamate CT-12)
Biodegradable: (Acylglutamate CT-12)
TOXICITY/HANDLING:
Nonskin irritation even for infants and eczema cases (Acylglutamate CT-12)

Myristic diethanolamide

SYNONYMS:
N,N-Bis (2-hydroxyethyl) myristamide
N,N-Bis (2-hydroxyethyl) tetradecanamide
Diethanolamine myristic acid condensate
Myristamide DEA (CTFA)
Myristoyl diethanolamide
Tetradecamide, N,N-bis
EMPIRICAL FORMULA:
$C_{18}H_{37}NO_3$
STRUCTURE:

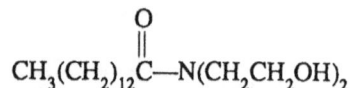

$$CH_3(CH_2)_{12}\overset{\overset{\displaystyle O}{\|}}{C}-N(CH_2CH_2OH)_2$$

CAS No.:
7545-23-5
TRADENAME EQUIVALENTS:
Condensate PM [Continental]
Monamid 150-MW [Mona]
Rewomid DLM/SE [Rewo Chemische Werke]

CATEGORY:
   Surfactant, detergent, emulsifier, wetting agent, thickener, foaming agent, foam
      stabilizer, dispersant, corrosion inhibitor, lubricant, superfatting agent
APPLICATIONS:
   Bath products: bubble bath (Condensate PM; Monamid 150-MW)
   Cleansers: hand cleanser (Monamid 150-MW)
   Cosmetic industry preparations: (Condensate PM; Monamid 150-MW; Rewomid
      DLM/SE); conditioners (Monamid 150-MW); shampoos (Condensate PM;
      Monamid 150-MW)
   Farm products: agricultural oils/sprays (Monamid 150-MW)
   Household detergents: (Condensate PM); carpet & upholstery shampoos (Monamid
      150-MW); dishwashing (Condensate PM; Monamid 150-MW); powdered deter-
      gents (Monamid 150-MW)
   Industrial applications: dyes and pigments (Monamid 150-MW); lubricating/cutting
      oils (Monamid 150-MW); petroleum industry (Monamid 150-MW); polishes and
      waxes (Condensate PM; Monamid 150-MW); textile/leather processing (Conden-
      sate PM; Monamid 150-MW)
   Industrial cleaners: (Condensate PM); bottle cleaners (Condensate PM); drycleaning
      compositions (Monamid 150-MW); metal processing surfactants (Monamid 150-
      MW); textile cleaning (Monamid 150-MW)
PROPERTIES:
Form:
   Solid (Condensate PM; Monamid 150-MW)
Color:
   White (Condensate PM)
   GVCS-33: 2 max. (Monamid 150-MW)
Odor:
   Typical (Condensate PM)
Composition:
   90–95% amide content (Condensate PM)
   100% active (Monamid 150-MW)
   100% conc. (Rewomid DLM/SE)
Solubility:
   Sol. in aromatic hydrocarbons (@ 10%) (Monamid 150-MW)
   Sol. in chlorinated hydrocarbons (@ 10%) (Monamid 150-MW)
   Sol. in ethyl alcohol (@ 10%) (Monamid 150-MW)
   Gels in natural fats and oils (@ 10%) (Monamid 150-MW)
   Disp. in water (@ 10%) (Monamid 150-MW); disp. in lows
      concs. and forms gels at higher levels (Condensate PM)
Ionic Nature:
   Nonionic (Monamid 150-MW; Rewomid DLM/SE)
Sp.gr.:
   0.97 (Condensate PM)

Myristic diethanolamide *(cont'd.)*

0.98 (45 C) (Monamid 150-MW)
Density:
   8.20 lb/gal (Monamid 150-MW)
M.P.:
   100–102 F (Condensate PM)
Acid No.:
   0–3 (Monamid 150-MW)
Alkali No.:
   30–48 (Condensate PM)
pH:
   9.0–9.7 (1% disp.) (Condensate PM)
   9.5–10.5 (10% sol'n.) (Monamid 150-MW)
Biodegradable: (Monamid 150-MW)
STD. PKGS.:
   100 lb net fiberboard drum (Monamid 150-MW)

Myristyl dimethyl benzyl ammonium chloride

SYNONYMS:
   Benzenemethanaminium, N,N-dimethyl-N-tetradecyl-, chloride
   N,N-Dimethyl-N-tetradecylbenzenemethanaminium chloride
   Myristalkonium chloride (CTFA)
   Tetradecyl dimethyl benzyl ammonium chloride
EMPIRICAL FORMULA:
   $C_{23}H_{42}N \cdot Cl$
STRUCTURE:

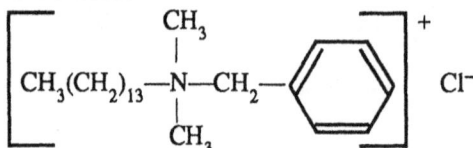

CAS No.:
   139-08-2
TRADENAME EQUIVALENTS:
   Arquad DM14B-90 [Akzo Chemie] (dihydrate)
   Barquat MS-100 [Lonza] (dihydrate)
   Barquat MX-50, MX-80 [Lonza]
   BTC 824 [Onyx]
   Cyncal 80% [Hilton-Davis]
   Dibactol [Hexcel] (dihydrate)
   Nissan Cation M_2-100 [Nippon Oil & Fats]

Myristyl dimethyl benzyl ammonium chloride

CATEGORY:
Surfactant, detergent, antimicrobial, bactericide, biocide, germicide, algicide, fungicide, disinfectant, dispersant, coagulant, antistat, softener

APPLICATIONS:
Farm products: (Cyncal 80%); herbicides (Nissan Cation M_2-100)
Food applications: (Cyncal 80%; Nissan Cation M_2-100)
Household detergents: hard surface cleaner (BTC 824)
Industrial applications: dyes and pigments (Nissan Cation M_2-100); paper mfg. (Nissan Cation M_2-100); petroleum industry (Nissan Cation M_2-100); plastics (Nissan Cation M_2-100); textile/leather processing (Cyncal 80%; Nissan Cation M_2-100); water treatment (BTC 824; Cyncal 80%; Nissan Cation M_2-100)
Industrial cleaners: (Cyncal 80%); beverage processing, food service (Cyncal 80%); hospital cleaners (Cyncal 80%); sanitizers/germicides (Barquat MX-50, MX-80; BTC 824)

PROPERTIES:
Form:
Liquid (Barquat MX-50, MX-80; BTC 824; Cyncal 80%)
Powder (Arquad DM14B-90; Barquat MS-100; Dibactol)
Crystalline powder (Nissan Cation M_2-100)
Color:
White (Arquad DM14B-90; Dibactol)
White or light yellow (Nissan Cation M_2-100)
Light yellow (Cyncal 80%)
Odor:
Mild, pleasant (Cyncal 80%)
Composition:
50% conc. (Barquat MX-50)
50% quaternary min. (BTC 824)
80% active (Cyncal 80%)
80% conc. (Barquat MX-80)
85% active min. (Nissan Cation M_2-100)
90% active (Arquad DM14B-90)
99–103% assay (Dibactol)
100% conc. (Barquat MS-100)
Solubility:
Sol. in alcohols (Dibactol); sol. in lower alcohols (Cyncal 80%)
Sol. in glycols (Cyncal 80%)
Sol. in ketones (Cyncal 80%)
Sol. in water (Cyncal 80%; Dibactol)
Ionic Nature:
Cationic (BTC 824; Cyncal 80%; Nissan Cation M_2-100)
M.W.:
359 (of active) (Cyncal 80%)

Myristyl dimethyl benzyl ammonium chloride (cont'd.)

404.06 (Dibactol)

Sp.gr.:

0.94 (Cyncal 80%)

0.96 (25/20 C) (BTC 824)

Density:

7.8 lb/gal (Cyncal 80%)

Flash Pt.:

120 F (BTC 824)

Storage Stability:

Crystallizes after prolonged storage at low temps. (returns to homogeneous state on warming to R.T.) (Cyncal 80%)

pH:

5.0–8.0 (1% aq. sol'n.) (Dibactol)

TOXICITY/HANDLING:

Skin irritant, severe eye irritant (Arquad DM14B-90)

Skin irritant; causes eye damage; corrosive; goggles and gloves should be worn; harmful or fatal if swallowed; avoid contamination of food (Cyncal 80%)

STORAGE/HANDLING:

Avoid contact with strong oxidizing agents, anionics; product will kill bacteria in biological disposal systems (Arquad DM14B-90)

STD. PKGS.:

10-kg paper box (Nissan Cation M_2-100)

50-kg net fiberboard kegs (Arquad DM14B-90)

Oleyl amine

SYNONYMS:
9-Octadecen-1-amine
Oleamine (CTFA)
EMPIRICAL FORMULA:
$C_{18}H_{35}N$
STRUCTURE:
$CH_3(CH_2)_7CH=CH(CH_2)_7CH_2NH_2$
CAS No.:
112-90-3
TRADENAME EQUIVALENTS:
Armeen O [Armak]
Armeen OD [Akzo Chemie]
Crodamine 1.O, 1.OD [Croda Universal Ltd.]
Kemamine P-989 [Humko Sheffield] (tech.)
Lilamin 172 [Lilachim S.A.]
Lilamin 172D [Lilachim S.A.] (distilled)
Nissan Amine OB [Nippon Oil & Fats]
Noram O [Ceca S.A.]
Radiamine 6172 [Oleofina S.A.]
Radiamine 6173 [Oleofina S.A.] (distilled)
CATEGORY:
Wetting agent, emulsifier, flotation agent, dispersant, intermediate, corrosion inhib-
itor, germicide, bactericide, lubricant, mold release agent, softener
APPLICATIONS:
Cosmetic industry preparations: (Radiamine 6172, 6173)
Farm products: (Nissan Amine OB)
Industrial applications: ceramics (Nissan Amine OB); construction (Lilamin 172;
Nissan Amine OB); dyes and pigments (Crodamine 1.O, 1.OD; Lilamin 172; Nissan
Amine OB; Radiamine 6172, 6173); industrial processing (Radiamine 6172, 6173);
lubricating/cutting oils (Armeen OD; Kemamine P-989; Lilamin 172; Nissan
Amine OB; Noram O); metalworking (Crodamine 1.O, 1.OD; Kemamine P-989);
ore flotation (Crodamine 1.O, 1.OD; Nissan Amine OB; Radiamine 6172, 6173);
petroleum industry (Kemamine P-989; Lilamin 172); plastics (Crodamine 1.O,
1.OD; Kemamine P-989); printing inks (Lilamin 172); rubber (Crodamine 1.O,
1.OD; Kemamine P-989; Lilamin 172; Nissan Amine OB; Radiamine 6172, 6173);
textile/leather processing (Crodamine 1.O, 1.OD; Nissan Amine OB; Noram O;
Radiamine 6172, 6173); water treatment (Nissan Amine OB)

Oleyl amine *(cont'd.)*

PROPERTIES:
Form:
    Liquid (Crodamine 1.OD; Kemamine P-989; Lilamin 172)
    Liquid in summer, solid in winter (Nissan Amine OB)
    Liquid/paste (Radiamine 6172, 6173)
    Paste (Armeen O, OD)
    Solid (Crodamine 1.O; Noram O)
Color:
    APHA 140 max. (Nissan Amine OB)
    Gardner 2 (Armeen OD)
    Gardner 3 max. (Kemamine P-989; Lilamin 172)
    Gardner 8 (Armeen O)
Composition:
    93% conc. (Kemamine P-989)
    95% active (Lilamin 172)
    98% active (Armeen OD)
    98% primary amine min. (Nissan Amine OB)
    100% conc. (Crodamine 1.O, 1.OD; Radiamine 6172, 6173)
Solubility:
    Sol. in acetone (Armeen O)
    Sol. in carbon tetrachloride (Armeen O)
    Sol. in chloroform (Armeen O)
    Sol. in ethanol (Armeen O)
    Sol. in isopropanol (Armeen O)
    Sol. in kerosene (Armeen O)
    Sol. in methanol (Armeen O)
    Sol. in common organic solvents (Kemamine P-989)
    Sol. in toluene (Armeen O)
    Insol. in water (Armeen OD)
    Sol. in white min. oil (Armeen O)
Ionic Nature:
    Cationic (Crodamine 1.O, 1.OD; Kemamine P-989; Lilamin 172; Radiamine 6172,
      6173)
M.W.:
    $\approx$ 270 (Lilamin 172)
Sp.gr.:
    0.79 (60 C) (Armeen OD)
    0.802 (60 C) (Lilamin 172)
    0.820 (38/4 C) (Armeen O)
Visc.:
    56.6 SSU (Armeen OD)
    57.0 SSU (Armeen O)
    3.20 cps (60 C) (Lilamin 172)

Oleyl amine *(cont'd.)*

M.P.:
8–18 C (Armeen OD)
19 C (Lilamin 172)
50–68 F (Armeen O)
Solidification Pt.:
30 C max. (Nissan Amine OB)
Flash Pt.:
154 C (COC) (Armeen OD)
155 C (OC) (Lilamin 172)
320 F (Armeen O)
Fire Pt.:
360 F (Armeen O)
Iodine No.:
60 min. (Nissan Amine OB)
70 min. (Kemamine P-989; Lilamin 172)
78 min. (Armeen OD)
TOXICITY/HANDLING:
Skin irritant, severe eye irritant (Armeen OD)
STORAGE/HANDLING:
Avoid contact with strong mineral acids, strong oxidizing agents (Armeen OD)
STD. PKGS.:
200 L. bung-type steel drums (Armeen OD)
13-kg can, 160-kg drum (Nissan Amine OB)

Oleyl amine acetate

TRADENAME EQUIVALENTS:
Armac OD [Akzo Chemie]
Noramac O [Diamond Shamrock Process]
CATEGORY:
Corrosion inhibitor, flotation agent, bactericide, emulsifier, anticaking agent, stabilizer, flocculation agent
APPLICATIONS:
Farm applications: soil stabilization (Noramac O)
Industrial applications: mining industry (Armac OD); ore flotation (Armac OD)
PROPERTIES:
Form:
Pasty/solid (Noramac O)
Solid (Armac OD)

Oleyl amine acetate *(cont'd.)*

Color:
   Gardner 13 max. (Armac OD)
Composition:
   95% active min. (Armac OD)
Ionic Nature:
   Cationic (Noramac O)
M.P.:
   55 C (Armac OD)
Flash Pt.:
   170 C (COC) (Armac OD)
STD. PKGS.:
   172-kg net open-head drums (Armac OD)

Oleyl imidazoline *(CTFA)*

SYNONYMS:
   2-(8-Heptadecenyl)-4,5-dihydro-1H-imidazole-1-ethanol
   Hydroxyethyl imidazoline, oleic hydrophobe
   1-Hydroxyethyl-2-oleyl imidazoline
   1H-Imidazole-1-ethanol, 2-(8-heptadecenyl)-4,5-dihydro-
   Oleic acid imidazoline
   Oleic fatty acid imidazoline
   Oleic hydroxyethyl imidazoline
   Oleic imidazoline
   Oleyl hydroxyethyl imidazoline
   Substituted imidazoline of oleic acid
EMPIRICAL FORMULA:
   $C_{22}H_{42}N_2O$
STRUCTURE:

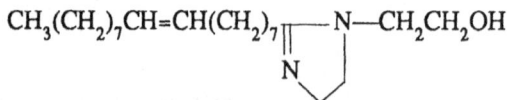

CAS No.:
   95-38-5; 21652-27-7; 27136-73-8
TRADENAME EQUIVALENTS:
   Alkazine O [Alkaril]
   Amine O [Ciba-Geigy]
   Crodazoline O [Croda]
   Finazoline OA [Finetex]
   Mazoline OA [Mazer]

Oleyl imidazoline (cont'd.)

TRADENAME EQUIVALENTS *(cont'd.):*
  Miramine OC [Miranol]
  Monazoline O [Mona]
  Nopcogen 22-O [Diamond Shamrock]
  Schercozoline O [Scher]
  Textamine O-1 [Henkel]
  Unamine O [Lonza]
  Varine O [Sherex]

CATEGORY:
  Emulsifier, wetting agent, dispersant, detergent, antistat, softener, corrosion inhibitor, intermediate, thickener, microbicide

APPLICATIONS:
  Automobile cleaners: car wax (Nopcogen 22-O)
  Cosmetic industry preparations: (Finazoline OA; Schercozoline O)
  Degreasers: (Alkazine O)
  Farm products: agricultural oils/sprays (Nopcogen 22-O; Schercozoline O); dairy cleaners (Textamine O-1); herbicides (Textamine O-1); bactericide/fungicide/pesticide (Schercozoline O; Textamine O-1)
  Industrial applications: construction (Alkazine O; Amine O; Textamine O-1); dyes and pigments (Textamine O-1); lubricating/cutting oils (Amine O; Miramine OC; Schercozoline O); metalworking (Finazoline OA; Miramine OC; Schercozoline O; Textamine O-1); paint mfg. (Amine O; Schercozoline O; Textamine O-1); paper mfg. (Nopcogen 22-O); petroleum industry (Textamine O-1); plastics (Amine O; Textamine O-1); polishes and waxes (Schercozoline O); surface treatment (Schercozoline O); textile/leather processing (Alkazine O; Amine O; Finazoline OA; Mazoline OA; Nopcogen 22-O; Schercozoline O; Textamine O-1)
  Industrial cleaners: (Miramine OC; Schercozoline O; Textamine O-1); acid cleaners (Alkazine O; Amine O; Nopcogen 22-O; Schercozoline O)

PROPERTIES:
Form:
  Liquid (Alkazine O; Amine O; Finazoline OA; Mazoline OA; Monazoline O; Schercozoline O; Textamine O-1; Unamine O; Varine O)
  Liquid to pasty solid (Miramine OC)
  Clear liquid
Color:
  Amber (Amine O; Miramine OC; Textamine O-1)
  Dark amber (Schercozoline O)
  Brown (Alkazine O)
  Gardner 6–7 (Crodazoline O)
Odor:
  Amine (Alkazine O)
  Ammoniacal (Miramine OC)

Oleyl imidazoline (cont'd.)

Composition:
85% active (Alkazine O)
90% conc. (Nopcogen 22-O)
90% imidazoline (Textamine O-1)
90% imidazoline min. (Crodazoline O; Schercozoline O)
99% active min. (Miramine OC)
100% active (Amine O; Varine O)
100% conc. (Finazoline OA; Mazoline OA; Monazoline O)

Solubility:
Sol. in aq. acids (Alkazine O); sol. in acidic sol'ns. (Unamine O)
Sol. in alcohols (Textamine O-1)
Sol. in chlorinated hydrocarbons (Textamine O-1)
Sol. in hydrocarbons (Amine O)
Sol. in oils (Mazoline OA; Miramine OC; Monazoline O; Textamine O-1)
Sol. in organic solvents (Alkazine O)
Sol. in polar, organic solvents (Amine O)
Disp. in water (Miramine OC; Textamine O-1); relatively insol. in water (Amine O)

Ionic Nature:
Cationic (Amine O; Finazoline OA; Mazoline OA; Miramine OC; Monazoline O; Nopcogen 22-O; Schercozoline O; Varine O)
Nonionic (Textamine O-1)

M.W.:
345 (Crodazoline O)
350 (Schercozoline O)

Sp.gr.:
0.930 (Miramine OC)

Density:
7.66 lb/gal (Textamine O-1)

F.P.:
5 C (Alkazine O)

Pour Pt.:
−12 C (Crodazoline O)

Acid No.:
< 1 (Textamine O-1)

Alkali No.:
160–170 (Schercozoline O)
168 (Textamine O-1)

Stability:
Good (Amine O)
Not stable in alkaline media (Alkazine O)

Storage Stability:
Good storage stability but may crystallize at low temps. or during aging (Textamine O-1)

pH:

10.5 (10% sol'n.) (Textamine O-1)

11.1 (10% aq. disp.) (Crodazoline O)

Surface Tension:

26.5 dynes/cm (1%/0.363%) (Crodazoline O)

Biodegradable: (Miramine OC; Textamine O-1)

TOXICITY/HANDLING:

Very irritating to skin and eyes (Amine O)

Severe skin and eye irritant; wear protective goggles and gloves (Miramine OC)

Strong base; irritating to skin and eyes in conc. form; avoid ingestion (Textamine O-1)

STORAGE/HANDLING:

If crystallization occurs, mild heating and agitation will reliquefy product; avoid prolonged heating over 165 C; exposure to water or humidity will gradually cause hydrolysis; store in closed containers (Textamine O-1)

STD. PKGS.:

55-gal steel drums (Textamine O-1)

Palmityl dimethyl amine

SYNONYMS:
N,N-dimethyl-1-hexadecanamine
Dimethyl palmitamine (CTFA)
Dimethyl palmitylamine
1-Hexadecanamine, N,N-dimethyl-
Hexadecyl dimethyl amine
Palmityl dimethyl tertiary amine

EMPIRICAL FORMULA:
$C_{18}H_{39}N$

STRUCTURE:

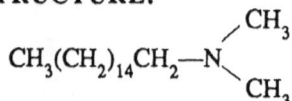

CAS No.:
112-69-6

TRADENAME EQUIVALENTS:
Adma 6 [Ethyl Corp.]
Armeen DM16D [Akzo Chemie]
Nissan Tertiary Amine PB [Nippon Oil & Fats]
Onamine 16 [Onyx]

Distilled:
Crodamine 3.A16D [Croda]
Lilamin 316D [Lilachim S.A.]

CATEGORY:
Intermediate, corrosion inhibitor, germicide, catalyst, curing agent, emulsifier, dispersant, extraction reagent

APPLICATIONS:
Farm products: herbicides (Crodamine 3.A16D)

Household detergents: surfactant intermediate (Armeen DM16D; Nissan Tertiary Amine PB)

Industrial applications: dyes and pigments (Crodamine 3.A16D; Lilamin 316D); lubricating/cutting oils (Lilamin 316D; Onamine 16; Nissan Tertiary Amine PB); metalworking (Crodamine 3.A16D); ore flotation (Crodamine 3.A16D); petroleum industry (Lilamin 316D); plastics (Crodamine 3.A16D; Nissan Tertiary Amine PB); polymers/polymerization (Lilamin 316D); printing inks (Lilamin 316D); rubber (Crodamine 3.A16D; Lilamin 316D); textile/leather processing (Crodamine 3.A16D)

PROPERTIES:
Form:
    Liquid (Adma 6; Armeen DM16D; Lilamin 316D; Nissan Tertiary Amine PB; Onamine 16)
    Semisolid (Nissan Tertiary Amine PB)
Color:
    Light yellow (Nissan Tertiary Amine PB)
    Yellow (Armeen DM16D)
    Gardner 2 max. (Lilamin 316D)
Odor:
    Amine (Armeen DM16D)
Composition:
    95% active (Lilamin 316D)
    95% tertiary amine min. (Nissan Tertiary Amine PB)
    98% active (Armeen DM16D)
    100% conc. (Adma 6; Onamine 16)
Solubility:
    Insol. in water (Armeen DM16D)
Ionic Nature:
    Cationic (Adma 6; Armeen DM16D; Crodamine 3.A16D; Lilamin 316D; Onamine 16)
M.W.:
    ≈ 280 (Lilamin 316D)
Sp.gr.:
    0.80 (Armeen DM16D); (25/20 C) (Onamine 16)
Visc.:
    6.75 cs (Armeen DM16D)
F.P.:
    8 C (Armeen DM16D)
B.P.:
    100–136 C (3 mm Hg) (Armeen DM16D)
Flash Pt.:
    150 C (COC) (Armeen DM16D)
    > 200 F (Onamine 16)
Iodine No.:
    1 max. (Nissan Tertiary Amine PB)
    2 max. (Lilamin 316D)
    3 max. (Armeen DM16D)
TOXICITY/HANDLING:
    Skin irritant, severe eye irritant (Armeen DM16D)
STD. PKGS.:
    13-kg can, 160-kg drum (Nissan Tertiary Amine PB)

Pareth-25-3 (CTFA)

SYNONYMS:

C_{12-15} alcohol (3EO) ethoxylate

Polyethylene glycol ether of a mixture of synthetic C_{12-15} fatty alcohols with avg. of 3 moles ethylene oxide

CAS No.:

68131-39-5 (generic)

RD No.: 977061-52-1

TRADENAME EQUIVALENTS:

Alkasurf LA-3 [Alkaril]

Dobanol 25-3 [Shell]

Neodol 25-3 [Shell]

Sterling Emulsifier #3 [Canada Packers]

Tergitol 25-L-3 [Union Carbide]

CATEGORY:

Emulsifier, surfactant, detergent intermediate, wetting agent

APPLICATIONS:

Degreasers: (Alkasurf LA-3)

Farm products: agricultural oils/sprays (Alkasurf LA-3)

Household detergents: dishwashing (Dobanol 25-3); liquid detergents (Neodol 25-3)

Industrial applications: dyes and pigments (Alkasurf LA-3; Tergitol 25-L-3); lubricating/cutting oils (Alkasurf LA-3); paint mfg. (Alkasurf LA-3); polishes and waxes (Alkasurf LA-3); textile/leather processing (Alkasurf LA-3; Tergitol 25-L-3)

Industrial cleaners: drycleaning compositions (Alkasurf LA-3); textile cleaning (Tergitol 25-L-3)

PROPERTIES:

Form:

Liquid (Sterling Emulsifier #3; Tergitol 25-L-3)

Clear liquid (Dobanol 25-3)

Translucent to opaque liquid (Alkasurf LA-3)

Clear to slightly hazy liquid (Neodol 25-3)

Color:

Colorless (Neodol 25-3)

Pt-Co 50 max. (Tergitol 25-L-3)

Odor:

Mild (Neodol 25-3)

Composition:

100% active (Alkasurf LA-3; Dobanol 25-3; Neodol 25-3; Sterling Emulsifier #3)

100% conc. (Tergitol 25-L-3)

Solubility:

Sol. in Aromatic 150 (@ 10%) (Tergitol 25-L-3)

Sol. in butyl Cellosolve (@ 10%) (Tergitol 25-L-3)

Sol. in isopropanol (@ 10%) (Tergitol 25-L-3)

Sol. in oil (Sterling Emulsifier #3)

Sol. in perchloroethylene (@ 10%) (Tergitol 25-L-3)
Insol. in water (Alkasurf LA-3)
Disp. in white min. oil (@ 10%) (Tergitol 25-L-3)
Sol. in xylene (@ 10%) (Tergitol 25-L-3)
Ionic Nature:
Nonionic (Alkasurf LA-3; Dobanol 25-3; Tergitol 25-L-3)
M.W.:
320–350 (Neodol 25-3)
341 avg. (Tergitol 25-L-3)
Sp.gr.:
0.9 (25/25 C) (Alkasurf LA-3)
0.925 (Neodol 25-3)
0.929 (20/20 C) (Tergitol 25-L-3)
Density:
0.93 kg/l (20 C) (Dobanol 25-3)
7.70 lb/gal (Neodol 25-3)
7.73 lb/gal (20 C) (Tergitol 25-L-3)
Visc.:
18 cs (40 C) (Dobanol 25-3)
19 cs (100 F) (Alkasurf LA-3; Neodol 25-3)
36 cs (20 C) (Tergitol 25-L-3)
M.P.:
–5–6 C (Neodol 25-3)
5–6 C (Dobanol 25-3)
Pour Pt.:
0 C (Neodol 25-3)
8 C (Tergitol 25-L-3)
Flash Pt.:
162 C (Dobanol 25-3)
260 F (COC) (Tergitol 25-L-3)
315 F (PMCC) (Neodol 25-3)
Cloud Pt.:
< 0 C (1% aq.) (Tergitol 25-L-3)
58–62 C (Alkasurf LA-3)
HLB:
7.7 (Tergitol 25-L-3)
7.8 (Alkasurf LA-3; Dobanol 25-3)
7.9 (Neodol 25-3)
Acid No.:
0.05–0.2 (Dobanol 25-3)
Hydroxyl No.:
165 (Tergitol 25-L-3)
166 ± 6 (Dobanol 25-3)

Pareth-25-3 *(cont'd.)*

167 (Neodol 25-3)
Stability:
Good (Alkasurf LA-3)
pH:
5–7 (1% sol'n. in 1:1 isopropanol/water) (Tergitol 25-L-3)
Biodegradable: (Tergitol 25-L-3); > 95% (Dobanol 25-3)
TOXICITY/HANDLING:
Severe eye irritant, and skin irritant on prolonged contact with conc. form (Neodol 25-3)
Defatting to skin (Tergitol 25-L-3)
STD. PKGS.:
1- and 5-gal containers, 55-gal drums (Tergitol 25-L-3)

Pareth-25-7 *(CTFA)*

SYNONYMS:
C_{12-15} alcohol (7EO) ethoxylate
Polyethylene glycol ether of a mixture of synthetic C_{12-15} fatty alcohols with avg. of 7 moles ethylene oxide
CAS No.:
68131-39-5 (generic)
RD No.: 977054-91-3
TRADENAME EQUIVALENTS:
Alkasurf LA-7 [Alkaril]
Dobanol 25-7 [Shell]
Neodol 25-7 [Shell]
Tergitol 25-L-7 [Union Carbide]
CATEGORY:
Detergent, wetting agent, emulsifier, desizing agent
APPLICATIONS:
Household detergents: (Alkasurf LA-7); heavy-duty cleaner (Neodol 25-7); laundry detergent (Alkasurf LA-7; Dobanol 25-7); powdered detergents (Dobanol 25-7)
Industrial applications: dyes and pigments (Alkasurf LA-7; Tergitol 25-L-7); electroplating (Alkasurf LA-7); lubricating/cutting oils (Alkasurf LA-7); paint mfg. (Alkasurf LA-7; Neodol 25-7); paper mfg. (Alkasurf LA-7; Neodol 25-7); printing inks (Alkasurf LA-7); textile/leather processing (Neodol 25-7; Tergitol 25-L-7)
Industrial cleaners: (Alkasurf LA-7); acid cleaners (Alkasurf LA-7); maintenance (Neodol 25-7); metal processing surfactants (Neodol 25-7); textile cleaning (Alkasurf LA-7; Tergitol 25-L-7)

PROPERTIES:
Form:
    Liquid (Alkasurf LA-7; Tergitol 25-L-7)
    Paste-like (Neodol 25-7)
    Paste/solid (Dobanol 25-7)
Color:
    Light (Alkasurf LA-7)
    Practically colorless (Neodol 25-7)
    White (Dobanol 25-7)
    Pt-Co 50 max. (Tergitol 25-L-7)
Odor:
    Mild (Neodol 25-7)
Composition:
    100% active (Alkasurf LA-7; Dobanol 25-7; Neodol 25-7)
    100% conc. (Tergitol 25-L-7)
Solubility:
    Sol. in Aromatic 150 (@ 10%) (Tergitol 25-L-7)
    Sol. in butyl Cellosolve (@ 10%) (Tergitol 25-L-7)
    Sol. in isopropanol (@ 10%) (Tergitol 25-L-7)
    Sol. in perchloroethylene (@ 10%) (Tergitol 25-L-7)
    Sol. in water (Alkasurf LA-7); (@ 10%) (Tergitol 25-L-7)
    Sol. in xylene (@ 10%) (Tergitol 25-L-7)
Ionic Nature:
    Nonionic (Alkasurf LA-7; Dobanol 25-7; Neodol 25-7; Tergitol 25-L-7)
M.W.:
    519 (Neodol 25-7)
    550 avg. (Tergitol 25-L-7)
Sp.gr.:
    0.958 (50/25 C) (Neodol 25-7)
    0.96 (50/25 C) (Alkasurf LA-7)
    0.985 (30/20 C) (Tergitol 25-L-7)
Density:
    0.97 kg/l (40 C) (Dobanol 25-7)
    8.12 lb/gal (100 F) (Neodol 25-7)
    8.19 lb/gal (30 C) (Tergitol 25-L-7)
Visc.:
    31 cs (40 C) (Dobanol 25-7)
    32 cps (100 F) (Alkasurf LA-7)
    34 cs (100 F) (Neodol 25-7)
    51 cps (38 C) (Tergitol 25-L-7)
M.P.:
    11–15 C (Alkasurf LA-7)
    12–17 C (Neodol 25-7)

Pareth-25-7 *(cont'd.)*

21–23 C (Dobanol 25-7)
Pour Pt.:
21 C (Neodol 25-7)
23 C (Tergitol 25-L-7)
70 F (Alkasurf LA-7)
Flash Pt.:
180 C (Dobanol 25-7)
340 F (COC) (Tergitol 25-L-7)
350 F (PMCC) (Neodol 25-7)
440 F (COC) (Alkasurf LA-7)
Cloud Pt.:
44–48 C (Alkasurf LA-7)
50 C (1% aq.) (Neodol 25-7; Tergitol 25-L-7)
51 ± 3 C (5% aq. sol'n.) (Dobanol 25-7)
HLB:
12.0 (Alkasurf LA-7; Dobanol 25-7)
12.2 (Neodol 25-7)
12.4 (Tergitol 25-L-7)
Acid No.:
0.05–0.2 (Dobanol 25-7)
Hydroxyl No.:
102 (Tergitol 25-L-7)
108 (Neodol 25-7)
109 ± 5 (Dobanol 25-7)
Stability:
Good (Alkasurf LA-7)
pH:
5.0–7.0 (1%) (Tergitol 25-L-7)
6.0 ± 0.5 (1% aq.) (Neodol 25-7)
6.8 (0.5% aq. sol'n.) (Dobanol 25-7)
Surface Tension:
28.7 dynes/cm (0.1%) (Tergitol 25-L-7)
Biodegradable: (Alkasurf LA-7; Dobanol 25-7; Tergitol 25-L-7)
TOXICITY/HANDLING:
Severe eye irritant, skin irritant on prolonged contact with conc. form (Neodol 25-7)
Eye irritant, skin irritant on repeated/prolonged contact (Dobanol 25-7)
Causes eye burns and skin irritation (Tergitol 25-L-7)
STORAGE/HANDLING:
Storage temperature should not exceed 50 C (Dobanol 25-7)
STD. PKGS.:
190-kg mild steel drums (Dobanol 25-7)
1- and 5-gal containers, 55-gal drums (Tergitol 25-L-7)

Pareth 25-12 (CTFA)

SYNONYMS:

C_{12-15} alcohol ethoxylate (12 EO)

Polyethylene glycol ether of a mixture of synthetic C_{12-15} fatty alcohols with an avg. of 12 moles of ethylene oxide

CAS No.:

68131-39-5 (generic)

RD No.: 977054-92-4

TRADENAME EQUIVALENTS:

Alkasurf LA-12 [Alkaril]

Neodol 25-12 [Shell]

Tergitol 25-L-12 [Union Carbide]

CATEGORY:

Detergent, wetting agent, emulsifier, desizing agent

APPLICATIONS:

Cosmetic industry preparations: (Alkasurf LA-12); topical cosmetics (Alkasurf LA-12)

Household detergents: (Alkasurf LA-12); heavy-duty cleaner (Neodol 25-12); laundry detergent (Alkasurf LA-12)

Industrial applications: dyes and pigments (Alkasurf LA-12; Tergitol 25-L-12); electroplating (Alkasurf LA-12); lubricating/cutting oils (Alkasurf LA-12); paint mfg. (Alkasurf LA-12; Neodol 25-12); paper mfg. (Alkasurf LA-12; Neodol 25-12); printing inks (Alkasurf LA-12); textile/leather processing (Alkasurf LA-12; Neodol 25-12; Tergitol 25-L-12)

Industrial cleaners: (Alkasurf LA-12) maintenance (Neodol 25-12); metal processing surfactants (Neodol 25-12); textile cleaning (Alkasurf LA-12; Tergitol 25-L-12)

PROPERTIES:

Form:

Paste-like (Neodol 25-12)

Solid (Alkasurf LA-12; Tergitol 25-L-12)

Color:

Practically colorless (Neodol 25-12)

White (Alkasurf LA-12)

Pt-Co 100 max. (Tergitol 25-L-12)

Odor:

Low (Alkasurf LA-12)

Mild (Neodol 25-12)

Composition:

100% active (Alkasurf LA-12; Neodol 25-12)

100% conc. (Tergitol 25-L-12)

Solubility:

Sol. in butyl Cellosolve (@ 10%) (Tergitol 25-L-12)

Sol. in isopropanol (@ 10%) (Tergitol 25-L-12)

Sol. in perchloroethylene (@ 10%) (Tergitol 25-L-12)

Pareth-25-12 *(cont'd.)*

Sol. in water (Alkasurf LA-12); (@ 10%) (Tergitol 25-L-12)
Sol. in xylene (@ 10%) (Tergitol 25-L-12)

Ionic Nature:
Nonionic (Alkasurf LA-12; Neodol 25-12; Tergitol 25-L-12)

M.W.:
729 (Neodol 25-12)
730 avg. (Tergitol 25-L-12)

Sp.gr.:
1.0 (50/25 C) (Alkasurf LA-12)
1.003 (50/25 C) (Neodol 25-12)
1.013 (30/20 C) (Tergitol 25-L-12)

Density:
8.37 lb/gal (40 C) (Tergitol 25-L-12)
8.40 lb/gal (100 F) (Neodol 25-12)

Visc.:
51 cs (38 C) (Tergitol 25-L-12)
53 cs (100 F) (Neodol 25-12)

M.P.:
30–33 C (Neodol 25-12)

Pour Pt.:
27 C (Neodol 25-12)
30 C (Tergitol 25-L-12)

Flash Pt.:
380 F (PMCC) (Neodol 25-12)
410 F (COC) (Tergitol 25-L-12)

Cloud Pt.:
90 C (1% aq.) (Tergitol 25-L-12)
97 C (1% aq.) (Neodol 25-12)

HLB:
14.2 (Tergitol 25-L-12)
14.4 (Alkasurf LA-12; Neodol 25-12)

Hydroxyl No.:
75 (Neodol 25-12)
77 (Tergitol 25-L-12)

pH:
5.0–7.0 (1%) (Tergitol 25-L-12)
6.0 ± 0.5 (1% aq.) (Neodol 25-12)

Surface Tension:
32.0 dynes/cm (0.1%) (Tergitol 25-L-12)

Biodegradable: (Tergitol 25-L-12)

TOXICITY/HANDLING:
Eye irritant (Tergitol 25-L-12)

Severe eye irritant, and skin irritant on prolonged contact with conc. form (Neodol 25-12)

STD. PKGS.:
1- and 5-gal containers, 55-gal drums (Tergitol 25-L-12)

PEG-1 lauryl ether

SYNONYMS:
2-(Dodecyloxy) ethanol
Ethanol, 2-(dodecyloxy)-
Ethylene glycol monolauryl ether
Laureth-1 (CTFA)
EMPIRICAL FORMULA:
$C_{14}H_{30}O_2$
STRUCTURE:
$CH_3(CH_2)_{10}CH_2OCH_2CH_2OH$
CAS No.:
4536-30-5; 9002-92-0 (generic)
RD No.: 977054-82-2
TRADENAME EQUIVALENTS:
Alkasurf LAN-1 [Alkaril]
Lipocol L-1 [Lipo]
Siponic L-1 [Alcolac]
CATEGORY:
Detergent intermediate, emulsifier, defoamer, wetting agent, solubilizer, conditioner, thickener, stabilizer, emollient
APPLICATIONS:
Bath products: bubble bath (Siponic L-1)
Cosmetic industry preparations: creams and lotions (Lipocol L-1); shampoos (Alkasurf LAN-1; Siponic L-1)
Industrial applications: dyes and pigments (Lipocol L-1); paper mfg. (Siponic L-1); polishes and waxes (Siponic L-1); polymers/polymerization (Siponic L-1); rubber (Siponic L-1)
Pharmaceutical applications: (Siponic L-1); antiperspirant/deodorant (Lipocol L-1); depilatories (Lipocol L-1)
PROPERTIES:
Form:
Liquid (Alkasurf LAN-1; Lipocol L-1; Siponic L-1)
Color:
Colorless (Lipocol L-1)

PEG-1 lauryl ether *(cont'd.)*

Composition:
100% active (Lipocol L-1; Siponic L-1)
Solubility:
Sol. in aromatic solvent (@ 10%) (Alkasurf LAN-1)
Sol. in min. oil (@ 10%) (Alkasurf LAN-1)
Sol. in min. spirits (@ 10%) (Alkasurf LAN-1)
Sol. in perchloroethylene (@ 10%) (Alkasurf LAN-1)
Insol. in water (@ 10%) (Alkasurf LAN-1)
Ionic Nature:
Nonionic (Lipocol L-1; Siponic L-1)
Density:
0.88 g/ml (Alkasurf LAN-1)
Cloud Pt.:
37–39 C (10% in 25% butyl Carbitol) (Alkasurf LAN-1)
HLB:
3.6 ± 1 (Lipocol L-1)
3.7 (Alkasurf LAN-1; Siponic L-1)
Acid No.:
2.0 max. (Lipocol L-1)
Hydroxyl No.:
231–243 (Lipocol L-1)
pH:
6.5 (1% sol'n.) (Siponic L-1)

POE (2) cetyl ether

SYNONYMS:
Ceteth-2 (CTFA)
PEG-2 cetyl ether
PEG 100 cetyl ether
POE (2) cetyl alcohol
STRUCTURE:
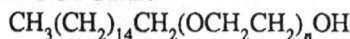
$CH_3(CH_2)_{14}CH_2(OCH_2CH_2)_nOH$
where avg. $n = 2$
CAS No.:
9004-95-9 (generic)
RD No.: 977054-67-3
TRADENAME EQUIVALENTS:
Brij 52 [ICI United States]
CE-55-2 [Hefti]

118

TRADENAME EQUIVALENTS *(cont'd.):*
Ethosperse CA2 [Glyco]
Hetoxol CA2 [Heterene]
Lipocol C-2 [Lipo]
Nikkol BC-2 [Nikko]
Simulsol 52 [Seppic]
Siponic C-20 [Alcolac]

CATEGORY:
Surfactant, detergent, emulsifier, coemulsifier, leveling agent, intermediate, defoamer, wetting agent, solubilizer, conditioning agent, antistat, thickener, stabilizer, opacifier

APPLICATIONS:
Bath products: bath oils (Hetoxol CA2)
Cosmetic industry preparations: (Ethosperse CA2); conditioners (Ethosperse CA2); creams and lotions (CE-55-2; Hetoxol CA2; Lipocol C-2; Nikkol BC-2; Siponic C-20); hair preparations (Ethosperse CA2; Nikkol BC-2); shampoos (Hetoxol CA2)
Household detergents: (Hetoxol CA2)
Industrial applications: dyes and pigments (Hetoxol CA2; Lipocol C-2); plastics (CE-55-2); silicone products (Hetoxol CA2); textile/leather processing (CE-55-2; Hetoxol CA2); waxes (Hetoxol CA2)
Industrial cleaners: textile cleaning (Hetoxol CA2)
Pharmaceutical applications: (CE-55-2); antiperspirant/deodorant (Ethosperse CA2; Lipocol C-2); depilatories (Ethosperse CA2; Lipocol C-2); ointments (Nikkol BC-2)

PROPERTIES:
Form:
Paste (Nikkol BC-2)
Solid (CE-55-2; Hetoxol CA2; Simulsol 52)
Solid wax (Lipocol C-2)
Color:
White (Ethosperse CA2; Lipocol C-2)
Composition:
100% active (Lipocol C-2; Siponic C-20)
100% conc. (CE-55-2; Nikkol BC-2; Simulsol 52)
Solubility:
Sol. in ethanol (Ethosperse CA2)
Sol. in isopropanol (Hetoxol CA2)
Sol. in min. oil (Hetoxol CA2); misc. in certain proportions (Ethosperse CA2)
Disp. in veg. oil (Ethosperse CA2)
Disp. hot in water (Ethosperse CA2)
Ionic Nature:
Nonionic (Brij 52; CE-55-2; Ethosperse CA2; Hetoxol CA2; Lipocol C-2; Nikkol BC-2; Simulsol 52; Siponic C-20)

POE (2) cetyl ether (cont'd.)

M.P.:
  29–33 C (Ethosperse CA2)
HLB:
  5.3 (Brij 52; Hetoxol CA2; Simulsol 52; Siponic C-20)
  5.3 ± 1 (Lipocol C-2)
  6.0 ± 1 (Ethosperse CA2)
  8.0 (Nikkol BC-2)
Acid No.:
  0.5 max. (Ethosperse CA2)
  1 max. (Lipocol C-2)
Hydroxyl No.:
  156–176 (Ethosperse CA2)
  160–180 (Hetoxol CA2; Lipocol C-2)
Stability:
  Chemically stable; resistant to acid hydrolysis and alkaline saponification (Ethosperse CA2)
TOXICITY/HANDLING:
  Low irritation to skin and eyes (Ethosperse CA2)

POE (6) cetyl ether

SYNONYMS:
  Ceteth-6 (CTFA)
  PEG-6 cetyl ether
  PEG 300 cetyl ether
STRUCTURE:
  $CH_3(CH_2)_{14}CH_2(OCH_2CH_2)_nOH$
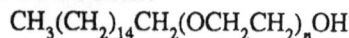
  where avg. n = 6
CAS No.:
  9004-95-9 (generic)
  RD No.: 977054-69-5
TRADENAME EQUIVALENTS:
  Nikkol BC-5.5 [Nikko]
CATEGORY:
  Emulsifier
APPLICATIONS:
  Cosmetic industry preparations: (Nikkol BC-5.5)
  Pharmaceutical applications: (Nikkol BC-5.5)

POE (6) cetyl ether *(cont'd.)*

PROPERTIES:
Form:
  Paste (Nikkol BC-5.5)
Composition:
  100% conc. (Nikkol BC-5.5)
Ionic Nature:
  Nonionic (Nikkol BC-5.5)
HLB:
  10.5 (Nikkol BC-5.5)

POE (10) cetyl ether

SYNONYMS:
  Ceteth-10 (CTFA)
  PEG-10 cetyl ether
  PEG 500 cetyl ether
  POE (10) cetyl alcohol
STRUCTURE:
  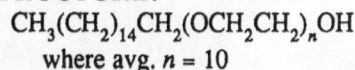
  $CH_3(CH_2)_{14}CH_2(OCH_2CH_2)_nOH$
    where avg. $n = 10$
CAS No.:
  9004-95-9 (generic)
  RD No.: 977054-70-8
TRADENAME EQUIVALENTS:
  Brij 56 [ICI United States]
  Hetoxol CA-10 [Heterene]
  Lipocol C-10 [Lipo]
  Nikkol BC-10TX [Nikko]
  Simulsol 56 [Seppic]
CATEGORY:
  Surfactant, detergent, emulsifier, defoamer, wetting agent, solubilizer, conditioning
    agent, leveling agent, intermediate
APPLICATIONS:
  Bath products: bath oils (Hetoxol CA-10)
  Cosmetic industry preparations: (Nikkol BC-10TX); creams and lotions (Hetoxol CA-
    10; Lipocol C-10); shampoos (Hetoxol CA-10)
  Household detergents: (Hetoxol CA-10)
  Industrial applications: dyes and pigments (Hetoxol CA-10; Lipocol C-10); silicone
    products (Hetoxol CA-10); textile/leather processing (Hetoxol CA-10); waxes
    (Hetoxol CA-10)

POE (10) cetyl ether (cont'd.)

Industrial cleaners: textile cleaning (Hetoxol CA-10)
Pharmaceutical applications: (Nikkol BC-10TX); antiperspirant/deodorant (Lipocol C-10); depilatories (Lipocol C-10)

PROPERTIES:

Form:
Solid (Hetoxol CA-10; Nikkol BC-10TX; Simulsol 56)
Waxy solid (Brij 56)
Solid wax (Lipocol C-10)

Color:
White (Brij 56; Lipocol C-10)
Gardner 1 max. (Hetoxol CA-10)

Composition:
100% active (Lipocol C-10)
100% conc. (Nikkol BC-10TX; Simulsol 56)

Solubility:
Sol. in alcohols (Brij 56)
Sol. in isopropanol (Hetoxol CA-10)
Insol. in min. oil (Hetoxol CA-10)
Sol. in water (Hetoxol CA-10)

Ionic Nature:
Nonionic (Brij 56; Lipocol C-10; Hetoxol CA-10; Nikkol BC-10TX; Simulsol 56)

Pour Pt.:
31 C (Brij 56)

Flash Pt.:
> 300 F (Brij 56)

Fire Pt.:
> 300 F (Brij 56)

HLB:
12.7 (Hetoxol CA-10)
12.9 (Brij 56; Simulsol 56)
12.9 ± 1 (Lipocol C-10)
13.5 (Nikkol BC-10TX)

Acid No.:
1 max. (Hetoxol CA-10; Lipocol C-10)

Hydroxyl No.:
75–90 (Hetoxol CA-10; Lipocol C-10)

Stability:
Acid and alkaline stable (Lipocol C-10)

POE (15) cetyl ether

SYNONYMS:
  Ceteth-15 (CTFA)
  PEG-15 cetyl ether
  PEG (15) cetyl ether
STRUCTURE:
  $CH_3(CH_2)_{14}CH_2(OCH_2CH_2)_nOH$
    where avg. $n = 15$
CAS No.:
  9004-95-9 (generic)
TRADENAME EQUIVALENTS:
  Nikkol BC-15TX [Nikko]
CATEGORY:
  Emulsifier
APPLICATIONS:
  Cosmetic industry preparations: (Nikkol BC-15TX)
  Pharmaceutical applications: (Nikkol BC-15TX)
PROPERTIES:
Form:
  Solid (Nikkol BC-15TX)
Composition:
  100% conc. (Nikkol BC-15TX)
Ionic Nature:
  Nonionic (Nikkol BC-15TX)
HLB:
  15.5 (Nikkol BC-15TX)

POE (20) cetyl ether

SYNONYMS:
  Ceteth-20 (CTFA)
  Cetomacrogol 1000
  Cetomacrogol 1000 BPC
  PEG-20 cetyl ether
  PEG 1000 cetyl ether
  POE (20) cetyl alcohol
STRUCTURE:
  $CH_3(CH_2)_{14}CH_2(OCH_2CH_2)_nOH$
    where avg. $n = 20$
CAS No.:
  9004-95-9 (generic)

POE (20) cetyl ether (cont'd.)

TRADENAME EQUIVALENTS:
Brij 58 [ICI United States]
CE-55-20 [Hefti]
Hetoxol CA-20 [Heterene]
Lipocol C-20 [Lipo]
Nikkol BC-20TX [Nikko]
Simulsol 58 [Seppic]

CATEGORY:
Surfactant, detergent, emulsifier, defoamer, wetting agent, solubilizer, conditioning agent, leveling agent, intermediate

APPLICATIONS:
Bath products: bath oils (Hetoxol CA-20)
Cosmetic industry preparations: (Nikkol BC-20TX); creams and lotions (Hetoxol CA-20; Lipocol C-20); shampoos (Hetoxol CA-20)
Household detergents: (Hetoxol CA-20)
Industrial applications: dyes and pigments (Hetoxol CA-20; Lipocol C-20); silicone products (Hetoxol CA-20); textile/leather processing (Hetoxol CA-20); waxes (CE-55-20; Hetoxol CA-20)
Industrial cleaners: textile cleaning (Hetoxol CA-20)
Pharmaceutical applications: (Nikkol BC-20TX); antiperspirant/deodorant (Lipocol C-20); depilatories (Lipocol C-20)

PROPERTIES:
Form:
Solid (CE-55-20; Hetoxol CA-20; Nikkol BC-20TX; Simulsol 58)
Waxy solid (Brij 58)
Solid wax (Lipocol C-20)
Color:
White (Brij 58; Lipocol C-20)
Composition:
100% active (Lipocol C-20)
100% conc. (CE-55-20; Nikkol BC-20TX; Simulsol 58)
Solubility:
Sol. in alcohols (Brij 58)
Sol. in isopropanol (Hetoxol CA-20)
Sol. in water (Brij 58; Hetoxol CA-20)
Ionic Nature:
Nonionic (Brij 58; CE-55-20; Lipocol C-20; Hetoxol CA-20; Nikkol BC-20TX; Simulsol 58)
Pour Pt.:
38 C (Brij 58)
Flash Pt.:
> 300 F (Brij 58)

Fire Pt.:
> 300 F (Brij 58)

HLB:
15.7 (Brij 58; Hetoxol CA-20; Simulsol 58)
16.0 (CE-55-20)
15.7 ± 1 (Lipocol C-20)
17.0 (Nikkol BC-20TX)

Acid No.:
2 max. (Lipocol C-20)

Hydroxyl No.:
45–60 (Hetoxol CA-20)
50–58 (Lipocol C-20)

Stability:
Acid and alkaline stable (Lipocol C-20)

POE (25) cetyl ether

SYNONYMS:
Ceteth-25 (CTFA)
PEG-25 cetyl ether
PEG (25) cetyl ether

STRUCTURE:
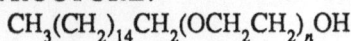
$CH_3(CH_2)_{14}CH_2(OCH_2CH_2)_nOH$
where avg. $n = 25$

CAS No.:
9004-95-9 (generic)

TRADENAME EQUIVALENTS:
Nikkol BC-25TX [Nikko]

CATEGORY:
Emulsifier

APPLICATIONS:
Cosmetic industry preparations: (Nikkol BC-25TX)
Pharmaceutical applications: (Nikkol BC-25TX)

PROPERTIES:

Form:
Solid (Nikkol BC-25TX)

Composition:
100% conc. (Nikkol BC-25TX)

Ionic Nature:
Nonionic (Nikkol BC-25TX)

POE (25) cetyl ether *(cont'd.)*

HLB:
   18.5 (Nikkol BC-25TX)

POE (30) cetyl ether

SYNONYMS:
   Ceteth-30 (CTFA)
   PEG-30 cetyl ether
   PEG (30) cetyl ether
STRUCTURE:
   $CH_3(CH_2)_{14}CH_2(OCH_2CH_2)_nOH$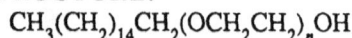
      where avg. $n = 30$
CAS No.:
   9004-95-9 (generic)
   RD No.: 977054-71-9
TRADENAME EQUIVALENTS:
   Nikkol BC-30TX [Nikko]
CATEGORY:
   Emulsifier
APPLICATIONS:
   Cosmetic industry preparations: (Nikkol BC-30TX)
   Pharmaceutical applications: (Nikkol BC-30TX)
PROPERTIES:
Form:
   Solid (Nikkol BC-30TX)
Composition:
   100% conc. (Nikkol BC-30TX)
Ionic Nature:
   Nonionic (Nikkol BC-30TX)
HLB:
   19.5 (Nikkol BC-30TX)

POE (11) cetyl/stearyl ether

SYNONYMS:
   Ceteareth-11 (CTFA)
   PEG-11 cetyl/stearyl ether

POE (11) cetyl/stearyl ether *(cont'd.)*

STRUCTURE:

R(OCH$_2$CH$_2$)$_n$OH

where R represents a blend of cetyl and stearyl radicals and
avg. $n = 11$

TRADENAME EQUIVALENTS:

Cremophor A11 [BASF AG]

CATEGORY:

Emulsifier

APPLICATIONS:

Cosmetic industry preparations: (Cremophor A11); creams and liquid emulsions (Cremophor A11)

Pharmaceutical applications: (Cremophor A11); ointments (Cremophor A11)

PROPERTIES:

Form:

Wax (Cremophor A11)

Color:

White (Cremophor A11)

Composition:

100% active (Cremophor A11)

Solubility:

Sol. in alcohol (Cremophor A11)

Sol. in water (Cremophor A11)

Ionic Nature: Nonionic

Density:

0.964–0.968 g/cm^3 (60 C) (Cremophor A11)

HLB:

12–14 (Cremophor A11)

Acid No.:

< 1 (Cremophor A11)

Iodine No.:

< 1 (Cremophor A11)

Saponification No.:

< 1 (Cremophor A11)

Hydroxyl No.:

70–80 (Cremophor A11)

Stability:

Stable to acids, bass and salts (Cremophor A11)

Ref. Index:

1.4464–1.4474 (60 C) (Cremophor A11)

pH:

6–7 (Cremophor A11)

POE (25) cetyl/stearyl ether

SYNONYMS:
Ceteareth-25 (CTFA)
PEG-25 cetyl/stearyl ether

STRUCTURE:
$R(OCH_2CH_2)_nOH$
> where R represents a blend of cetyl and stearyl radicals and
> avg. $n = 25$

TRADENAME EQUIVALENTS:
Cremophor A25 [BASF AG]

CATEGORY:
Emulsifier

APPLICATIONS:
Cosmetic industry preparations: (Cremophor A25); creams and liquid emulsions (Cremophor A25)
Pharmaceutical applications: (Cremophor A25); ointments (Cremophor A25)

PROPERTIES:
Form:
Powder (Cremophor A25)
Color:
White (Cremophor A25)
Composition:
100% active (Cremophor A25)
Solubility:
Sol. in alcohol (Cremophor A25)
Sol. in water (Cremophor A25)
Ionic Nature: Nonionic
Density:
1.020–1.028 g/cm^3 (60 C) (Cremophor A25)
HLB:
15–17 (Cremophor A25)
Acid No.:
< 1 (Cremophor A25)
Iodine No.:
< 1 (Cremophor A25)
Saponification No.:
< 3 (Cremophor A25)
Hydroxyl No.:
35–45 (Cremophor A25)
Stability:
Stable to acids, bases and salts (Cremophor A25)
Ref. Index:
1.4512–1.4520 (60 C) (Cremophor A25)

pH:
   5–7 (Cremophor A25)

POE (27) cetyl/stearyl ether

SYNONYMS:
   Ceteareth-27 (CTFA)
   PEG-27 cetyl/stearyl ether
   PEG (27) cetyl stearyl ether
STRUCTURE:
   $R(OCH_2CH_2)_n OH$
      where R represents a blend of cetyl and stearyl radicals and
         avg. $n = 27$
CAS No.:
   RD No.: 977063-71-0
TRADENAME EQUIVALENTS:
   Plurafac A-38 [BASF Wyandotte]
CATEGORY:
   Detergent
APPLICATIONS:
   Household detergents: (Plurafac A-38); all-purpose cleaner (Plurafac A-38); heavy-
      duty cleaner (Plurafac A-38); light-duty cleaners (Plurafac A-38); liquid detergents
      (Plurafac A-38); powdered detergents (Plurafac A-38)
   Industrial cleaners: (Plurafac A-38)
PROPERTIES:
Form:
   Solid (Plurafac A-38)
Color:
   White (Plurafac A-38)
Composition:
   100% conc. (Plurafac A-38)
Solubility:
   Sol. in acetone (Plurafac A-38)
   Sol. in butyl Cellosolve (Plurafac A-38)
   Sol. in chloroform (Plurafac A-38)
   Sol. in ethanol (Plurafac A-38)
   Sol. in isopropanol (Plurafac A-38)
   Sol. in MEK (Plurafac A-38)
   Sol. in methanol (Plurafac A-38)
   Sol. in perchloroethylene (Plurafac A-38)

POE (27) cetyl/stearyl ether (cont'd.)

Sol. in toluene (Plurafac A-38)
Sol. in water (Plurafac A-38)
Sol. in xylene (Plurafac A-38)
Ionic Nature: Nonionic
M.P.:
114 F (Plurafac A-38)
Flash Pt.:
480 F (Plurafac A-38)
Fire Pt.:
570 F (Plurafac A-38)
Cloud Pt.:
> 100 C (1%) (Plurafac A-38)
HLB:
19 (Plurafac A-38)
pH:
6.0–7.0 (Plurafac A-38)
Surface Tension:
43.6 dynes/cm (0.1%) (Plurafac A-38)
Biodegradable: (Plurafac A-38)
TOXICITY/HANDLING:
Mild skin irritant (Plurafac A-38)

POE (30) cetyl/stearyl ether

SYNONYMS:
Ceteareth-30 (CTFA)
PEG-30 cetyl/stearyl ether
PEG (30) cetyl/stearyl ether
POE (30) cetyl/stearyl alcohol
STRUCTURE:
$R(OCH_2CH_2)_nOH$
where   R represents a blend of cetyl and stearyl radicals and
avg. $n = 30$
CAS No.:
RD No.: 977063-72-1
TRADENAME EQUIVALENTS:
Eumulgin B3 [Henkel]
Hetoxol CS30 [Heterene]
Incropol CS-30 [Croda Surfactants]
Siponic E-15 [Alcolac]
Standamul B-3 [Henkel]

POE (30) cetyl/stearyl ether (cont'd.)

CATEGORY:

Solubilizer, coupling agent, emulsifier, detergent, leveling agent, antistat, dye assistant, intermediate, lubricant, conditioning agent

APPLICATIONS:

Automobile cleaners: car shampoo/conditioner (Standamul B-3)

Bath products: bath oils (Hetoxol CS30)

Cosmetic industry preparations: (Eumulgin B3; Incropol CS-30); creams and lotions (Hetoxol CS30); hair preparations (Standamul B-3); perfumery/essential oils (Eumulgin B3; Standamul B-3); shampoos (Hetoxol CS30; Siponic E-15)

Household detergents: (Hetoxol CS30; Incropol CS-30; Standamul B-3)

Industrial applications: (Incropol CS-30); dyes and pigments (Hetoxol CS30); paper mfg. (Siponic E-15; Standamul B-3); polishes and waxes (Siponic E-15; Standamul B-3); polymers/polymerization (Siponic E-15); rubber (Siponic E-15); silicone products (Hetoxol CS30); textile/leather processing (Hetoxol CS30; Incropol CS-30; Siponic E-15; Standamul B-3); waxes and oils (Hetoxol CS30)

Industrial cleaners: (Incropol CS-30; Standamul B-3); textile cleaning (Hetoxol CS30)

Pharmaceutical applications: (Eumulgin B3; Siponic E-15); deodorant sticks (Standamul B-3)

PROPERTIES:

Form:

Solid (Eumulgin B3; Hetoxol CS30; Incropol CS-30)

Wax (Siponic E-15)

Waxy solid (Standamul B-3)

Color:

Gardner 1 max. (Incropol CS-30)

Composition:

99.5–100.0% solids (Standamul B-3)

100% active (Incropol CS-30; Siponic E-15)

100% conc. (Eumulgin B3)

Solubility:

Sol. in isopropanol (Hetoxol CS30)

Sol. in water (Hetoxol CS30)

Ionic Nature:

Nonionic (Eumulgin B3; Hetoxol CS30; Incropol CS-30; Siponic E-15; Standamul B-3)

Sp.gr.:

1.023 (70 C) (Standamul B-3)

Solidification Pt.:

43–46 C (Standamul B-3)

Cloud Pt.:

> 95 C (1% NaCl) (Siponic E-15)

HLB:

15.0 (Standamul B-3)

POE (30) cetyl/stearyl ether (cont'd.)

16.7 (Hetoxol CS30; Siponic E-15)
Hydroxyl No.:
34–38 (Incropol CS-30)
35–50 (Standamul B-3)
40–52 (Hetoxol CS30)
pH:
5.5–7.5 (3%) (Incropol CS-30)
6.5 (1% sol'n.) (Siponic E-15)
Biodegradable: (Incropol CS-30)
TOXICITY/HANDLING:
Although nontoxic, ingestion should be avoided (Standamul B-3)
STORAGE/HANDLING:
Store for prolonged periods in a cool, dry place in sealed containers at 30 C (Standamul B-3)
STD. PKGS.:
450 lb net steel drums; 110 lb net fiber drums (Standamul B-3)

POE (50) cetyl/stearyl ether

SYNONYMS:
Ceteareth-50 (CTFA)
PEG-50 cetyl/stearyl ether
PEG (50) cetyl stearyl ether
POE (50) cetyl/stearyl alcohol
STRUCTURE:
$R(OCH_2CH_2)_nOH$
where R represents a blend of cetyl and stearyl radicals and
avg. $n = 50$
TRADENAME EQUIVALENTS:
Hetoxol CS50, CS50 Special [Heterene]
Incropol CS-50 [Croda Surfactants]
CATEGORY:
Detergent, emulsifier, leveling agent, intermediate, lubricant, coupling agent, solubilizer, antistat
APPLICATIONS:
Bath products: bath oils (Hetoxol CS50, CS50 Special)
Cosmetic industry preparations: (Incropol CS-50); creams and lotions (Hetoxol CS50, CS50 Special); shampoos (Hetoxol CS50, CS50 Special)
Household detergents: (Hetoxol CS50, CS50 Special; Incropol CS-50)
Industrial applications: (Incropol CS-50); dyes and pigments (Hetoxol CS50, CS50

POE (50) cetyl/stearyl ether *(cont'd.)*

Special); silicone products (Hetoxol CS50, CS50 Special); textile/leather processing (Hetoxol CS50, CS50 Special; Incropol CS-50); waxes and oils (Hetoxol CS50, CS50 Special)

Industrial cleaners: (Incropol CS-50); textile cleaning (Hetoxol CS50, CS50 Special)

PROPERTIES:
Form:
Flake (Hetoxol CS50, CS50 Special; Incropol CS-50)
Color:
Gardner 1 max. (Incropol CS-50)
Gardner 2 max. (Hetoxol CS50, CS50 Special)
Composition:
100% active (Incropol CS-50)
Solubility:
Sol. in isopropanol (Hetoxol CS50, CS50 Special)
Insol. in min. oil (Hetoxol CS50, CS50 Special)
Sol. in water (Hetoxol CS50); sol. in water (gels) (Hetoxol CS50 Special)
Ionic Nature:
Nonionic (Hetoxol CS50, CS50 Special; Incropol CS-50)
Setting Pt.:
44–48 C (Hetoxol CS50 Special)
Acid No.:
2.0 max. (Hetoxol CS50)
Hydroxyl No.:
18–24 (Hetoxol CS50 Special)
18–27 (Incropol CS-50)
20–40 (Hetoxol CS50)
pH:
6.0–8.0 (3%) (Incropol CS-50)
Biodegradable: (Incropol CS-50)

POE (55) cetyl/stearyl ether

SYNONYMS:
Ceteareth-55 (CTFA)
PEG-55 cetyl/stearyl ether
PEG (55) cetyl stearyl ether
STRUCTURE:
$R(OCH_2CH_2)_nOH$
where R represents a blend of cetyl and stearyl radicals and
avg. $n = 55$

133

POE (55) cetyl/stearyl ether *(cont'd.)*

TRADENAME EQUIVALENTS:
Plurafac A-39 [BASF Wyandotte]
CATEGORY:
Detergent
APPLICATIONS:
Household detergents: (Plurafac A-39); all-purpose cleaner (Plurafac A-39); heavy-duty cleaner (Plurafac A-39); light-duty cleaners (Plurafac A-39); liquid detergents (Plurafac A-39); powdered detergents (Plurafac A-39)
Industrial cleaners: (Plurafac A-39)
PROPERTIES:
Form:
Solid (Plurafac A-39)
Color:
White (Plurafac A-39)
Composition:
100% conc. (Plurafac A-39)
Solubility:
Sol. in acetone (< 10%) (Plurafac A-39)
Sol. in chloroform (Plurafac A-39)
Sol. in ethanol (< 10%) (Plurafac A-39)
Sol. in MEK (< 10%) (Plurafac A-39)
Sol. in methanol (< 10%) (Plurafac A-39)
Sol. in toluene (< 10%) (Plurafac A-39)
Sol. in water (Plurafac A-39)
Sol. in xylene (< 10%) (Plurafac A-39)
Ionic Nature: Nonionic
M.P.:
132 F (Plurafac A-39)
Flash Pt.:
480 F (Plurafac A-39)
Fire Pt.:
570 F (Plurafac A-39)
Cloud Pt.:
> 100 C (1%) (Plurafac A-39)
HLB:
20 (Plurafac A-39)
pH:
6.0–7.0 (Plurafac A-39)
Surface Tension:
53.2 dynes/cm (0.1%) (Plurafac A-39)
Biodegradable: (Plurafac A-39)
TOXICITY/HANDLING:
Mild skin irritant (Plurafac A-39)

POE (2) coconut amine

SYNONYMS:

Coco amine + EO (2 moles)

PEG 100 coconut amine

PEG-2 cocamine (CTFA)

POE (2) coco amine

STRUCTURE:

$$R-N \begin{cases} (CH_2CH_2O)_xH \\ (CH_2CH_2O)_yH \end{cases}$$

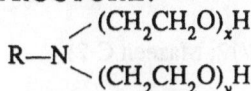

where R represents the coconut radical and
avg. $(x + y) = 2$

CAS No.:

61791-14-8 (generic)

RD No.: 977061-61-2

TRADENAME EQUIVALENTS:

Accomeen C2 [Armstrong]

Alkaminox C-2 [Alkaril]

Chemeen C-2 [Chemax]

Crodamet 1.C2 [Croda Chem. Ltd.]

Ethomeen C/12 [Armak]

Hetoxamine C-2 [Heterene]

Mazeen C-2 [Mazer]

Teric 12M2 [ICI Australia Ltd.]

CATEGORY:

Emulsifier, antistat, surfactant, corrosion inhibitor, dye leveler, wetting agent, rewetting agent, lubricant, dispersant, desizing agent, softener, stabilizer

APPLICATIONS:

Cosmetic industry preparations: (Mazeen C-2)

Farm products: agricultural oils/sprays (Hetoxamine C-2); herbicides (Mazeen C-2); insecticides/pesticides (Mazeen C-2)

Industrial applications: construction (Teric 12M2); dyes and pigments (Teric 12M2); lubricating/cutting oils (Mazeen C-2; Teric 12M2); metalworking (Teric 12M2); paper mfg. (Teric 12M2); polishes and waxes (Hetoxamine C-2); printing inks (Mazeen C-2); steam generating/circulating systems (Alkaminox C-2); textile/leather processing (Ethomeen C/12; Hetoxamine C-2; Mazeen C-2; Teric 12M2)

PROPERTIES:

Form:

Liquid (Accomeen C2; Chemeen C-2; Hetoxamine C-2; Mazeen C-2; Teric 12M2)

Clear liquid (Alkaminox C-2; Ethomeen C/12)

Color:

Amber (Alkaminox C-2)

Gardner 4 (Accomeen C2)

Gardner 6 max. (Ethomeen C/12)

135

POE (2) coconut amine (cont'd.)

Gardner 11 (Mazeen C-2)

Odor:

Amine (Accomeen C2)

Composition:

95% tertiary amine min. (Ethomeen C/12)

99% active (Accomeen C2)

100% active (Teric 12M2)

100% conc. (Chemeen C-2; Crodamet 1.C2; Ethomeen C/12; Mazeen C-2)

Solubility:

Sol. in acetone (Ethomeen C/12; Mazeen C-2)

Sol. in benzene (Ethomeen C/12; Mazeen C-2; Teric 12M2)

Sol. in carbon tetrachloride (Ethomeen C/12)

Sol. in ethanol (Teric 12M2)

Sol. in ethyl acetate (Teric 12M2)

Sol. in isopropanol (Ethomeen C/12; Hetoxamine C-2; Mazeen C-2)

Sol. in kerosene (Teric 12M2)

Sol. in min. acid (Alkaminox C-2)

Sol. in min. oil (Hetoxamine C-2; Mazeen C-2; Teric 12M2)

Sol. in olein (Teric 12M2)

Sol. in organic solvents (Accomeen C2; Alkaminox C-2)

Sol. in paraffin oil (Teric 12M2)

Sol. in perchlorethylene (Teric 12M2)

Sol. in Stoddard solvent (Ethomeen C/12)

Sol. in veg. oil (Teric 12M2)

Insol. in water (Accomeen C2; Alkaminox C-2); forms gel in water (Ethomeen C/12; Hetoxamine C-2; Mazeen C-2)

Ionic Nature:

Nonionic (Accomeen C2; Teric 12M2)

Cationic (Crodamet 1.C2; Ethomeen C/12; Hetoxamine C-2; Mazeen C-2)

M.W.:

285 (Hetoxamine C-2; Mazeen C-2)

290 (Chemeen C-2)

Sp.gr.:

0.87 (Accomeen C2; Alkaminox C-2; Ethomeen C/12)

0.874 (Mazeen C-2)

0.913 (Teric 12M2)

Density:

7.25 lb/gal (Accomeen C2)

Visc.:

166 cps (20 C) (Teric 12M2)

M.P.:

-1 ± 2 C (Teric 12M2)

POE (2) coconut amine *(cont'd.)*

HLB:
   6.4 (Ethomeen C/12)
   11.4 (Teric 12M2)

Stability:
   Good in hard or saline waters and in reasonable concs. of acids and alkalis (Teric 12M2)

pH:
   8–10 (1% aq.) (Teric 12M2)

Surface Tension:
   28 dynes/cm (0.1% sol'n.) (Mazeen C-2)
   28.2 dynes/cm (0.01%, 20 C) (Teric 12M2)

TOXICITY/HANDLING:
   Corrosive (Accomeen C2)
   Skin irritant, severe eye irritant (Ethomeen C/12)
   May cause skin and eye irritation; spillages are slippery (Teric 12M2)

POE (5) coconut amine

SYNONYMS:
   PEG-5 cocamine (CTFA)
   PEG (5) coconut amine

STRUCTURE:

$$R-N \begin{cases} (CH_2CH_2O)_xH \\ (CH_2CH_2O)_yH \end{cases}$$

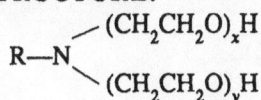

   where   R represents the coconut radical and
             avg. $(x + y) = 5$

CAS No.:
   61791-14-8 (generic)
   977066-78-6

TRADENAME EQUIVALENTS:
   Accomeen C5 [Armstrong]
   Chemeen C-5 [Chemax]
   Crodamet 1.C5 [Croda Chem. Ltd.]
   Ethomeen C/15 [Armak]
   Hetoxamine C5 [Heterene]
   Mazeen C-5 [Mazer]
   Teric 12M5 [ICI Australia Ltd.]

CATEGORY:
   Emulsifier, surfactant, antistat, desizing agent, softener, dye leveler, wetting agent, rewetting agent, lubricant, dispersant, water repellent, stabilizer

POE (5) coconut amine *(cont'd.)*

APPLICATIONS:
Cosmetic industry preparations: (Mazeen C-5)
Farm products: agricultural oils/sprays (Hetoxamine C5); herbicides (Mazeen C-5); insecticides/pesticides (Mazeen C-5)
Industrial applications: construction (Teric 12M5); dyes and pigments (Chemeen C-5; Teric 12M5); lubricating/cutting oils (Mazeen C-5; Teric 12M5); metalworking (Teric 12M5); paper mfg. (Teric 12M5); polishes and waxes (Hetoxamine C5); printing inks (Mazeen C-5); textile/leather processing (Chemeen C-5; Ethomeen C/ 15; Hetoxamine C5; Mazeen C-5; Teric 12M5)
Industrial cleaners: metal processing surfactants (Hetoxamine C5)

PROPERTIES:
Form:
Liquid (Accomeen C5; Chemeen C-5; Hetoxamine C5; Teric 12M5)
Clear liquid (Ethomeen C/15)
Color:
Gardner 4 (Accomeen C5)
Gardner 6 max. (Ethomeen C/15)
Odor:
Amine (Accomeen C5)
Composition:
99% active (Accomeen C5)
100% active (Teric 12M5)
100% conc. (Chemeen C-5; Crodamet 1.C5; Mazeen C-5)
Solubility:
Sol. in acetone (Mazeen C-5); (> 25 C) (Ethomeen C/15)
Sol. in benzene (Ethomeen C/15; Mazeen C-5; Teric 12M5)
Sol. in carbon tetrachloride (> 25 C) (Ethomeen C/15)
Sol. in dioxane (Ethomeen C/15)
Sol. in ethanol (Teric 12M5)
Sol. in ethyl acetate (Teric 12M5)
Sol. in isopropanol (Ethomeen C/15; Hetoxamine C5; Mazeen C-5)
Disp. in kerosene (Teric 12M5)
Sol. in min. oil (Hetoxamine C5; Mazeen C-5; Teric 12M5)
Sol. in olein (Teric 12M5)
Sol. in organic solvents (Accomeen C5)
Sol. in paraffin oil (Teric 12M5)
Sol. in perchloroethylene (Teric 12M5)
Sol. in Stoddard solvent (> 10 C) (Ethomeen C/15)
Sol. in veg. oil (Teric 12M5)
Sol. in water (Accomeen C5; Hetoxamine C5; Mazeen C-5); sol. cloudy (Ethomeen C/ 15); disp. (Teric 12M5)
Ionic Nature:
Nonionic (Accomeen C5; Teric 12M5)

Cationic (Crodamet 1.C5; Ethomeen C/15; Hetoxamine C5; Mazeen C-5)
M.W.:
   425 (Chemeen C-5; Hetoxamine C5; Mazeen C-5)
Sp.gr.:
   0.971 (Teric 12M5)
   0.976 (Mazeen C-5)
   0.98 (Accomeen C5; Ethomeen C/15)
Density:
   8.15 lb/gal (Accomeen C5)
Visc.:
   311 cps (20 C) (Teric 12M5)
M.P.:
   -9 ± 2 C (Teric 12M5)
HLB:
   12.4 (Teric 12M5)
   13.9 (Ethomeen C/15)
Stability:
   Good in hard or saline waters and in reasonable concs. of acids and alkalis (Teric 12M5)
pH:
   8–10 (1% aq.) (Teric 12M5)
Surface Tension:
   29.2 dynes/cm (0.01%, 20 C) (Teric 12M5)
   33 dynes/cm (0.1% sol'n.) (Ethomeen C/15; Mazeen C-5)
TOXICITY/HANDLING:
   Skin irritant, severe eye irritant (Ethomeen C/15)
   May cause skin and eye irritation; spillages are slippery (Teric 12M5)

POE (10) coconut amine

SYNONYMS:
   PEG-10 cocamine (CTFA)
   PEG 500 coconut amine
STRUCTURE:

$$R-N \begin{cases} (CH_2CH_2O)_xH \\ (CH_2CH_2O)_yH \end{cases}$$

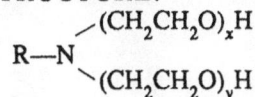

   where   R represents the coconut radical and
           avg. $(x + y) = 10$
CAS No.:
   61791-14-8 (generic)
   RD No.: 977063-21-0

POE (10) coconut amine (cont'd.)

TRADENAME EQUIVALENTS:
Accomeen C10 [Armstrong]
Chemeen C-10 [Chemax]
Crodamet 1.C10 [Croda Chem. Ltd.]
Ethomeen C/20 [Armak]
Mazeen C10 [Mazer]
Trymeen CAM-10 [Emery]

CATEGORY:
Emulsifier, coemulsifier, surfactant, antistat, dye leveler, wetting agent, rewetting agent, lubricant, dispersant

APPLICATIONS:
Cosmetic industry preparations: (Mazeen C10)
Farm products: herbicides (Mazeen C10); insecticides/pesticides (Mazeen C10)
Industrial applications: lubricating/cutting oils (Mazeen C10; Trymeen CAM-10); metalworking (Chemeen C-10); printing inks (Mazeen C10); textile/leather processing (Chemeen C-10; Ethomeen C/20; Mazeen C10; Trymeen CAM-10)

PROPERTIES:

Form:
Liquid (Accomeen C10; Chemeen C-10; Mazeen C10; Trymeen CAM-10)
Clear liquid (Ethomeen C/20)

Color:
Gardner 6 (Accomeen C10)
Gardner 8 (Trymeen CAM-10)
Gardner 11 (Mazeen C10)
Gardner 11 max. (Ethomeen C/20)

Odor:
Amine (Accomeen C10)
Characteristic
Mild
Pleasant
Typical

Composition:
99% active (Accomeen C10)
100% active (Trymeen CAM-10)
100% conc. (Chemeen C-10; Crodamet 1.C10; Mazeen C10)

Solubility:
Sol. in acetone (Ethomeen C/20; Mazeen C10)
Sol. in benzene (Ethomeen C/20; Mazeen C10)
Disp. in butyl stearate (@ 5%) (Trymeen CAM-10)
Sol. in carbon tetrachloride (> 25 C) (Ethomeen C/20)
Sol. in dioxane (Ethomeen C/20)
Sol. in glycerol trioleate (@ 5%) (Trymeen CAM-10)
Sol. in isopropanol (Ethomeen C/20; Mazeen C10)

POE (10) coconut amine (cont'd.)

Sol. in organic solvents (Accomeen C10)
Sol. in perchloroethylene (@ 5%) (Trymeen CAM-10)
Sol. in Stoddard solvent (> 75 C) (Ethomeen C/20); disp. (@ 5%) (Trymeen CAM-10)
Sol. in water (Accomeen C10; Mazeen C10); sol. (@ 5%) (Trymeen CAM-10); sol. cloudy (Ethomeen C/20)

Ionic Nature:
Nonionic (Accomeen C10)
Cationic (Crodamet 1.C10; Ethomeen C/20; Mazeen C10; Trymeen CAM-10)

M.W.:
645 (Chemeen C-10; Mazeen C10)

Sp.gr.:
1.0 (Accomeen C10)
1.017 (Mazeen C10)
1.02 (Ethomeen C/20)

Density:
8.3 lb/gal (Accomeen C10)
8.4 lb/gal (Trymeen CAM-10)

Visc.:
260 cs (Trymeen CAM-10)

Cloud Pt.:
> 100 C (Trymeen CAM-10)

HLB:
13.8 (Trymeen CAM-10)

Surface Tension:
39 dynes/cm (0.1% sol'n.) (Ethomeen C/20; Mazeen C10)

TOXICITY/HANDLING:
Corrosive (Accomeen C10)
Skin irritant, severe eye irritant (Ethomeen C/20)

POE (15) coconut amine

SYNONYMS:
PEG-15 cocamine (CTFA)
PEG (15) coconut amine

STRUCTURE:

$$R-N \begin{cases} (CH_2CH_2O)_xH \\ (CH_2CH_2O)_yH \end{cases}$$

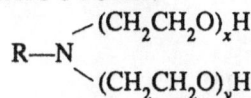

where R represents the coconut radical and
avg. $(x + y) = 15$

141

POE (15) coconut amine (cont'd.)

CAS No.:
8051-52-3; 61791-14-8 (generic)

TRADENAME EQUIVALENTS:
Accomeen C15 [Armstrong]
Chemeen C-15 [Chemax]
Crodamet 1.C15 [Croda Chem. Ltd.]
Ethomeen C/25 [Armak]
Hetoxamine C15 [Heterene]
Mazeen C15 [Mazer]
Teric 12M15 [ICI Australia Ltd.]
Trymeen CAM-15 [Emery]

CATEGORY:
Emulsifier, coemulsifier, surfactant, antistat, dye leveler, wetting agent, rewetting
agent, lubricant, dispersant, desizing agent, softener, water repellent, stabilizer

APPLICATIONS:
Cosmetic industry preparations: (Mazeen C15)
Farm products: agricultural oils/sprays (Hetoxamine C15); herbicides (Mazeen C15);
insecticides/pesticides (Mazeen C15)
Industrial applications: construction (Teric 12M15); dyes and pigments (Teric
12M15); lubricating/cutting oils (Mazeen C15; Teric 12M15; Trymeen CAM-15);
metalworking (Chemeen C-15; Teric 12M15); paper mfg. (Teric 12M15); polishes
and waxes (Hetoxamine C15); printing inks (Mazeen C15); textile/leather processing (Accomeen C15; Chemeen C-15; Ethomeen C/25; Hetoxamine C15; Mazeen
C15; Teric 12M15; Trymeen CAM-15)
Industrial cleaners: metal processing surfactants (Hetoxamine C15)

PROPERTIES:

Form:
Liquid (Accomeen C15; Chemeen C-15; Hetoxamine C15; Mazeen C15; Trymeen
CAM-15)
Clear liquid (Ethomeen C/25)

Color:
Gardner 6 (Trymeen CAM-15)
Gardner 6 max. (Accomeen C15)
Gardner 9 (Mazeen C15)
Gardner 12 max. (Ethomeen C/25)

Odor:
Mild amine (Accomeen C15)

Composition:
99% active (Accomeen C15)
100% active (Teric 12M15; Trymeen CAM-15)
100% conc. (Chemeen C-15; Crodamet 1.C15; Mazeen C15)

Solubility:
Sol. in acetone (Ethomeen C/25; Mazeen C15)

Sol. in benzene (Ethomeen C/25; Mazeen C15; Teric 12M15)
Sol. in carbon tetrachloride (> 25 C) (Ethomeen C/25)
Sol. in dioxane (Ethomeen C/25)
Sol. in ethanol (Teric 12M15)
Sol. in ethyl acetate (Teric 12M15)
Sol. in glycerol trioleate (@ 5%) (Trymeen CAM-15)
Sol. in isopropanol (Ethomeen C/25; Hetoxamine C15; Mazeen C15)
Partly sol. in kerosene (Teric 12M15)
Partly sol. in min. oil (Teric 12M15)
Sol. in olein (Teric 12M15)
Partly sol. in paraffin oil (Teric 12M15)
Sol. in perchloroethylene (Teric 12M15); sol. (@ 5%) (Trymeen CAM-15)
Sol. in Stoddard solvent (> 80 C) (Ethomeen C/25)
Sol. in veg. oil (Teric 12M15)
Sol. in water (Ethomeen C/25; Hetoxamine C15; Mazeen C15; Teric 12M15); sol. (@ 5%) (Trymeen CAM-15)
Ionic Nature:
Nonionic (Accomeen C15; Teric 12M15)
Cationic (Crodamet 1.C15; Ethomeen C/25; Hetoxamine C15; Mazeen C15; Trymeen CAM-15)
M.W.:
860 (Hetoxamine C15; Mazeen C15)
890 (Chemeen C-15)
Sp.gr.:
1.038 (Teric 12M15)
1.04 (Accomeen C15; Ethomeen C/25)
1.042 (Mazeen C15)
Density:
8.6 lb/gal (Trymeen CAM-15)
8.7 lb/gal (Accomeen C15)
Visc.:
120 cs (Trymeen CAM-15)
343 cps (20 C) (Teric 12M15)
M.P.:
−11 ± 2 C (Teric 12M15)
Cloud Pt.:
> 100 C (Trymeen CAM-15); (1% in hard water) (Teric 12M15)
HLB:
15.4 (Trymeen CAM-15)
15.7 (Teric 12M15)
Stability:
Good in hard or saline waters and in reasonable concs. of acids and alkalis (Teric 12M15)

POE (15) coconut amine (cont'd.)

pH:
  8–10 (1% aq.) (Teric 12M15)
Surface Tension:
  39.6 dynes/cm (0.01%, 20 C) (Teric 12M15)
  41 dynes/cm (0.1% sol'n.) (Ethomeen C/25; Mazeen C15)
TOXICITY/HANDLING:
  Corrosive (Accomeen C15)
  Skin irritant, severe eye irritant (Ethomeen C/25)
  May cause skin and eye irritation; spillages are slippery (Teric 12M15)

POE (4) dioleate

SYNONYMS:
  PEG-4 dioleate (CTFA)
  PEG 200 dioleate
STRUCTURE:

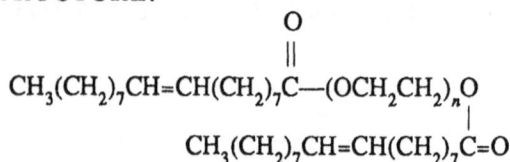

  where avg. $n = 4$
CAS No.:
  9005-07-6 (generic); 52688-97-0 (generic)
  RD No.: 977065-57-8
TRADENAME EQUIVALENTS:
  Alkamuls 200DO [Alkaril]
  Chemax PEG 200 DO [Chemax]
  Cithrol 2DO [Croda Chem. Ltd.]
  Kessco PEG 200 Dioleate [Armak]
  Mapeg 200DO [Mazer]

CATEGORY:
  Emulsifier, coemulsifier, surfactant, dispersant, defoamer, lubricant, mold release
  agent, softener, thickener, solubilizer
APPLICATIONS:
  Bath products: bath oils (Kessco PEG 200 Dioleate)
  Cosmetic industry preparations: (Kessco PEG 200 Dioleate; Mapeg 200DO); hair
    preparations (Kessco PEG 200 Dioleate); perfumery (Kessco PEG 200 Dioleate);
    shampoos (Kessco PEG 200 Dioleate)
  Farm products: (Kessco PEG 200 Dioleate)

Food applications: (Kessco PEG 200 Dioleate)
Industrial applications: (Alkamuls 200DO); metalworking lubricants (Mapeg 200DO); mold release (Chemax PEG 200 DO); paper mfg. (Alkamuls 200DO); plastics (Kessco PEG 200 Dioleate); textile/leather processing (Alkamuls 200DO; Mapeg 200DO)
Pharmaceutical applications: (Kessco PEG 200 Dioleate; Mapeg 200DO)
PROPERTIES:
Form:
Liquid (Alkamuls 200DO; Chemax PEG 200 DO; Cithrol 2DO; Kessco PEG 200 Dioleate; Mapeg 200DO)
Color:
Light amber (Kessco PEG 200 Dioleate)
Amber (Alkamuls 200DO)
Yellow (Mapeg 200DO)
Composition:
97% conc. (Cithrol 2DO)
100% conc. (Mapeg 200DO)
Solubility:
Sol. in acetone (Kessco PEG 200 Dioleate)
Sol. in aromatic solvent (@ 10%) (Alkamuls 200DO)
Sol. in carbon tetrachloride (Kessco PEG 200 Dioleate)
Sol. in ethyl acetate (Kessco PEG 200 Dioleate)
Sol. in isopropanol (Kessco PEG 200 Dioleate; Mapeg 200DO)
Sol. in isopropyl myristate (Kessco PEG 200 Dioleate)
Sol. in kerosene (Kessco PEG 200 Dioleate)
Sol. in min. oil (Mapeg 200DO); sol. (@ 10%) (Alkamuls 200DO)
Sol. in min. spirits (@ 10%) (Alkamuls 200DO)
Sol. in naphtha (Kessco PEG 200 Dioleate)
Sol. in oil (Chemax PEG 200 DO)
Sol. in peanut oil (Kessco PEG 200 Dioleate)
Sol. in perchloroethylene (@ 10%) (Alkamuls 200DO)
Sol. in soybean oil (Mapeg 200DO)
Sol. in toluol (Kessco PEG 200 Dioleate; Mapeg 200DO)
Disp. in water (Kessco PEG 200 Dioleate; Mapeg 200DO); disp. (@ 10%) (Alkamuls 200DO)
Sol. in white oil (Kessco PEG 200 Dioleate)
Ionic Nature:
Nonionic (Alkamuls 200DO; Cithrol 2DO; Kessco PEG 200 Dioleate)
Sp.gr.:
0.942 (Kessco PEG 200 Dioleate)
0.95 (Mapeg 200DO)
Density:
0.95 g/ml (Alkamuls 200DO)

POE (4) dioleate (cont'd.)

7.9 lb/gal (Kessco PEG 200 Dioleate)
F.P.:
> −15 C (Kessco PEG 200 Dioleate)
M.P.:
< −10 C (Mapeg 200DO)
Flash Pt.:
545 F (COC) (Kessco PEG 200 Dioleate)
Fire Pt.:
625 F (Kessco PEG 200 Dioleate)
HLB:
5.4 (Chemax PEG 200 DO)
6.0 (Alkamuls 200DO; Kessco PEG 200 Dioleate; Mapeg 200DO)
Acid No.:
10.0 max. (Kessco PEG 200 Dioleate; Mapeg 200DO)
Iodine No.:
70 (Kessco PEG 200 Dioleate)
70 max. (Mapeg 200DO)
Saponification No.:
147–157 (Alkamuls 200DO)
148–158 (Kessco PEG 200 Dioleate; Mapeg 200DO)
pH:
5.0 (3% disp.) (Kessco PEG 200 Dioleate)

POE (8) dioleate

SYNONYMS:
PEG-8 dioleate (CTFA)
PEG 400 dioleate
STRUCTURE:

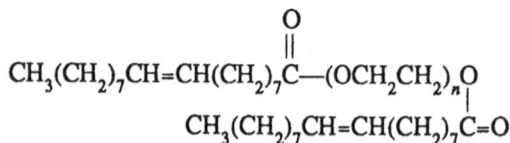

where avg. $n = 8$
CAS No.:
9005-07-6 (generic); 52688-97-0 (generic)
RD No.: 977051-75-4
TRADENAME EQUIVALENTS:
Alkamuls 400DO [Alkaril]
Chemax PEG 400 DO [Chemax]

POE (8) dioleate (cont'd.)

TRADENAME EQUIVALENTS *(cont'd.):*
Emerest 2648 [Emery]
Industrol DO-9 [BASF Wyandotte]
Kessco PEG 400 Dioleate [Armak]
Lipopeg 4DO [Lipo]
Lonzest PEG 4DO [Lonza]
Mapeg 400DO [Mazer]
Nonisol 210 [Ciba-Geigy]
Pegosperse 400 DO [Glyco]
Radiasurf 7443 [Oleofina]

CATEGORY:
Emulsifier, solubilizer, anticorrosive, antifog aid, antistat, defoamer, detergent, dispersant, emollient, lubricant, opacifier, o/w emulgent, plasticizer, rust inhibitor, scouring and detergent aid, softener, surfactant, thickener, viscosity modifier, wetting aid

APPLICATIONS:
Bath products: bath oils (Kessco PEG 400 Dioleate; Lipopeg 4DO)
Cosmetic industry preparations: (Alkamuls 400DO; Kessco PEG 400 Dioleate; Lipopeg 4DO; Mapeg 400DO; Radiasurf 7443); hair rinses (Kessco PEG 400 Dioleate); personal care products (Kessco PEG 400 Dioleate); shampoos (Kessco PEG 400 Dioleate)
Degreasers: (Lonzest PEG 4DO)
Farm products: (Alkamuls 400DO; Emerest 2648; Kessco PEG 400 Dioleate; Lonzest PEG 4DO); insecticides (Lonzest PEG 4DO; Radiasurf 7443)
Food applications: (Kessco PEG 400 Dioleate)
Industrial applications: (Nonisol 210); adhesives (Lonzest PEG 4DO); dyes and pigments (Radiasurf 7443); glass processing (Radiasurf 7443); lubricating/ cutting oils (Alkamuls 400DO; Emerest 2648; Lonzest PEG 4DO; Radiasurf 7443); metalworking (Alkamuls 400DO; Emerest 2648; Lonzest PEG 4DO; Kessco PEG 400 Dioleate; Mapeg 400DO); paint mfg. (Alkamuls 400DO; Emerest 2648; Radiasurf 7443); petroleum industry (Lonzest PEG 4DO); plastics (Kessco PEG 400 Dioleate; Radiasurf 7443); polishes and waxes (Radiasurf 7443); printing inks (Radiasurf 7443); textile/leather processing (Alkamuls 400DO; Emerest 2648; Kessco PEG 400 Dioleate; Lonzest PEG 4DO; Mapeg 400DO; Nonisol 210; Radiasurf 7443)
Industrial cleaners: drycleaning compositions (Lonzest PEG 4DO)
Pharmaceutical applications: (Kessco PEG 400 Dioleate; Mapeg 400DO; Radiasurf 7443)

PROPERTIES:
Form:
Liquid (Alkamuls 400DO; Chemax PEG400DO; Emerest 2648; Industrol DO-9; Kessco PEG 400 Dioleate; Lipopeg 4DO; Lonzest PEG 4DO; Mapeg 400DO; Nonisol 210; Pegosperse 400DO; Radiasurf 7443)

147

POE (8) dioleate (cont'd.)

Color:
    Light amber (Kessco PEG 400 Dioleate)
    Amber (Alkamuls 400DO; Lipopeg 4DO; Pegosperse 400 DO; Radiasurf 7443)
    Light yellow (Nonisol 210)
    Yellow (Lonzest PEG 4DO; Mapeg 400DO)
    Gardner 3 (Emerest 2648)
    Gardner 4 (Industrol DO9)

Composition:
    100% active (Alkamuls 400-DO; Emerest 2648)
    100% conc. (Lonzest PEG-4DO; Mapeg 400DO; Nonisol 210; Pegosperse 400 DO)

Solubility:
    Sol. in acetone (Kessco PEG 400 Dioleate)
    Sol. in aromatic solvent (Alkamuls 400-DO)
    Sol. in benzene (Radiasurf 7443)
    Disp. in butyl stearate (Emerest 2648)
    Sol. in carbon tetrachloride (Kessco PEG 400 Dioleate)
    Sol. in ethanol (Pegosperse 400DO)
    Sol. in ethyl acetate (Kessco PEG 400 Dioleate; Pegosperse 400DO)
    Sol. in glycerol trioleate (Emerest 2648)
    Sol. in hexane (Radiasurf 7443)
    Sol. in isopropanol (Kessco PEG 400 Dioleate; Lonzest PEG 4DO; Mapeg 400DO;
        Radiasurf 7443)
    Sol. in isopropyl myristate (Kessco PEG 400 Dioleate)
    Partly sol. in kerosene (Kessco PEG 400 Dioleate)
    Sol. in methanol (Pegosperse 400DO)
    Sol. in min. oil (Alkamuls 400-DO; Emerest 2648; Lonzest PEG 4DO; Pegosperse
        400DO; Radiasurf 7443); partly sol. in min. oil (Mapeg 400DO)
    Sol. in min. spirits (Alkamuls 400-DO)
    Sol. in naphtha (Kessco PEG 400 Dioleate; Pegosperse 400DO)
    Sol. in peanut oil (Kessco PEG 400 Dioleate)
    Sol. in perchloroethylene (Alkamuls 400-DO)
    Disp. in sodium sulfate (5%) (Kessco PEG 400 Dioleate)
    Sol. in soybean oil (Mapeg 400DO)
    Disp. in Stoddard solvent (Emerest 2648)
    Sol. in toluol (Kessco PEG 400 Dioleate; Lonzest PEG 4DO; Mapeg 400DO;
        Pegosperse 400DO)
    Sol. in trichlorethylene (Radiasurf 7443)
    Sol. in veg. oils (Pegosperse 400DO; Radiasurf 7443)
    Disp. in water (Alkamuls 400DO; Emerest 2648; Kessco PEG 400 Dioleate; Lonzest
        PEG 4DO; Mapeg 400DO; Pegosperse 400DO; Radiasurf 7443)
    Partly sol. in white oil (Kessco PEG 400 Dioleate)
    Sol. in xylene (Emerest 2648)

Ionic Nature: Nonionic

M.W.:
911 avg. (Radiasurf 7443)
Density:
8.1 lb/gal (Emerest 2648, Kessco PEG 400 Dioleate)
Sp.gr.:
0.962 (Radiasurf 7443)
0.97 (Industrol DO-9; Pegosperse 400DO)
0.977 (Kessco PEG 400 Dioleate)
0.98 (Lonzest PEG 4DO; Mapeg 400DO)
M.P.:
0 C (Lonzest PEG 4DO)
< 7 C (Mapeg 400DO)
Pour Pt.:
6 C (Emerest 2648; Industrol DO-9)
Visc.:
47.00 cps @ 37.8 C (Radiasurf 7443)
45 cst @ 100 F (Emerest 2648)
HLB:
7.2 (Alkamuls 400DO)
7.2 ± 0.5 (Pegosperse 400DO)
7.2 ± 1 (Lipopeg 4-DO)
7.4 (Radiasurf 7443)
8.5 (Chemax PEG-400 DO; Industrol DO9; Kessco PEG 400 Dioleate; Mapeg 400DO)
8.8 (Emerest 2648)
9.9 (Lonzest PEG 4DO)

Acid No.:
< 5 (Radiasurf 7443)
5–10 (Lonzest PEG-4DO)
10 max. (Industrol DO-9; Kessco PEG 400 Dioleate; Lipopeg 4DO; Mapeg 400DO;
   Pegosperse 400DO)

Iodine No.:
45–55 (Lonzest PEG-4DO)
50–60 (Radiasurf 7443)
52–57 (Pegosperse 400DO)
55 max. (Kessco PEG 400 Dioleate; Mapeg 400DO)

Saponification No.:
105–115 (Alkamuls 400DO)
113–122 (Kessco PEG 400 Dioleate)
113–128 (Lipopeg 4DO)
114–122 (Mapeg 400DO)
115–125 (Pegosperse 400DO)
120–130 (Lonzest PEG 4DO, Radiasurf 7443)

POE (8) dioleate *(cont'd.)*

Ref. Index:
  1.4655 (Radiasurf 7443)
pH:
  4.5–6.5 (5% aq. disp.) (Pegosperse 400DO)
  5.0 (3% disp.) (Kessco PEG 400 Dioleate)
  5.0–7.0 (5% aq. disp.) (Industrol DO-9)
Flash Pt.:
  248 C (Radiasurf 7443)
  515 F (Emerest 2648)
  520 F (Kessco PEG 400 Dioleate)
Fire Pt.:
  635 F (Kessco PEG 400 Dioleate)
Cloud Pt.:
  –3 C (Radiasurf 7443)
  < 25 C (Emerest 2648; Industrol DO-9)
STD. PKGS.:
  450 lb steel drums (Kessco PEG 400 Dioleate)

POE (12) dioleate

SYNONYMS:
  PEG-12 dioleate (CTFA)
  PEG 600 dioleate
STRUCTURE:

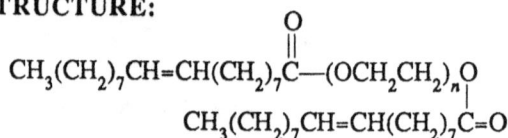

  where avg. $n = 12$
CAS No.:
  9005-07-6 (generic); 52688-97-0 (generic)
TRADENAME EQUIVALENTS:
  Alkamuls 600-DO [Alkaril]
  Chemax PEG 600 DO [Chemax]
  Cithrol 6DO [Croda Chem. Ltd.]
  CPH-213-N [C.P. Hall Co.]
  Emerest 2665 [Emery]
  Industrol DO-13 [BASF Wyandotte]
  Kessco PEG 600 Dioleate [Armak]
  Mapeg 600 DO [Mazer]
  Marlipal FS [Chemische Werke Huls AG]

150

CATEGORY:
Emulsifier, wetting agent, surfactant, thickener, opacifier, dispersant, cosolvent, solubilizer, lubricant, superfatting agent

APPLICATIONS:
Bath products: bath oils (Kessco PEG 600 Dioleate)

Cosmetic industry preparations: (Cithrol 6DO; Kessco PEG 600 Dioleate; Mapeg 600 DO; Marlipal FS); creams and lotions (Alkamuls 600-DO); hair preparations (Kessco PEG 600 Dioleate); perfumery (Kessco PEG 600 Dioleate); shampoos (Alkamuls 600-DO; Kessco PEG 600 Dioleate; Marlipal FS)

Farm products: (Kessco PEG 600 Dioleate); agricultural oils/sprays (Alkamuls 600-DO; Emerest 2665); insecticides/pesticides (Emerest 2665)

Food applications: (Kessco PEG 600 Dioleate)

Household detergents: liquid detergents (Marlipal FS); paste detergents (Marlipal FS)

Industrial applications: (Chemax PEG 600 DO; Cithrol 6DO); lubricating/cutting oils (Emerest 2665; Mapeg 600 DO); metalworking (Emerest 2665; Mapeg 600 DO); paint mfg. (Emerest 2665); plastics (Kessco PEG 600 Dioleate); textile/leather processing (Alkamuls 600-DO; Mapeg 600 DO)

Pharmaceutical applications: (Mapeg 600 DO)

PROPERTIES:

Form:
Liquid (Alkamuls 600-DO; Cithrol 6DO; CPH-213-N; Emerest 2665; Industrol DO-13; Kessco PEG 600 Dioleate; Mapeg 600 DO; Marlipal FS)

Semiliquid (Chemax PEG 600 DO)

Color:
Light amber (Kessco PEG 600 Dioleate)

Amber (Alkamuls 600-DO)

Yellow (Mapeg 600 DO)

Gardner 3 (Emerest 2665)

Gardner 4 max. (Industrol DO-13)

Composition:
97% conc. (Cithrol 6DO)

100% active (Alkamuls 600-DO; Industrol DO-13; Marlipal FS)

100% conc. (CPH-213-N; Mapeg 600 DO)

Solubility:
Sol. in acetone (Kessco PEG 600 Dioleate)

Sol. in butyl stearate (@ 5%) (Emerest 2665)

Sol. in carbon tetrachloride (Kessco PEG 600 Dioleate)

Sol. in ethyl acetate (Kessco PEG 600 Dioleate)

Sol. in glycerol trioleate (@ 5%) (Emerest 2665)

Sol. in isopropanol (Kessco PEG 600 Dioleate; Mapeg 600 DO)

Sol. in isopropyl myristate (Kessco PEG 600 Dioleate)

Disp. in min. oil (@ 5%) (Emerest 2665)

Partly sol. in naphtha (Kessco PEG 600 Dioleate)

POE (12) dioleate (cont'd.)

Sol. in peanut oil (Kessco PEG 600 Dioleate)
Sol. in soybean oil (Mapeg 600 DO)
Sol. in Stoddard solvent (@ 5%) (Emerest 2665)
Sol. in toluol (Kessco PEG 600 Dioleate; Mapeg 600 DO)
Disp. in water (Industrol DO-13; Kessco PEG 600 Dioleate; Mapeg 600 DO); disp. (@ 5%) (Emerest 2665)
Sol. in xylene (@ 5%) (Emerest 2665)

Ionic Nature:
Nonionic (Alkamuls 600-DO; Cithrol 6DO; CPH-213-N; Industrol DO-13; Marlipal FS)

M.W.:
1160 (Industrol DO-13)

Sp.gr.:
1.0 (Industrol DO-13)
1.001 (Kessco PEG 600 Dioleate)
1.01–1.03 (Alkamuls 600-DO)

Density:
8.3 lb/gal (Emerest 2665; Kessco PEG 600 Dioleate)

Visc.:
64 cSt (100 F) (Emerest 2665)

F.P.:
19 C (Kessco PEG 600 Dioleate)

Pour Pt.:
19 C (Emerest 2665)
20 C (Industrol DO-13)

Solidification Pt.:
15 C (Alkamuls 600-DO)

Flash Pt.:
495 F (COC) (Kessco PEG 600 Dioleate)
530 F (Emerest 2665)

Fire Pt.:
615 F (Kessco PEG 600 Dioleate)

Cloud Pt.:
< 25 C (Emerest 2665); (1% aq.) (Industrol DO-13)

HLB:
10.3 (Emerest 2665)
10.5 (Chemax PEG 600 DO; CPH-213-N; Kessco PEG 600 Dioleate)
10.6 (Industrol DO-13)

Acid No.:
10 max. (Industrol DO-13; Kessco PEG 600 Dioleate)

Iodine No.:
45 max. (Kessco PEG 600 Dioleate)

Saponification No.:
92–102 (Kessco PEG 600 Dioleate)
pH:
5.0 (3% disp.) (Kessco PEG 600 Dioleate)
5.0–7.0 (5% aq.) (Industrol DO-13)
STD. PKGS.:
55-gal (450 lb net) steel drums (Industrol DO-13)

POE (4) distearate

SYNONYMS:
PEG-4 distearate (CTFA)
PEG 200 distearate
STRUCTURE:

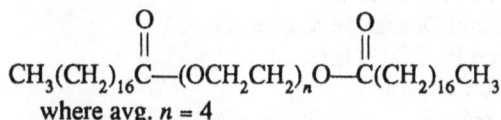

where avg. $n = 4$

CAS No.:
9005-08-7 (generic)
RD No.: 977065-58-9
TRADENAME EQUIVALENTS:
Alkamuls 200-DS [Alkaril]
Cithrol 2DS [Croda Chem. Ltd.]
Kessco PEG 200 Distearate [Armak]
Mapeg 200DS [Mazer]
CATEGORY:
Emulsifier, surfactant, lubricant, softener, opacifier, antistat, thickener, solubilizer, dispersant
APPLICATIONS:
Bath products: bath oils (Kessco PEG 200 Distearate)
Cosmetic industry preparations: (Alkamuls 200-DS; Kessco PEG 200 Distearate; Mapeg 200DS); hair preparations (Kessco PEG 200 Distearate); perfumery (Kessco PEG 200 Distearate)
Farm products: (Kessco PEG 200 Distearate)
Food applications: (Kessco PEG 200 Distearate)
Industrial applications: lubricating/cutting oils (Mapeg 200DS); metalworking (Mapeg 200DS); plastics (Kessco PEG 200 Distearate); textile/leather processing (Alkamuls 200-DS; Cithrol 2DS; Mapeg 200DS)
Pharmaceutical applications: (Kessco PEG 200 Distearate; Mapeg 200DS)

POE (4) distearate *(cont'd.)*

PROPERTIES:
Form:
  Paste (Cithrol 2DS)
  Soft solid (Kessco PEG 200 Distearate)
  Solid (Alkamuls 200-DS; Mapeg 200DS)
Color:
  White (Mapeg 200DS)
  White to cream (Kessco PEG 200 Distearate)
  Cream (Alkamuls 200-DS)
Composition:
  97% conc. (Cithrol 2DS)
  100% conc. (Mapeg 200DS)
Solubility:
  Sol. in acetone (Kessco PEG 200 Distearate)
  Sol. in aromatic solvent (@ 10%) (Alkamuls 200-DS)
  Sol. in carbon tetrachloride (Kessco PEG 200 Distearate)
  Sol. in ethyl acetate (Kessco PEG 200 Distearate)
  Sol. in isopropanol (Kessco PEG 200 Distearate; Mapeg 200DS)
  Sol. in isopropyl myristate (Kessco PEG 200 Distearate)
  Sol. in kerosene (Kessco PEG 200 Distearate)
  Sol. in min. oil (Mapeg 200DS); sol. (@ 10%) (Alkamuls 200-DS)
  Sol. in min. spirits (@ 10%) (Alkamuls 200-DS)
  Sol. in naphtha (Kessco PEG 200 Distearate)
  Sol. in peanut oil (Kessco PEG 200 Distearate)
  Sol. in perchloroethylene (@ 10%) (Alkamuls 200-DS)
  Sol. in soybean oil (Mapeg 200DS)
  Sol. in toluol (Kessco PEG 200 Distearate; Mapeg 200DS)
  Disp. in water (@ 10%) (Alkamuls 200-DS); disp. hot (Kessco PEG 200 Distearate;
    Mapeg 200DS)
  Sol. in white oil (Kessco PEG 200 Distearate)
Ionic Nature:
  Nonionic (Cithrol 2DS; Kessco PEG 200 Distearate; Mapeg 200DS)
Sp.gr.:
  0.9060 (65 C) (Kessco PEG 200 Distearate)
M.P.:
  34 C (Kessco PEG 200 Distearate; Mapeg 200DS)
Flash Pt.:
  475 F (COC) (Kessco PEG 200 Distearate)
Fire Pt.:
  525 F (Kessco PEG 200 Distearate)
HLB:
  4.7 (Mapeg 200DS)
  5.0 (Alkamuls 200-DS; Kessco PEG 200 Distearate)

5.2 (Cithrol 2DS)
Acid No.:
10 max. (Kessco PEG 200 Distearate; Mapeg 200DS)
Iodine No.:
0.5 max. (Kessco PEG 200 Distearate)
1.0 max. (Mapeg 200DS)
Saponification No.:
153–162 (Kessco PEG 200 Distearate)
155–165 (Mapeg 200DS)
160–170 (Alkamuls 200-DS)
pH:
5.0 (3% disp.) (Kessco PEG 200 Distearate)

POE (8) distearate

SYNONYMS:
PEG-8 distearate (CTFA)
PEG 400 distearate
STRUCTURE:

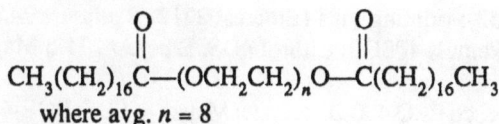

$$CH_3(CH_2)_{16}C\!-\!(OCH_2CH_2)_nO\!-\!C(CH_2)_{16}CH_3$$
where avg. $n = 8$
CAS No.
9005-08-7 (generic)
RD No. 977053-29-4
TRADENAME EQUIVALENTS:
Alkamuls 400DS [Alkaril]
Cithrol 4DS [Croda]
Cyclochem PEG-400DS [Cyclo]
Emerest 2642 [Emery]
Emerest 2712 [Emery]
Kessco PEG 400 Distearate [Armak]
Kessco PEG 400 DS-356 [Armak]
Lipal 400DS [PVO]
Lipopeg 4DS [Lipo]
Mapeg 400DS [Mazer]
Pegosperse 400DS [Glyco]
PGE-400-DS [Hefti Ltd.]
Radiasurf 7453 [Oleofina]
Scher PEG 400 Distearate [Scher]

POE (8) distearate (cont'd.)

CATEGORY:
Emulsifier, thickener, lubricant, softener, antifog aid, antistat, chemical intermediate, defoamer, dispersant, o/w emulgent, opacifier, plasticizer, rust inhibitor, scouring and detergent aid, solubilizer, surfactant, viscosity modifier, wetting agent

APPLICATIONS:
Bath products: bath oils (Kessco PEG 400 Distearate; Lipopeg 4DS)

Cleansers: body cleansers (Emerest 2712; Lipal 400DS)

Cosmetic industry preparations: (Alkamuls 400DS; Emerest 2642; Emerest 2712; Kessco PEG 400 Distearate; Lipal 400DS; Lipopeg 4DS; Mapeg 400DS; PGE-400-DS; Radiasurf 7453; Scher PEG 400 Distearate); hair rinses (Emerest 2712; Kessco PEG 400 Distearate; Lipal 400DS); makeup (Emerest 2712; Lipal 400DS); perfumery (Kessco PEG 400 Distearate); shampoos (Emerest 2712; Kessco PEG 400 Distearate; Lipal 400DS); shaving preparations (Emerest 2712; Lipal 400DS)

Farm products: (Kessco PEG 400 Distearate); insecticides/pesticides (Emerest 2712; Radiasurf 7453)

Food applications: (Kessco PEG 400 Distearate)

Industrial applications: (Alkamuls 400DS; Emerest 2642); dyes and pigments (Emerest 2712; Radiasurf 7453); glass processing (Emerest 2712; Radiasurf 7453); lubricating/cutting oils (Emerest 2712; Radiasurf 7453); metalworking (Mapeg 400DS); paint mfg. (Emerest 2712; Radiasurf 7453); plastics (Emerest 2712; Kessco PEG 400 Distearate; PGE-400-DS; Radiasurf 7453); polishes and waxes (Emerest 2712; Radiasurf 7453); printing inks (Emerest 2712; Radiasurf 7453); textile/leather processing (Alkamuls 400DS; Cithrol 4DS; Emerest 2712; Mapeg 400DS; Radiasurf 7453)

Pharmaceutical applications: (Kessco PEG 400 Distearate; Mapeg 400DS; PGE-400-DS; Radiasurf 7453); antiseptic soaps (Emerest 2712; Lipal 400DS)

PROPERTIES:
Form:
Paste (Kessco PEG 400 DS-356; Radiasurf 7453)

Solid (Alkamuls 400DS; Cithrol 4DS; Emerest 2642; Lipal 400DS; Mapeg 400DS; PGE-400-DS)

Soft solid (Cyclochem PEG-400DS; Emerest 2712; Kessco PEG 400 Distearate; Pegosperse 400DS; Scher PEG 400 Distearate)

Soft wax (Lipopeg 4DS)

Color:
White (Cyclochem PEG 400DS; Kessco PEG 400 Distearate; Mapeg 400DS; Radiasurf 7453; Scher PEG 400 Distearate)

Cream (Alkamuls 400DS; Kessco PEG 400 Distearate; Lipopeg 4DS; Pegosperse 400DS)

Gardner 2 (Emerest 2642)

Garder 4 (Lipal 400DS)

Odor:
Mild, typical (Scher PEG 400 Distearate)

Composition:
97% conc. (Cithrol 4DS)
100% active (Alkamuls 400-DS; Emerest 2642, 2712)
100% conc. (Mapeg 400DS; Pegosperse 400DS; PGE-400-DS)
Solubility:
Sol. in acetate esters (Scher PEG 400 Distearate)
Sol. in most alcohols (Scher PEG 400 Distearate)
Sol. in aliphatic hydrocarbons (Scher PEG 400 Distearate)
Sol. in aromatic solvent (Alkamuls 400DS); aromatic hydrocarbons (Scher PEG 400
    Distearate)
Sol. in benzene (Radiasurf 7453)
Sol. in butyl stearate (Emerest 2642)
Sol. in carbon tetrachloride (Kessco PEG 400 Distearate)
Sol. in ethanol (Pegosperse 400DS)
Sol. in ethyl acetate (Kessco PEG 400 Distearate; Pegosperse 400DS)
Sol. in glycerol trioleate (Emerest 2642)
Sol. in glycols (Scher PEG 400 Distearate)
Sol. in glycol ethers (Scher PEG 400 Distearate)
Sol. in isopropanol (Kessco PEG 400 Distearate; Lipal 400DS; Mapeg 400DS)
Sol. in isopropyl myristate (Kessco PEG 400 Distearate)
Partly sol. in kerosene (Kessco PEG 400 Distearate)
Sol. in ketones (Scher PEG 400 Distearate)
Sol. in methanol (Pegosperse 400DS)
Sol. in min. oil (Alkamuls 400DS; Emerest 2642; Lipal 400DS; Pegosperse 40DS);
    partly sol. (Mapeg 400DS; Scher PEG 400 Distearate)
Sol. in min. spirits (Alkamuls 400DS)
Sol. in naphtha (Kessco PEG 400 Distearate; Pegosperse 400DS)
Sol. in peanut oil (Kessco PEG 400 Distearate; Lipal 400DS)
Sol. in perchloroethylene (Alkamuls 400DS)
Sol. in soybean oil (Mapeg 400DS)
Sol. in Stoddard solvent (Emerest 2642)
Sol. in toluol (Kessco PEG 400 Distearate; Mapeg 400DS; Pegosperse 400DS)
Sol. in trichlorethylene (Radiasurf 7453)
Sol. in veg. oil (Pegosperse 400DS); partly sol. (Scher PEG 400 Distearate)
Disp. in water (Alkamuls 400DS; Emerest 2642; Kessco PEG 400 Distearate; Lipal
    400DS; Pegosperse 400DS; Scher PEG 400 Distearate)
Sol. in white oil (Kessco PEG 400 Distearate)
Sol. in xylene (Emerest 2642)
Ionic Nature: Nonionic
M.W.:
926 (Scher PEG 400 Distearate)
Sp.gr.:
0.920 (98.9 C) (Radiasurf 7453)

POE (8) distearate (cont'd.)

0.9390 (65 C) (Kessco PEG 400 Distearate)
0.950 (Scher PEG 400 Distearate)
0.98 (Pegosperse 400DS)

Visc.:
52 cSt (100 F) (Emerest 2642)
9.85 cps (98.9 C) (Radiasurf 7453)

M.P.:
29–37 C (Pegosperse 400DS)
33 C (Cyclochem PEG 400DS)
36 C (Emerest 2642; Kessco PEG 400 Distearate; Mapeg 400DS)
39 C (Radiasurf 7453)

Flash Pt.:
> 170 C (Scher PEG 400 Distearate)
242 C (Radiasurf 7453)
470 F (Emerest 2642)
500 F (Kessco PEG 400 Distearate)

Fire Pt.:
545 F (Kessco PEG 400 Distearate)

Cloud Pt.:
< 25 C (Emerest 2642)

HLB:
7.2 ± 1 (Lipal 400DS)
7.5 (Emerest 2642)
7.7 (Radiasurf 7453)
7.8 (Alkamuls 400DS)
7.8 ± 0.5 (Pegosperse 400DS)
8.0 (Kessco PEG 400 Distearate; PGE-400-DS)
8.0 ± 1 (Lipopeg 4-DS)
8.1 (Emerest 2712; Mapeg 400DS)
8.8 (Scher PEG 400 Distearate)

Acid No.:
< 5 (Radiasurf 7453)
< 10 (Pegosperse 400DS)
10 max. (Cyclochem PEG 400DS; Kessco PEG 400 Distearate; Lipopeg 4-DS; Mapeg 400DS; Scher PEG 400 Distearate)

Iodine No.:
0.5 max. (Kessco PEG 400 Distearate)
< 1 (Pegosperse 400DS)
1. max. (Mapeg 400DS; Radiasurf 7453; Scher PEG 400 Distearate)
3 (Lipal 400DS)

Saponification No.:
113–128 (Lipopeg 4DS)
115–124 (Kessco PEG 400 Distearate)

115–125 (Pegosperse 400DS; Scher PEG 400 Distearate)
116–125 (Mapeg 400DS)
120–130 (Alkamuls 400DS; Radiasurf 7453)
122–132 (Lipal 400DS)
123 (Cyclochem PEG400DS)

Stability:

Stable over a wide pH range (Lipal 400DS)

pH:

4.0–6.5 (5% aq. disp.) (Pegosperse 400DS)
5.0 (3% aq. disp.) (Kessco PEG 400 Distearate)

STD. PKGS.:

190-kg net bung drums or bulk (Radiasurf 7453)
450-lb steel drums (Kessco PEG 400 Distearate)

POE (12) distearate

SYNONYMS:

PEG-12 distearate (CTFA)
PEG 600 distearate

STRUCTURE:

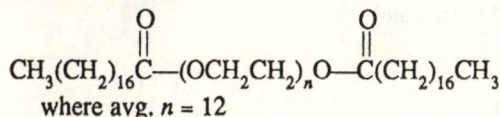

$$CH_3(CH_2)_{16}\overset{O}{\overset{\|}{C}}-(OCH_2CH_2)_nO-\overset{O}{\overset{\|}{C}}(CH_2)_{16}CH_3$$

where avg. $n = 12$

CAS No.:

9005-08-7 (generic)
RD No.: 977055-07-4

TRADENAME EQUIVALENTS:

Alkamuls 600-DS [Alkaril]
Cithrol 6DS [Croda Chem. Ltd.]
Kessco PEG 600 Distearate [Armak]
Mapeg 600DS [Mazer]
PGE-600-DS [Hefti Ltd.]
Radiasurf 7454 [Oleofina S.A.]
Scher PEG-600 Distearate [Scher]

CATEGORY:

Emulsifier, surfactant, lubricant, softener, opacifier, antistat, thickener, viscosity modifier, solubilizer, dispersant, plasticizer, wetting aid, o/w emulgent, scouring and detergent aid, defoamer, rust inhibitor, antifog aid

159

POE (12) distearate (cont'd.)

APPLICATIONS:

Bath products: bath oils (Kessco PEG 600 Distearate)

Cosmetic industry preparations: (Alkamuls 600-DS; Kessco PEG 600 Distearate; Mapeg 600DS; PGE-600-DS; Radiasurf 7454; Scher PEG-600 Distearate); creams and lotions (Scher PEG-600 Distearate); hair preparations (Kessco PEG 600 Distearate); perfumery (Kessco PEG 600 Distearate)

Farm products: (Kessco PEG 600 Distearate); insecticides/pesticides (Radiasurf 7454)

Food applications: (Kessco PEG 600 Distearate)

Industrial applications: (PGE-600-DS); dyes and pigments (Radiasurf 7454); glass fiber (Radiasurf 7454); lubricating/cutting oils (Mapeg 600DS; Radiasurf 7454); metalworking (Mapeg 600DS); paint mfg. (Radiasurf 7454); plastics (Kessco PEG 600 Distearate; PGE-600-DS; Radiasurf 7454); polishes and waxes (Radiasurf 7454); printing inks (Radiasurf 7454); textile/leather processing (Alkamuls 600-DS; Cithrol 2DS; Mapeg 600DS; Radiasurf 7454)

Pharmaceutical applications: (Kessco PEG 600 Distearate; Mapeg 600DS; PGE-600-DS; Radiasurf 7454)

PROPERTIES:

Form:

Soft solid (Kessco PEG 600 Distearate)

Solid (Alkamuls 600-DS; Cithrol 6DS; Mapeg 600DS; Radiasurf 7454; Scher PEG-600 Distearate)

Flake (Mapeg 600DS; PGE-600-DS)

Color:

White (Mapeg 600DS; Radiasurf 7454; Scher PEG-600 Distearate)

White to cream (Kessco PEG 600 Distearate)

Cream (Alkamuls 600-DS)

Odor:

Mild (Scher PEG-600 Distearate)

Composition:

97% conc. (Cithrol 2DS)

100% conc. (Mapeg 600DS; PGE-600-DS)

Solubility:

Sol. in acetate esters (Scher PEG-600 Distearate)

Sol. in acetone (Kessco PEG 600 Distearate)

Sol. in alcohols (Scher PEG-600 Distearate)

Sol. in aliphatic hydrocarbons (Scher PEG-600 Distearate)

Sol. in aromatic hydrocarbons (Scher PEG-600 Distearate)

Sol. in aromatic solvent (@ 10%) (Alkamuls 600-DS)

Sol. in benzene (@ 10%) (Radiasurf 7454)

Sol. in carbon tetrachloride (Kessco PEG 600 Distearate)

Sol. in ethyl acetate (Kessco PEG 600 Distearate)

Sol. in glycol ethers (Scher PEG-600 Distearate)

Sol. in glycols (Scher PEG-600 Distearate)

Sol. cloudy in hexane (@ 10%) (Radiasurf 7454)
Sol. in isopropanol (Kessco PEG 600 Distearate; Mapeg 600DS); sol. cloudy (@ 10%)
    (Radiasurf 7454)
Sol. in isopropyl myristate (Kessco PEG 600 Distearate)
Partly sol. in kerosene (Kessco PEG 600 Distearate)
Sol. in ketones (Scher PEG-600 Distearate)
Partly sol. in min. oil (Scher PEG-600 Distearate); sol. cloudy (@ 10%) (Radiasurf
    7454); disp. (@ 10%) (Alkamuls 600-DS)
Sol. in min. spirits (@ 10%) (Alkamuls 600-DS)
Partly sol. in naphtha (Kessco PEG 600 Distearate)
Sol. in peanut oil (Kessco PEG 600 Distearate)
Sol. in perchloroethylene (@ 10%) (Alkamuls 600-DS)
Sol. in soybean oil (Mapeg 600DS)
Sol. in toluol (Kessco PEG 600 Distearate; Mapeg 600DS)
Sol. in trichlorethylene (@ 10%) (Radiasurf 7454)
Partly sol. in veg. oil (Scher PEG-600 Distearate)
Disp. in water (Scher PEG-600 Distearate); disp. (@ 10%) (Alkamuls 600-DS); disp.
    hot (Kessco PEG 600 Distearate; Mapeg 600DS)
Ionic Nature:
Nonionic (Cithrol 2DS; Kessco PEG 600 Distearate; Mapeg 600DS; PGE-600-DS)
M.W.:
1101 avg. (Radiasurf 7454)
1104 avg. (Scher PEG-600 Distearate)
Sp.gr.:
0.940 (98.9 C) (Radiasurf 7454)
0.9670 (65 C) (Kessco PEG 600 Distearate)
0.978 (Scher PEG-600 Distearate)
Visc.:
12.20 cps (98.9 C) (Radiasurf 7454)
M.P.:
39 C (Kessco PEG 600 Distearate)
40 C (Radiasurf 7454; Scher PEG-600 Distearate)
41 C (Mapeg 600DS)
Flash Pt.:
> 170 C (OC) (Scher PEG-600 Distearate)
249 C (Radiasurf 7454)
490 F (COC) (Kessco PEG 600 Distearate)
Fire Pt.:
530 F (Kessco PEG 600 Distearate)
HLB:
9.7 (Radiasurf 7454)
10.5 (PGE-600-DS)
10.6 (Alkamuls 600-DS; Kessco PEG 600 Distearate; Mapeg 600DS)

POE (12) distearate *(cont'd.)*

10.9 (Scher PEG-600 Distearate)
Acid No.:
  < 5 (Radiasurf 7454)
  10 max. (Kessco PEG 600 Distearate; Mapeg 600DS; Scher PEG-600 Distearate)
Iodine No.:
  0.5 max. (Kessco PEG 600 Distearate)
  1.0 max. (Mapeg 600DS; Radiasurf 7454; Scher PEG-600 Distearate)
Saponification No.:
  92–102 (Scher PEG-600 Distearate)
  93–102 (Kessco PEG 600 Distearate)
  94–104 (Mapeg 600DS)
  96–106 (Alkamuls 600-DS)
  100–110 (Radiasurf 7454)
pH:
  5.0 (3% disp.) (Kessco PEG 600 Distearate)
STD. PKGS.:
  25-kg net multiply paper bags or bulk (Radiasurf 7454)

POE (20) glyceryl monostearate

SYNONYMS:
  PEG-20 glyceryl stearate (CTFA)
  PEG 1000 glyceryl monostearate
  POE (20) glycerol monostearate
STRUCTURE:

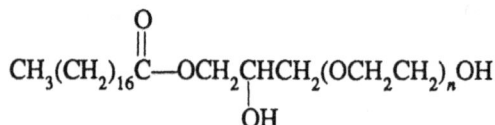

$$CH_3(CH_2)_{16}\overset{\displaystyle O}{\overset{\|}{C}}\!\!-\!\!OCH_2\underset{\underset{\textstyle OH}{|}}{CH}CH_2(OCH_2CH_2)_n OH$$

where avg. $n = 20$
CAS No.:
  RD No.: 977055-13-2
TRADENAME EQUIVALENTS:
  Aldosperse MS-20, MS-20 FG [Glyco]
  Capmul EMG [Capital City Products]
  Cutina E-24 [Henkel Canada]
  Durfax EOM, EOM K [Durkee Foods]
  Mazol 80MGK [Mazer]
  Radiasurf 7000 [Oleofina S.A.]

POE (20) glyceryl monostearate (cont'd.)

TRADENAME EQUIVALENTS *(cont'd.):*
Tagat S2 [Goldschmidt]
Varonic LI42 [Sherex]

CATEGORY:
Emulsifier, solubilizer, dispersant, antistat, conditioner, lubricant, mold release agent, softener, chemical intermediate, detergent, wetting agent

APPLICATIONS:
Cosmetic industry preparations: (Radiasurf 7000); creams and lotions (Cutina E-24; Varonic LI42); ointments (Cutina E-24); perfumery (Tagat S2); shampoos (Radiasurf 7000)

Food applications: (Capmul EMG; Durfax EOM; Mazol 80MGK); flavors (Tagat S2); food emulsifying (Aldosperse MS-20 FG; Durfax EOM, EOM K; Mazol 80MGK)

Household detergents: (Radiasurf 7000)

Industrial applications: dyes and pigments (Radiasurf 7000); industrial processing (Radiasurf 7000); lubricating/cutting oils (Radiasurf 7000); paint mfg. (Radiasurf 7000); plastics (Radiasurf 7000); printing inks (Radiasurf 7000); textile/leather processing (Durfax EOM; Radiasurf 7000)

Pharmaceutical applications: (Radiasurf 7000); vitamin oils (Tagat S2)

PROPERTIES:

Form:
Liquid (Radiasurf 7000; Tagat S2)
Paste (Cutina E-24; Durfax EOM K)
Soft solid (Aldosperse MS-20)
Solid (Aldosperse MS-20 FG; Capmul EMG; Mazol 80MGK; Varonic LI42)

Color:
White (Aldosperse MS-20)

Composition:
100% active (Durfax EOM K)
100% conc. (Aldosperse MS-20 FG; Capmul EMG; Mazol 80MGK; Tagat S2; Varonic LI42)

Solubility:
Sol. in benzene (@ 10%) (Radiasurf 7000)
Sol. cloudy in hexane (@ 10%) (Radiasurf 7000)
Sol. in isopropanol (@ 10%) (Radiasurf 7000)
Sol. in min. oil (@ 10%) (Radiasurf 7000)
Sol. in trichlorethylene (@ 10%) (Radiasurf 7000)
Sol. in veg. oil (@ 10%) (Radiasurf 7000)
Disp. in water (@ 10%) (Radiasurf 7000)

Ionic Nature:
Nonionic (Aldosperse MS-20 FG; Capmul EMG; Cutina E-24; Durfax EOM K; Mazol 80MGK; Tagat S2; Varonic LI42)

M.W.:
1395 avg. (Radiasurf 7000)

163

POE (20) glyceryl monostearate (cont'd.)

Sp.gr.:
1.032 (37.8 C) (Radiasurf 7000)
Visc.:
113.30 cps (37.8 C) (Radiasurf 7000)
M.P.:
30–40 C (Aldosperse MS-20)
80–85 F (Durfax EOM K)
Flash Pt.:
244 C (COC) (Radiasurf 7000)
Cloud Pt.:
27.5 C (Radiasurf 7000)
HLB:
13.0 (Aldosperse MS-20 FG; Radiasurf 7000; Varonic LI42)
13.1 (Aldosperse MS-20; Capmul EMG; Durfax EOM K; Mazol 80MGK)
13.5 (Durfax EOM)
15.0 (Tagat S2)
Saponification No.:
65–75 (Aldosperse MS-20; Durfax EOM)
Hydroxyl No.:
65–80 (Durfax EOM)
Ref. Index:
1.4672 (Radiasurf 7000)

POE (5) hydrogenated castor oil

SYNONYMS:
PEG-5 hydrogenated castor oil (CTFA)
PEG (5) hydrogenated castor oil
CAS No.:
61788-85-0 (generic)
RD No.: 977065-66-9
TRADENAME EQUIVALENTS:
Chemax HCO-5 [Chemax]
Nikkol HCO-5 [Nikko]
Trylox HCO-5 [Emery]
CATEGORY:
Emulsifier, coemulsifier, lubricant, softener, hydrotrope
APPLICATIONS:
Cosmetic industry preparations: (Nikkol HCO-5)
Industrial applications: textile/leather processing (Trylox HCO-5)
Pharmaceutical applications: (Nikkol HCO-5)

POE (5) hydrogenated castor oil *(cont'd.)*

PROPERTIES:
Form:
   Liquid (Chemax HCO-5; Nikkol HCO-5; Trylox HCO-5)
Color:
   Gardner 1 (Trylox HCO-5)
Composition:
   100% active (Trylox HCO-5)
   100% conc. (Nikkol HCO-5)
Solubility:
   Disp. in butyl stearate (@ 5%) (Trylox HCO-5)
   Sol. in glycerol trioleate (@ 5%) (Trylox HCO-5)
   Sol. in min. oil (@ 5%) (Trylox HCO-5)
   Sol. in oil (Chemax HCO-5)
   Disp. in perchloroethylene (@ 5%) (Trylox HCO-5)
   Sol. in Stoddard solvent (@ 5%) (Trylox HCO-5)
   Disp. in water (@ 5%) (Trylox HCO-5)
Ionic Nature:
   Nonionic (Nikkol HCO-5; Trylox HCO-5)
Density:
   7.9 lb/gal (Trylox HCO-5)
Visc.:
   1200 cs (Trylox HCO-5)
Cloud Pt.:
   < 25 C (Trylox HCO-5)
HLB:
   3.8 (Chemax HCO-5; Trylox HCO-5)
   6.0 (Nikkol HCO-5)

POE (7) hydrogenated castor oil

SYNONYMS:
   PEG-7 hydrogenated castor oil (CTFA)
   PEG (7) hydrogenated castor oil
CAS No.:
   61788-85-0 (generic)
TRADENAME EQUIVALENTS:
   Arlacel 989 [ICI United States]
CATEGORY:
   Emulsifier
APPLICATIONS:
   Cosmetic industry preparations: creams and lotions (Arlacel 989)

POE (7) hydrogenated castor oil *(cont'd.)*

PROPERTIES:
Form:
   Solid (Arlacel 989)
Composition:
   100% conc. (Arlacel 989)

POE (25) hydrogenated castor oil

SYNONYMS:
   PEG-25 castor oil, hydrogenated
   PEG-25 hydrogenated castor oil (CTFA)
   PEG (25) hydrogenated castor oil
CAS No.:
   61788-85-0 (generic)
   RD No.: 977055-14-3
TRADENAME EQUIVALENTS:
   Arlatone G [ICI United States]
   Chemax HCO-25 [Chemax]
   Industrol COH-25 [BASF Wyandotte]
   Mapeg CO-25H [Mazer]
   Peganate COH25 [GAF]
   Trylox HCO-25 [Emery]
CATEGORY:
   Emulsifier, coemulsifier, surfactant
APPLICATIONS:
   Cosmetic industry preparations: gels (Arlatone G; Mapeg CO-25H)
   Industrial applications: dye carrier systems (Trylox HCO-25); lanolin systems (Chemax HCO-25; Trylox HCO-25); solvent systems (Chemax HCO-25; Trylox HCO-25); textile/leather processing (Trylox HCO-25)
PROPERTIES:
Form:
   Liquid (Chemax HCO-25; Industrol COH-25; Mapeg CO-25H; Peganate COH25; Trylox HCO-25)
   Liquid to soft paste (Arlatone G)
Color:
   Yellow (Arlatone G)
   Gardner 1 (Trylox HCO-25)
   Gardner 3 max. (Industrol COH-25)
Composition:
   100% active (Arlatone G; Industrol COH-25; Trylox HCO-25)
   100% conc. (Mapeg CO-25H; Peganate COH25)

Solubility:
Disp. in butyl stearate (@ 5%) (Trylox HCO-25)
Sol. in ethanol (Arlatone G)
Sol. in glycerol trioleate (@ 5%) (Trylox HCO-25)
Sol. in isopropanol (Arlatone G)
Sol. in min. oil (@ 5%) (Trylox HCO-25)
Disp. in perchloroethylene (@ 5%) (Trylox HCO-25)
Disp. in Stoddard solvent (@ 5%) (Trylox HCO-25)
Sol. in water (Arlatone G; Industrol COH-25); disp. (@ 5%) (Trylox HCO-25)
Ionic Nature:
Nonionic (Arlatone G; Industrol COH-25; Mapeg CO-25H; Peganate COH25; Trylox HCO-25)
M.W.:
2140 (theoret.) (Industrol COH-25)
Sp. gr.:
1.0 (Arlatone G)
1.03 (Industrol COH-25)
Density:
8.6 lb/gal (Trylox HCO-25)
Visc.:
1100 cs (Trylox HCO-25)
Pour Pt.:
5 C (Industrol COH-25)
Flash Pt.:
> 300 F (Arlatone G)
Fire Pt.:
> 300 F (Arlatone G)
Cloud Pt.:
25 C (Trylox HCO-25)
HLB:
10.8 (Arlatone G; Chemax HCO-25; Mapeg CO-25H; Trylox HCO-25)
11.0 (Industrol COH-25)
Acid No.:
2 max. (Industrol COH-25)
pH:
6.0–7.5 (5% aq.) (Industrol COH-25)
STD. PKGS.:
55-gal (450 lb net) steel drums (Industrol COH-25)

POE (2) isostearyl ether

SYNONYMS:
Isosteareth-2 (CTFA)
PEG-2 isostearyl ether
PEG 100 isostearyl ether

STRUCTURE:
$C_{18}H_{37}(OCH_2CH_2)_nOH$
where avg. $n = 2$

CAS No.:
52292-17-8 (generic)
RD. No.: 977064-96-2

TRADENAME EQUIVALENTS:
Arosurf 66-E2 [Sherex]

CATEGORY:
Emulsifier, coupling agent, stabilizer

APPLICATIONS:
Bath products: bath oils (Arosurf 66-E2)
Cosmetic industry preparations: (Arosurf 66-E2); conditioners (Arosurf 66-E2); creams and lotions (Arosurf 66-E2); hair preparations (Arosurf 66-E2); shampoos (Arosurf 66-E2); shaving preparations (Arosurf 66-E2); skin preparations (Arosurf 66-E2); toiletries (Arosurf 66-E2)

PROPERTIES:

Form:
Liquid (Arosurf 66-E2)

Color:
Gardner 1 (Arosurf 66-E2)

Ionic Nature:
Nonionic (Arosurf 66-E2)

M.P.:
−5 C (Arosurf 66-E2)

HLB:
4.6 (Arosurf 66-E2)

Stability:
Good (Arosurf 66-E2)

pH:
7 (1% sol'n. in DW) (Arosurf 66-E2)

TOXICITY/HANDLING:
Nonirritating to skin and eyes; nontoxic orally (Arosurf 66-E2)

SYNONYMS:
Isosteareth-10 (CTFA)
PEG-10 isostearyl ether
PEG 500 isostearyl ether
STRUCTURE:
$C_{18}H_{37}(OCH_2CH_2)_nOH$
where avg. $n = 10$
CAS No.:
52292-17-8 (generic)
RD. No.: 977059-85-0
TRADENAME EQUIVALENTS:
Arosurf 66-E10 [Sherex]
CATEGORY:
Emulsifier, detergent, coupling agent, stabilizer
APPLICATIONS:
Bath products: bath oils (Arosurf 66-E10)
Cosmetic industry preparations: (Arosurf 66-E10); conditioners (Arosurf 66-E10); creams and lotions (Arosurf 66-E10); hair preparations (Arosurf 66-E10); shampoos (Arosurf 66-E10); shaving preparations (Arosurf 66-E10); skin preparations (Arosurf 66-E10); toiletries (Arosurf 66-E10)
PROPERTIES:
Form:
Semisolid (Arosurf 66-E10)
Color:
Gardner 1 (Arosurf 66-E10)
Ionic Nature:
Nonionic (Arosurf 66-E10)
M.P.:
22 C (Arosurf 66-E10)
HLB:
12.0 (Arosurf 66-E10)
Stability:
Good (Arosurf 66-E10)
pH:
7 (1% sol'n. in DW) (Arosurf 66-E10)
TOXICITY/HANDLING:
Toxic orally and irritating to eyes in solid form; 5% sol'n. found to be nontoxic orally and nonirritating to skin and eyes (Arosurf 66-E10)

169

POE (20) isostearyl ether

SYNONYMS:
Isosteareth-20 (CTFA)
PEG-20 isostearyl ether
PEG 1000 isostearyl ether

STRUCTURE:
$C_{18}H_{37}(OCH_2CH_2)_nOH$
where avg. $n = 20$

CAS No.:
52292-17-8 (generic)
RD. No.: 977064-97-3

TRADENAME EQUIVALENTS:
Arosurf 66-E20 [Sherex]

CATEGORY:
Emulsifier, detergent, coupling agent, stabilizer

APPLICATIONS:
Bath products: bath oils (Arosurf 66-E20)
Cosmetic industry preparations: (Arosurf 66-E20); conditioners (Arosurf 66-E20); creams and lotions (Arosurf 66-E20); hair preparations (Arosurf 66-E20); shampoos (Arosurf 66-E20); shaving preparations (Arosurf 66-E20); skin preparations (Arosurf 66-E20); toiletries (Arosurf 66-E20)

PROPERTIES:
Form:
Soft solid (Arosurf 66-E20)
Color:
Gardner 1 (Arosurf 66-E20)
Ionic Nature:
Nonionic (Arosurf 66-E20)
M.P.:
35 C (Arosurf 66-E20)
HLB:
15.0 (Arosurf 66-E20)
Stability:
Good (Arosurf 66-E20)
pH:
7 (1% sol'n. in DW) (Arosurf 66-E20)

TOXICITY/HANDLING:
Toxic orally in solid form; 5% sol'n. found to be nontoxic orally and nonirritating to skin and eyes (Arosurf 66-E20)

SYNONYMS:
Laureth-4 (CTFA)
PEG-4 lauryl ether
PEG 200 lauryl ether

EMPIRICAL FORMULA:
$C_{20}H_{42}O_5$

STRUCTURE:
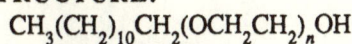
$CH_3(CH_2)_{10}CH_2(OCH_2CH_2)_nOH$
where avg. $n = 4$

CAS No.:
5274-68-0; 9002-92-0 (generic)
RD No.: 977054-83-3

TRADENAME EQUIVALENTS:
Brij 30, 30 SP [ICI United States]
Chemal LA-4 [Chemax]
Dehydol LS-4 [Henkel]
Emthox 5882 [Emery]
Ethosperse LA-4 [Glyco]
Hetoxol L-4N [Heterene]
LA-55-4 [Hefti Ltd.]
Lipal 4LA [PVO Int'l.]
Lipocol L-4 [Lipo]
Macol LA-4 [Mazer]
Simulsol P4 [Seppic]
Siponic L4 [Alcolac]
Volpo L4 [Croda]

CATEGORY:
Surfactant, emulsifier, detergent, solubilizer, coupling agent, raw material, intermediate, leveling agent, dispersant, stabilizer, wetting agent, defoamer, conditioner, thickener

APPLICATIONS:
Bath products: bubble bath (Siponic L4); bath oils (Emthox 5882; Hetoxol L-4N; LA-55-4)

Cleansers: germicidal skin cleanser (Lipal 4LA)

Cosmetic industry preparations: (Chemal LA-4; Ethosperse LA-4; Macol LA-4); creams and lotions (Emthox 5882; Hetoxol L-4N; LA-55-4; Lipal 4LA; Lipocol L-4; Siponic L4); hair preparations (Lipal 4LA); makeup (LA-55-4; Lipal 4LA); shampoos (Hetoxol L-4N; Siponic L4); shaving preparations (Lipal 4LA)

Household detergents: (Chemal LA-4; Dehydol LS-4; Hetoxol L-4N; Macol LA-4); dishwashing (Dehydol LS-4)

Industrial applications: (Ethosperse LA-4; Macol LA-4); dyes and pigments (Hetoxol L-4N; LA-55-4; Lipocol L-4); lubricating/cutting oils (Macol LA-4); metalworking

    (Macol LA-4); mold release (Chemal LA-4); paper mfg. (Siponic L4); polishes and waxes (Chemal LA-4; LA-55-4; Siponic L4); polymers/polymerization (LA-55-4; Siponic L4); rubber (Siponic L4); silicone products (Chemal LA-4; Hetoxol L-4N); textile/leather processing (Hetoxol L-4N; Macol LA-4)

Industrial cleaners: textile cleaning (Hetoxol L-4N)

Pharmaceutical applications: (Ethosperse LA-4; Siponic L4); antiperspirant/ deodorant (LA-55-4; Lipal 4LA; Lipocol L-4); depilatories (Lipocol L-4); ointments (LA-55-4; Lipal 4LA)

PROPERTIES:

Form:

    Liquid (Brij 30, 30SP; Chemal LA-4; Dehydol LS-4; Emthox 5882; Ethosperse LA-4; Hetoxol L-4N; LA-55-4; Lipal 4LA; Lipocol L-4; Macol LA-4; Simulsol P4; Siponic L4; Volpo L4)

Color:

    Colorless (Lipocol L-4; Macol LA-4)

    Colorless to light yellow (Brij 30, 30SP)

    Pale straw (Ethosperse LA-4)

    Gardner 1 (Lipal 4LA)

Composition:

    97% conc. (Volpo L4)

    99–100% conc. (Dehydol LS-4)

    100% active (Ethosperse LA-4; Lipal 4LA; Lipocol L-4; Siponic L4)

    100% conc. (Emthox 5882; LA-55-4; Macol LA-4; Simulsol P4)

Solubility:

    Sol. in acetone (Ethosperse LA-4)

    Sol. in alcohols (Brij 30, 30SP)

    Sol. in cottonseed oil (Brij 30, 30SP)

    Sol. in ethanol (Ethosperse LA-4)

    Sol. in ethyl acetate (Ethosperse LA-4)

    Sol. in isopropanol (Hetoxol L-4N; Lipal 4LA; Macol LA-4)

    Sol. in methanol (Ethosperse LA-4)

    Sol. in min. oil (Ethosperse LA-4; Hetoxol L-4N; Macol LA-4); disp. (Lipal 4LA)

    Sol. in peanut oil (Lipal 4LA)

    Sol. in propylene glycol (Brij 30, 30SP; Lipal 4LA)

    Sol. in toluol (Ethosperse LA-4)

    Misc. with veg. oil in certain proportions (Ethosperse LA-4)

    Misc. with water (Ethosperse LA-4); disp. (Lipal 4LA)

Ionic Nature:

    Nonionic (Brij 30, 30SP; Dehydol LS-4; Emthox 5882; Ethosperse LA-4; Hetoxol L-4N; LA-55-4; Lipal 4LA; Lipocol L-4; Macol LA-4; Simulsol P4; Siponic L4; Volpo L4)

Sp.gr.:

    0.95 (Brij 30, 30SP; Ethosperse LA-4)

Visc.:
  30 cps (Brij 30, 30SP; Ethosperse LA-4)
Flash Pt.:
  > 300 F (Brij 30, 30SP)
Fire Pt.:
  > 300 F (Brij 30, 30SP)
Cloud Pt.:
  52–53 C (1% sol'n.) (Lipal 4LA)
HLB:
  9.4 (Siponic L4)
  9.5 (Brij 30SP; Ethosperse LA-4; LA-55-4)
  9.7 (Brij 30; Emthox 5882; Hetoxol L-4N; Macol LA-4)
  9.7 ± 1 (Lipal 4LA; Lipocol L-4)
Acid No.:
  0.5 max. (Lipocol L-4)
  2.0 max. (Ethosperse LA-4)
Hydroxyl No.:
  145–160 (Ethosperse LA-4; Lipocol L-4)
  145–165 (Hetoxol L-4N; Macol LA-4)
  150–165 (Chemal LA-4)
  150–170 (Lipal 4LA)
Stability:
  Stable over a wide pH range (Lipal 4LA)
  Stable to hydrolysis by strong acids and alkalis (Macol LA-4)
pH:
  6.5 (1% sol'n.) (Siponic L4)

POE (12) lauryl ether

SYNONYMS:
  Laureth-12 (CTFA)
  PEG-12 lauryl ether
  PEG 600 lauryl ether
STRUCTURE:
  $CH_3(CH_2)_{10}CH_2(OCH_2CH_2)_nOH$
    where avg. $n = 12$
CAS No.:
  9002-92-0 (generic)
  RD No.: 977054-84-4

POE (12) lauryl ether (cont'd.)

TRADENAME EQUIVALENTS:
Chemal LA-12 [Chemax]
Cyclogol EL [Cyclo]
Ethosperse LA-12 [Glyco]
Hetoxol L-12 [Heterene]
Incropol L-12 [Croda Surfactants]
Lipal 12LA [PVO]
Lipocol L-12 [Lipo]
Macol LA-12 [Mazer]
Siponic L-12 [Alcolac]

CATEGORY:
Emulsiifier, wetting agent, lubricant, detergent, coupling agent, antistat, solubilizer, dispersant, defoamer, conditioning agent, stabilizer

APPLICATIONS:
Bath products: bath oils (Cyclogol EL)
Cleansers: germicidal skin cleanser (Lipal 12LA)
Cosmetic industry preparations: (Chemal LA-12; Ethosperse LA-12; Incropol L-12; Macol LA-12); creams and lotions (Lipal 12LA; Lipocol L-12; Siponic L-12); hair preparations (Lipal 12LA); makeup (Lipal 12LA); personal care products (Hetoxol L-12); perfumery (Cyclogol EL); shaving preparations (Lipal 12LA)
Farm products: microbicides (Cyclogol EL)
Household detergents: (Chemal LA-12; Incropol L-12; Macol LA-12)
Industrial applications: (Ethosperse LA-12; Macol LA-12); metalworking (Macol LA-12); mold release (Chemal LA-12); paper mfg. (Siponic L-12); pigment dispersions (Lipocol L-12); polishes and waxes (Chemal LA-12; Siponic L-12); polymers/polymerization (Siponic L-12); rubber (Siponic L-12);l textile/leather processing (Incropol L-12)
Industrial cleaners: (Incropol L-12); textile cleaning (Macol LA-12)
Pharmaceutical applications: (Ethosperse LA-12; Siponic L-12); antiperspirant/deodorant (Lipal 12LA; Lipocol L-12); depilatories (Lipocol L-12); ointments (Lipal 12LA)

PROPERTIES:
Form:
Liquid (Chemal LA-12)
Turbid liquid to soft paste (Ethosperse LA-12)
Paste (Cyclogol EL; Incropol L-12)
Semisolid (Hetoxol L-12)
Solid (Lipal 12LA; Macol LA-12)
Solid wax (Lipocol L-12)
Wax (Siponic L-12)

Color:
White (Cyclogol EL; Lipocol L-12; Macol LA-12)
Gardner 1 (Lipal 12LA)

Gardner 1 max. (Hetoxol L-12; Incropol L-12)

Composition:

100% active (Ethosperse LA-12; Incropol L-12; Lipal 12LA; Lipocol L-12; Siponic L-12)

100% conc. (Macol LA-12)

Solubility:

Sol. in acetone (Ethosperse LA-12)

Sol. in ethanol (Ethosperse LA-12)

Sol. in ethyl acetate (Ethosperse LA-12)

Sol. in isopropanol (Hetoxol L-12; Lipal 12LA; Macol LA-12)

Sol. in methanol (Ethosperse LA-12)

Sol. in propylene glycol (Lipal 12LA)

Sol. in toluol (Ethosperse LA-12)

Miscible hot with veg. oil (Ethosperse LA-12)

Sol. in water (Hetoxol L-12; Lipal 12LA; Macol LA-12); miscible with water in certain proportions (Ethosperse LA-12)

Ionic Nature:

Nonionic (Cyclogol EL; Ethosperse LA-12; Incropol L-12; Lipal 12LA; Lipocol L-12; Macol LA-12)

Sp.gr.:

1.10 (Ethosperse LA-12)

Visc.:

1000 cps (Ethosperse LA-12)

M.P.:

28–33 C (Cyclogol EL)

Cloud Pt.:

65–68 C (1% sol'n.) (Lipal 12LA)

89 C (1% NaCl) (Siponic L-12)

HLB:

4.8 (Ethosperse LA-12)

14.5 (Macol LA-12)

14.5 ± 1 (Lipocol L-12)

14.6 (Hetoxol L-12; Siponic L-12)

14.8 ± 1 (Lipal 12LA)

Acid No.:

1.0 max. (Hetoxol L-12; Lipocol L-12)

2.0 max. (Cyclogol EL; Ethosperse LA-12)

Hydroxyl No.:

70–80 (Macol LA-12)

72–82 (Ethosperse LA-12)

72–87 (Chemal LA-12; Lipal 12LA; Lipocol L-12)

74–81 (Incropol L-12)

74–82 (Hetoxol L-12)

POE (12) lauryl ether *(cont'd.)*

Stability:
    Stable over a wide pH range (Lipal 12LA)
    Acid and alkaline stable (Lipocol L-12)
    Stable to hydrolysis by strong acids and alkalies (Macol LA-12)
pH:
    5.5–7.5 (3%) (Incropol L-12)
    6.5 (1% sol'n.) (Siponic L-12)

POE (15) lauryl ether

SYNONYMS:
    Laureth-15 (CTFA)
    PEG-15 lauryl ether
    PEG (15) lauryl ether
STRUCTURE:
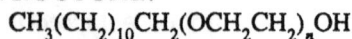
    $CH_3(CH_2)_{10}CH_2(OCH_2CH_2)_nOH$
    where avg. $n = 15$
CAS No.:
    9002-92-0 (generic)
    RD No.: 977069-29-6
TRADENAME EQUIVALENTS:
    Alkasurf LAN-15 [Alkaril]
CATEGORY:
    Coemulsifier
APPLICATIONS:
    Industrial applications: polymers/polymerization (Alkasurf LAN-15)
PROPERTIES:
Form:
    Solid (Alkasurf LAN-15)
Color:
    White (Alkasurf LAN-15)
Solubility:
    Sol. in aromatic solvent (@ 10%) (Alkasurf LAN-15)
    Disp. in min. spirits (@ 10%) (Alkasurf LAN-15)
    Sol. in perchloroethylene (@ 10%) (Alkasurf LAN-15)
    Sol. in water (@ 10%) (Alkasurf LAN-15)
Cloud Pt.:
    85–88 C (1% in 5% sodium chloride) (Alkasurf LAN-15)
HLB:
    15.5 (Alkasurf LAN-15)

POE (23) lauryl ether

SYNONYMS:
Laureth-23 (CTFA)
PEG (23) lauryl ether
PEG-23 lauryl ether
POE (23) lauryl alcohol

STRUCTURE:
$CH_3(CH_2)_{10}CH_2(OCH_2CH_2)_nOH$
where avg. $n = 23$

CAS No.:
9002-92-0 (generic)
RD No.: 977054-85-5

TRADENAME EQUIVALENTS:
Alkasurf LAN-23 [Alkaril]
Brij 35, 35SP [ICI United States]
Chemal LA-23 [Chemax]
Ethosperse LA23 [Glyco]
Hetoxol L-23N [Heterene]
Incropol L-23 [Croda Surfactants]
LA-55-23 [Hefti Ltd.]
Lipal 23LA [PVO]
Lipocol L-23 [Lipo]
Macol LA-23 [Mazer]
Siponic L-25 [Alcolac]
Trycol LAL-23 [Emery]

CATEGORY:
Emulsifier, surfactant, detergent, wetting agent, leveling agent, intermediate, solubilizer, dispersant, coupling agent, defoamer, conditioning agent, stabilizer, thickener, lubricant

APPLICATIONS:
Bath products: bath oils (Hetoxol L-23N)
Cleansers: germicidal skin cleanser (Lipal 23LA)
Cosmetic industry preparations: (Chemal LA-23; Incropol L-23; Macol LA-23); conditioners (Ethosperse LA23); creams and lotions (Hetoxol L-23N; LA-55-23; Lipal 23LA; Lipocol L-23; Siponic L-25); hair preparations (Ethosperse LA23; Lipal 23LA); makeup (Lipal 23LA); perfumery (Ethosperse LA23); shampoos (Ethosperse LA23; Hetoxol L-23N); shaving preparations (Lipal 23LA); topical cosmetics (Alkasurf LAN-23)
Household detergents: (Chemal LA-23; Hetoxol L-23N; Incropol L-23; Macol LA-23)
Industrial applications: (Incropol L-23; Macol LA-23); dyes and pigments (Hetoxol L-23N; Lipocol L-23; Trycol LAL-23); metalworking (Macol LA-23); mold release (Chemal LA-23; Trycol LAL-23); paper mfg. (Siponic L-25); polishes and waxes (Chemal LA-23; LA-55-23; Siponic L-25; Trycol LAL-23); polymers/polymerization (Siponic L-25); rubber (Siponic L-25); textile/leather processing (Hetoxol L-

177

POE (23) lauryl ether *(cont'd.)*

23N; Incropol L-23; LA-55-23; Macol LA-23; Trycol LAL-23)

Industrial cleaners: textile cleaning (Hetoxol L-23N)

Pharmaceutical applications: (Siponic L-25); antiperspirant/deodorant (Ethosperse LA23; Lipal 23LA; Lipocol L-23); depilatories (Ethosperse LA23; Lipocol L-23); ointments (LA-55-23; Lipal 23LA)

PROPERTIES:

Form:

Solid (Alkasurf LAN-23; Chemal LA-23; Ethosperse LA23; Hetoxol L-23N; Incropol L-23; LA-55-23; Lipal 23LA; Macol LA-23; Trycol LAL-23)

Wax (Siponic L-25)

Waxy solid (Brij 35, 35SP; Lipocol L-23)

Color:

Almost colorless (Alkasurf LAN-23)

White (Brij 35, 35SP; Ethosperse LA23; Lipocol L-23; Macol LA-23)

< Gardner 1 (Trycol LAL-23)

Gardner 1 (Lipal 23LA)

Odor:

Characteristic (Alkasurf LAN-23)

Composition:

100% active (Alkasurf LAN-23; Incropol L-23; Lipal 23LA; Lipocol L-23; Siponic L-25; Trycol LAL-23)

100% conc. (LA-55-23; Macol LA-23)

Solubility:

Sol. in alcohol (Brij 35, 35SP)

Sol. hot in ethanol (Ethosperse LA23)

Sol. in isopropanol (Hetoxol L-23N; Lipal 23LA; Macol LA-23)

Disp. in perchloroethylene (@ 5%) (Trycol LAL-23)

Sol. in propylene glycol (Brij 35, 35SP; Lipal 23LA)

Sol. in water (Brij 35, 35SP; Ethosperse LA23; Hetoxol L-23N; Lipal 23LA; Macol LA-23); (@ 5%) (Trycol LAL-23)

Ionic Nature: Nonionic

Sp.gr.:

1.05 (Brij 35, 35SP)

M.P.:

30–45 C (Ethosperse LA23)

32 C (Macol LA-23)

46 C (Trycol LAL-23)

Pour Pt.:

33 C (Brij 35, 35SP)

Flash Pt.:

> 300 F (Brij 35, 35SP)

Fire Pt.:

> 300 F (Brij 35, 35SP)

Cloud Pt.:
   75–78 C (1% sol'n.) (Lipal 23LA)
   > 95 C (1% NaCl) (Siponic L-25)
   > 100 C (Trycol LAL-23)
HLB:
   16.9 (Brij 35, 35SP; Hetoxol L-23N; Macol LA-23; Siponic L-25; Trycol LAL-23)
   16.9 ± 1 (Lipal 23LA; Lipocol L-23)
   17.0 (LA-55-23)
   17.0 ± 1 (Ethosperse LA23)
Acid No.:
   0.3 max. (Ethosperse LA23)
   0.5 max. (Lipocol L-23)
Hydroxyl No.:
   40–55 (Chemal LA-23; Hetoxol L-23N; Lipal 23LA; Macol LA-23)
   42–52 (Lipocol L-23)
   45–52 (Ethosperse LA23)
Stability:
   Stable over a wide pH range (Lipal 23LA)
   Stable to hydrolysis by acids and alkalis (Trycol LAL-23)
   Stable to hydrolysis by strong acids and alkalis (Macol LA-23)
   Chemically stable; resistant to acid hydrolysis and alkaline saponification (Ethosperse
      LA23)
pH:
   6.5 (1% sol'n.) (Siponic L-25)
Biodegradable: (Alkasurf LAN-23; Incropol L-23)
STORAGE/HANDLING:
   Conc. mineral acids will react chemically with product (Trycol LAL-23)

POE (8) monococoate

SYNONYMS:
   PEG-8 cocoate (CTFA)
   PEG 400 monococoate
STRUCTURE:

$$RC\!\!-\!\!(OCH_2CH_2)_nOH$$
with O double-bonded to C

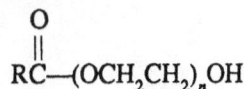

   where   RCO⁻ represents the coconut acid radicals and
              avg. $n = 8$

POE (8) monococoate (cont'd.)

CAS No.:
61791-29-5 (generic)
RD No.: 977055-01-8
TRADENAME EQUIVALENTS:
Alkamuls 400 MC [Alkaril]
Nopalcol 4-C, 4-CH [Diamond Shamrock]
Pegosperse 400 MC [Glyco]
CATEGORY:
Emulsifier, wetting agent, binding agent, defoamer, dispersant, lubricant, plasticizer, softener, thickening agent
APPLICATIONS:
Cosmetic industry preparations (Alkamuls 400MC; Nopalcol 4-C); personal care products (Nopalcol 4-C); shampoos (Nopalcol 4-C)
Farm products (Nopalcol 4-C)
Industrial applications: construction (Nopalcol 4-C); dyes and pigments (Nopalcol 4-C); paper mfg. (Nopalcol 4-C); textile/leather processing (Nopalcol 4-C)
Industrial cleaners: drycleaning compositions (Nopalcol 4-C)
PROPERTIES:
Form:
Liquid (Alkamuls 400MC; Nopalcol 4-C, 4-CH; Pegosperse 400MC)
Color:
Yellow (Pegosperse 400MC)
Composition:
99% conc. (Nopalcol 4-C, 4-CH)
100% active (Alkamuls 400MC)
100% conc. (Pegosperse 400MC)
Solubility:
Sol. in acetone, ethanol, ethyl acetate, methanol, toluol, veg. oil; sol. hot in min. oil; miscible with water and naphtha (Pegosperse 400MC)
Ionic Nature: Nonionic
Sp.gr.:
1.03 (Pegosperse 400MC)
Solidification Pt.:
< 6 C (Pegosperse 400MC)
HLB:
13.9 (Alkamuls 400MC; Nopalcol 4-C, 4-CH)
13.9 ± 0.5 (Pegosperse 400MC)
Acid No.:
< 4 (Pegosperse 400MC)
Iodine No.:
< 5 (Pegosperse 400MC)
Saponification No.:
88–100 (Pegosperse 400MC)

Stability:
Stable to acid and alkali at room temp.; may hydrolyze at higher temperatures or in conc. acid and alkaline sol'ns. (Nopalcol 4-C, 4-CH)
pH:
4–6 (5% aq. disp.) (Pegosperse 400MC)

POE (4) monolaurate

SYNONYMS:
PEG-4 laurate (CTFA)
PEG 200 monolaurate
STRUCTURE:

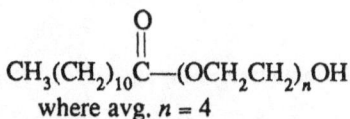

$$CH_3(CH_2)_{10}\overset{\displaystyle O}{\overset{\|}{C}}-(OCH_2CH_2)_nOH$$
where avg. $n = 4$

CAS No.:
9004-81-3 (generic)
RD No: 977055-22-3
TRADENAME EQUIVALENTS:
Alkamuls 200ML [Alkaril]
Cithrol 2ML [Croda]
CPH-27-N [C.P. Hall]
Emerest 2620 [Emery]
Hodag 20-L [Hodag]
Kessco PEG 200 Monolaurate [Armak]
Mapeg 200ML [Mazer]
Nopalcol 2-L [Diamond Shamrock]
Pegosperse 200ML [Glyco]
Radiasurf 7422 [Oleofina]
CATEGORY:
Emulsifier, coupling agent, solubilizer, wetting agent, antifog aid, antistat, chemical intermediate, defoamer, detergent, dispersant, emulsifier, lubricant, opacifier, o/w emulgent, plasticizer, rust inhibitor, scouring aid, softening agent, surfactant, thickening agent, visc. modifier
APPLICATIONS:
Bath products: (Kessco PEG 200 Monolaurate)
Cosmetic industry preparations (Cithrol 2ML; Hodag 20-L; Kessco PEG 200 Monolaurate; Mapeg 200ML; Radiasurf 7422); hair rinses (Kessco PEG 200 Monolaurate); shampoos (Kessco PEG 200 Monolaurate)

POE (4) monolaurate (cont'd.)

Farm products: insecticides (Radiasurf 7422)

Food applications (Kessco PEG 200 Monolaurate)

Industrial applications: dyes (Radiasurf 7422); glass processing (Cithrol 2ML); lubricating/cutting oils (Alkamuls 200ML; Radiasurf 7422); metalworking (Alkamuls 200ML; Cithrol 2ML); paint mfg.(Radiasurf 7422); plastics (Kessco PEG 200 Monolaurate; Radiasurf 7422); polishes and waxes (Radiasurf 7422); printing inks (Radiasurf 7422); textile/leather processing (Alkamuls 200ML; Mapeg 200ML; Radiasurf 7422)

Industrial cleaners: metal processing surfactants (Mapeg 200ML)

Pharmaceutical applications (Cithrol 2ML; Hodag 20-L; Kessco PEG 200 Monolaurate; Mapeg 200ML; Radiasurf 7422)

PROPERTIES:

Form:

Liquid (Alkamuls 200ML; CPH-27-N; Hodag 20-L; Kessco PEG 200 Monolaurate; Mapeg 200ML; Nopalcol 2-L; Pegosperse 200ML; Radiasurf 7422)

Color:

White (Radiasurf 7422)

Yellow (Alkamuls 200ML; Kessco PEG 200 Monolaurate; Mapeg 200ML; Pegosperse 200ML)

Composition:

100% active (Emerest 2620)

100% conc. (CPH-27-N; Hodag 20-L; Mapeg 200ML; Nopalcol 2-L; Pegosperse 200ML)

Solubility:

Sol. in acetone (Kessco PEG 200 Monolaurate; Pegosperse 200ML)

Disp. in aromatic solvents (Alkamuls 200ML)

Sol. in benzene (Radiasurf 7422)

Sol. in carbon tetrachloride (Kessco PEG 200 Monolaurate)

Sol. in ethanol (Pegosperse 200ML)

Sol. in ethyl acetate (Kessco PEG 200 Monolaurate; Pegosperse 200ML)

Sol. in hexane (cloudy) (Radiasurf 7422)

Sol. in isopropanol (Kessco PEG 200 Monolaurate; Mapeg 200ML; Radiasurf 7422)

Sol. in isopropyl myristate (Kessco PEG 200 Monolaurate)

Sol. in methanol (Pegosperse 200ML)

Sol. in min. oils (Radiasurf 7422); partly sol. (Mapeg 200ML); disp. (Alkamuls 200ML); miscible (Pegosperse 200ML)

Disp. in min. spirits (Alkamuls 200ML)

Miscible with naphtha (Pegosperse 200ML)

Disp. in perchloroethylene (Alkamuls 200ML)

Partly sol. in soybean oil (Mapeg 200ML)

Sol. in toluol (Kessco PEG 200 Monolaurate; Mapeg 200ML; Pegosperse 200ML)

Sol. in trichloroethylene (Radiasurf 7422)

Sol. in veg. oils (Radiasurf 7422); miscible (Pegosperse 200ML)

POE (4) monolaurate *(cont'd.)*

Disp. in water (Alkamuls 200ML; Kessco PEG 200 Monolaurate; Mapeg 200ML; Radiasurf 7422); miscible (Pegosperse 200ML)

Ionic Nature:
Nonionic (Alkamuls 200ML; Cithrol 2ML; CPH-27-N; Emerest 2620; Hodag 20-L; Kessco PEG-200 Monolaurate; Mapeg 200ML; Nopalcol 2-L; Pegosperse 200ML; Radiasurf 7422)

Density:
8.2 lb/gal (Kessco PEG 200 Monolaurate)
1.00 g/ml (Alkamuls 200ML)

Sp.gr.:
0.983 @ 37.8 C (Radiasurf 7422)
0.985 (Kessco PEG 200 Monolaurate)
0.99 (Mapeg 200ML; Pegosperse 200 ML)

HLB:
8.6 ± 0.5 (Pegosperse 200ML)
8.8 (Cithrol 2ML)
9.6 (Radiasurf 7422)
9.8 (Alkamuls 200ML; Kessco PEG 200 Monolaurate; Mapeg 200ML)
10.0 (Hodag 20-L)
10.5 (Nopalcol 2-L)

Acid No.:
< 3 (Radiasurf 7422)
< 5 (Kessco PEG 200 Monolaurate; Mapeg 200ML; Pegosperse 200ML)

Iodine No.:
1 max. (Radiasurf 7422)
< 9 (Pegosperse 200ML)
9.5 max. (Kessco PEG 200 Monolaurate)
10 max. (Mapeg 200 ML)

Saponification No.:
132–142 (Kessco PEG 200 Monolaurate)
139–159 (Mapeg 200ML)
140–150 (Radiasurf 7422)
142–152 (Alkamuls 200ML)
149–159 (Pegosperse 200ML)
158–168 (Cithrol 2ML)

pH:
4.0–6.5 (5% aq. disp.) (Pegosperse 200ML)
4.5 (3% disp. @ 20 C) (Kessco PEG 200 Monolaurate)

Flash Pt.:
188 C (Radiasurf 7422)
385 F (Kessco PEG 200 Monolaurate)

POE (6) monolaurate

SYNONYMS:
  PEG-6 laurate (CTFA)
  PEG 300 monolaurate
STRUCTURE:

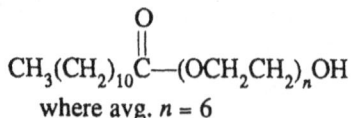

  where avg. $n = 6$
CAS No.:
  9004-81-3 (generic)
  RD No.: 977065-75-0
TRADENAME EQUIVALENTS:
  Chemax E-200ML [Chemax]
  Emerest 2630 [Emery]
  Kessco PEG 300 Monolaurate [Armak]
CATEGORY:
  Emulsifier, coemulsifier, viscosity control agent, thickener, defoaming agent, surfactant, solubilizer
APPLICATIONS:
  Bath products: bath oils (Kessco PEG 300 Monolaurate)
  Cosmetic industry preparations: (Kessco PEG 300 Monolaurate); hair preparations (Kessco PEG 300 Monolaurate); perfumery (Kessco PEG 300 Monolaurate); shampoos (Kessco PEG 300 Monolaurate)
  Farm products: (Kessco PEG 300 Monolaurate)
  Food applications: (Kessco PEG 300 Monolaurate)
  Industrial applications: (Kessco PEG 300 Monolaurate); lubricating/cutting oils (Emerest 2630); plastics (Kessco PEG 300 Monolaurate); plastisols (Emerest 2630); textile/leather processing (Emerest 2630); water-based coatings (Chemax E-200ML)
  Pharmaceutical applications: (Kessco PEG 300 Monolaurate)
PROPERTIES:
Form:
  Liquid (Chemax E-200ML; Emerest 2630; Kessco PEG 300 Monolaurate)
Color:
  Light yellow (Kessco PEG 300 Monolaurate)
  Gardner 2 (Emerest 2630)
Solubility:
  Sol. in acetone (Kessco PEG 300 Monolaurate)
  Disp. in butyl stearate (@ 5%) (Emerest 2630)
  Sol. in carbon tetrachloride (Kessco PEG 300 Monolaurate)
  Sol. in ethyl acetate (Kessco PEG 300 Monolaurate)
  Disp. in glycerol trioleate (@ 5%) (Emerest 2630)
  Sol. in isopropanol (Kessco PEG 300 Monolaurate)

Partly sol. in isopropyl myristate (Kessco PEG 300 Monolaurate)
Disp. in min. oil (@ 5%) (Emerest 2630)
Disp. in Stoddard solvent (@ 5%) (Emerest 2630)
Sol. in toluol (Kessco PEG 300 Monolaurate)
Disp. in water (Kessco PEG 300 Monolaurate); disp. (@ 5%) (Emerest 2630)
Sol. in xylene (@ 5%) (Emerest 2630)

Ionic Nature:
Nonionic (Kessco PEG 300 Monolaurate)
Sp.gr.:
1.011 (Kessco PEG 300 Monolaurate)
Density:
8.4 lb/gal (Emerest 2630; Kessco PEG 300 Monolaurate)
Visc.:
37 cSt (100 F) (Emerest 2630)
F.P.:
< 8 C (Kessco PEG 300 Monolaurate)
Pour Pt.:
9 C (Emerest 2630)
Flash Pt.:
445 F (COC) (Kessco PEG 300 Monolaurate)
475 F (Emerest 2630)
Fire Pt.:
500 F (Kessco PEG 300 Monolaurate)
Cloud Pt.:
< 25 C (Emerest 2630)
HLB:
9.8 (Chemax E-200ML)
11.4 (Kessco PEG 300 Monolaurate)
12.1 (Emerest 2630)
Acid No.:
5.0 max. (Kessco PEG 300 Monolaurate)
Iodine No.:
7.5 max. (Kessco PEG 300 Monolaurate)
Saponification No.:
104–114 (Kessco PEG 300 Monolaurate)
pH:
4.5 (3% disp., 20 C) (Kessco PEG 300 Monolaurate)

POE (8) monolaurate

SYNONYMS:
  PEG-8 laurate (CTFA)
  PEG-400 monolaurate
STRUCTURE:

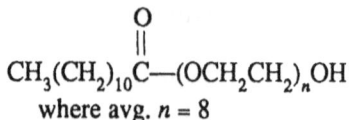

$$CH_3(CH_2)_{10}C-(OCH_2CH_2)_nOH$$
  where avg. $n = 8$

CAS No.:
  9004-81-3 (generic)
TRADENAME EQUIVALENTS:
  Alkamuls 400ML [Alkaril]
  Chemax E-400-ML [Chemax]
  Cithrol 4ML [Croda]
  Collemul L4 [Allied Colloids Ltd.]
  Cyclochem PEG 400ML [Cyclo]
  Emerest 2650 [Emery]
  Emerest 2705 [Emery]
  Gradonic 400 ML [Graden]
  Kessco PEG 400 Monolaurate [Armak]
  Lipopeg 4-L [Lipo]
  Lonzest PEG-4-L [Lonza]
  Mapeg 400ML [Mazer]
  Nonisol 100 [Ciba-Geigy]
  Nopalcol 4-L [Diamond Shamrock]
  Pegosperse 400ML [Glyco]
  PGE-400-ML[Hefti]
  Radiasurf 7423 [Oleofina]
  Scher PEG 400 Monolaurate [Scher]
CATEGORY:
  Emulsifier, wetting agent, antifog aid, antistat, binding agent, defoamer, detergent,
    dispersant, dye assistant, intermediate, leveling agent, lubricant, o/w emulgent,
    opacifier, penetrant, plasticizer, rust inhibitor, scouring and detergent aid, solubi-
    lizer, spreading agent, stabilizer, surfactant, thickener, viscosity modifier
APPLICATIONS:
  Bath products: bath oils (Emerest 2705; Kessco PEG 400 Monolaurate; Lipopeg 4-L;
    PGE-400-ML; Scher PEG 400 Monolaurate)
  Cleansers: hand cleanser (Nonisol 100)
  Cosmetic industry preparations: (Alkamuls 400ML; Kessco PEG 400 Monolaurate;
    Lipopeg 4-L; Lonzest PEG 4-L; Mapeg 400ML; Nonisol 100; PGE-400-ML;
    Radiasurf 7423); creams and lotions (Scher PEG 400 Monolaurate); hair rinses
    (Emerest 2705; Kessco PEG 400 Monolaurate); perfumery (Emerest 2705; Kessco
    PEG 400 Monolaurate; PGE-400-ML; Scher PEG 400 Monolaurate); shampoos

(Alkamuls 400ML; Emerest 2705; Kessco PEG 400 Monolaurate)

Farm products: (Alkamuls 400ML; Emerest 2705; Kessco PEG 400 Monolaurate); animal feed (Lonzest PEG-4-L); insecticides/pesticides (Cithrol 4ML; Gradonic 400-ML; Radiasurf 7423)

Food applications: (Emerest 2705; Kessco PEG 400 Monolaurate); food containers (Cithrol 4ML)

Household detergents: (Collemul L4)

Industrial applications: construction (Cithrol 4ML; Nopalcol 4-L; PGE-400-ML); dyes and pigments (Cithrol 4ML; Emerest 2650, 2705; Gradonic 400-ML; Lonzest PEG-4-L; Nopalcol 4-L; PGE-400-ML; Radiasurf 7423); glass processing (Gradonic 400-ML; Radiasurf 7423); lubricating/cutting oils (Gradonic 400-ML; Radiasurf 7423); metalworking (Emerest 2705; Mapeg 400 ML); paint mfg. (Chemax E-400-ML; Emerest 2650, 2705; Gradonic 400-ML; Lonzest PEG-4-L; Radiasurf 7423); paper mfg. (Cithrol 4ML; Lonzest PEG-4-L; Nopalcol 4-L; PGE-400-ML); plastics (Chemax E-400-ML; Cithrol 4ML; Emerest 2650, 2705; Gradonic 400-ML; Kessco PEG 400 Monolaurate; Lonzest PEG-4-L; Radiasurf 7423); polishes and waxes (Gradonic 400-ML; Lonzest PEG-4-L; Radiasurf 7423); printing inks (Gradonic 400-ML; Lonzest PEG-4-L; Radiasurf 7423); rubber (Lonzest PEG-4-L); textile/leather processing (Alkamuls 400ML; Cithrol 4ML; Emerest 2650, 2705; Gradonic 400-ML; Lonzest PEG-4-L; Mapeg 400 ML; Nonisol 100; Nopalcol 4-L; PGE-400-ML; Radiasurf 7423)

Pharmaceutical applications: (Kessco PEG 400 Monolaurate; Lonzest PEG 4-L; Mapeg 400ML; Radiasurf 7423); deodorant (Lonzest PEG-4-L)

PROPERTIES:
Form:
Clear liquid (Alkamuls 400ML; Scher PEG 400 Monolaurate)
Liquid (Cithrol 4 ML; Collemul L4; Cyclochem PEG 400 ML; Emerest 2650, 2705; Gradonic 400-ML; Kessco PEG 400 Monolaurate; Lipopeg 4-L; Lonzest PEG-4-L; Mapeg 400ML; Nonisol 100; Nopalcol 4-L; Pegosperse 400ML; PGE-400-ML; Radiasurf 7423)

Color:
White (Radiasurf 7423)
Amber (Alkamuls 400ML)
Light straw (Cyclochem PEG 400ML; Scher PEG 400 Monolaurate)
Straw (Pegosperse 400 ML)
Light yellow (Kessco PEG 400 Monolaurate; Mapeg 400 ML)
Yellow (Lipopeg 4-L; Lonzest PEG 4-L; Nonisol 100)
Gardner 1 (Emerest 2650)

Odor:
Mild, typical (Scher PEG 400 Monolaurate)

Composition:
100% active (Alkamuls 400ML; Emerest 2650; Lipopeg 4-L)
100% conc. (Lonzest PEG-4-L; Nonisol 100; Pegosperse 400ML)

POE (8) monolaurate (cont'd.)

Solubility:
Sol. in acetate esters (Scher PEG 400 Monolaurate)
Sol. in acetone (Kessco PEG 400 Monolaurate; Pegosperse 400ML)
Sol. in most alcohols (Scher PEG 400 Monolaurate)
Sol. in aliphatic hydrocarbons (Scher PEG 400 Monolaurate)
Sol. in aromatic hydrocarbons (Scher PEG 400 Monolaurate)
Sol. in benzene (Radiasurf 7423)
Disp. in butyl stearate (Emerest 2650)
Sol. in carbon tetrachloride (Kessco PEG 400 Monolaurate)
Sol. in ethanol (Pegosperse 400ML)
Sol. in ethyl acetate (Kessco PEG 400 Monolaurate; Pegosperse 400ML)
Sol. in glycol ethers (Scher PEG 400 Monolaurate)
Sol. in glycols (Scher PEG 400 Monolaurate)
Disp. in glycerol trioleate (Emerest 2650)
Sol. in hexane (Radiasurf 7423)
Sol. in isopropyl alcohol (Kessco PEG 400 Monolaurate; Lonzest PEG-4-L; Mapeg 400ML)
Sol. in ketones (Scher PEG 400 Monolaurate)
Sol. in methanol (Pegosperse 400ML)
Sol. in min. oil (Pegosperse 400ML; Radiasurf 7423)
Miscible with naphtha (Pegosperse 400ML)
Disp. in perchloroethylene (Emerest 2650)
Disp. in Stoddard solvent (Emerest 2650)
Sol. in toluol (Kessco PEG 400 Monolaurate; Lonzest PEG-4-L; Mapeg 400ML)
Sol. in trichlorethylene (Radiasurf 7423)
Sol. in veg. oil (Pegosperse 400ML; Radiasurf 7423)
Sol. in water (Cithrol 4ML; Emerest 2650; Kessco PEG 400 Monolaurate; Lonzest PEG-4-L; Mapeg 400ML; Nonisol 100; Radiasurf 7423); miscible with water (Pegosperse 400ML)

Ionic Nature:
Nonionic

M.W.:
400 (Cithrol 4ML)
592 avg. (Scher PEG 400 Monolaurate)
610 avg. (Radiasurf 7423)

Sp.gr.:
1.01 (Mapeg 400ML)
1.02–1.04 (Alkamuls 400ML)
1.023 (Radiasurf 7423)
1.028 (Kessco PEG 400 Monolaurate)
1.03 (Lonzest PEG 4-L; Pegosperse 400ML; Scher PEG 400 Monolaurate)

Density:
8.6 lb/gal (Emerest 2650; Kessco PEG 400 Monolaurate)

Visc.:
   41.70 cps (37.8 C) (Radiasurf 7423)
   100 cs (Emerest 2650)
F.P.:
   12 C (Kessco PEG 400 Monolaurate)
M.P.:
   10 C (Cyclochem PEG 400 ML)
   12 C (Mapeg 400 ML)
Solidification Pt.:
   < 7 C (Pegosperse 400ML)
Flash Pt.:
   > 170 C (OC) (Scher PEG 400 Monolaurate)
   238 C (COC) (Radiasurf 7423)
   475 F (Kessco PEG 400 Monolaurate)
Fire Pt.:
   535 F (Kessco PEG 400 Monolaurate)
Cloud Pt.:
   4 C (Radiasurf 7423)
   6 C (Scher PEG 400 Monolaurate)
   33 C (Emerest 2650)
   50 F (Alkamuls 400 ML)
HLB:
   12.8 (Alkamuls 400ML)
   13.0 ± 1 (Lipopeg 4-L)
   13.1 (Chemax E-400-ML; Emerest 2650; Kessco PEG 400 Monolaurate; Lonzest
      PEG-4-L; Mapeg 400ML)
   13.2 (Radiasurf 7423)
   13.5 (Scher PEG 400 Monolaurate)
   13.9 ± 0.5 (Pegosperse 400ML)
   19 (Cithrol 4ML)
Acid No.:
   3 max. (Pegosperse 400ML; Radiasurf 7423)
   5 max. (Cyclochem PEG 400ML; Kessco PEG 400 Monolaurate; Lipopeg 4-L; Mapeg
      400ML; Scher PEG 400 Monolaurate)
   5–10 (Lonzest PEG-4-L)
Iodine No.:
   1 max. (Radiasurf 7423)
   4 max. (Pegosperse 400ML)
   6.5 max. (Kessco PEG 400 Monolaurate; Scher PEG 400 Monolaurate)
   7–8 (Lonzest PEG-4-L)
   8 max. (Mapeg 400ML)
Saponification No.:
   86–96 (Kessco PEG 400 Monolaurate)

POE (8) monolaurate *(cont'd.)*

88–98 (Scher PEG 400 Monolaurate)
89–96 (Mapeg 400ML)
90 (Cyclochem PEG 400ML)
90–100 (Lipopeg 4-L; Lonzest PEG-4-L; Pegosperse 400ML; Radiasurf 7423)
92–98 (Cithrol 4ML)

Ref. Index:

1.4575 (Scher PEG 400 Monolaurate)
1.4596 (Radiasurf 7423)

pH:

4.0–6.0 (5% aq. disp.) (Pegosperse 400ML)

Surface Tension:

31.00 dynes/cm (0.1% aq.) (Radiasurf 7423)

STD. PKGS.:

450-lb steel drums (Kessco PEG 400 Monolaurate)
190-kg net bung drums or bulk (Radiasurf 7423)

POE (12) monolaurate

SYNONYMS:

PEG-12 laurate (CTFA)
PEG 600 monolaurate

STRUCTURE:

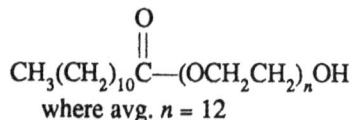

$$CH_3(CH_2)_{10}\overset{\overset{\textstyle O}{\|}}{C}—(OCH_2CH_2)_nOH$$

where avg. $n = 12$

CAS No.

9004-81-3 (generic)

TRADENAME EQUIVALENTS:

Alkamuls 600ML [Alkaril]
Cithrol 6ML [Croda]
CPH-43-N [C.P. Hall]
Emerest 2661 [Emery]
Kessco PEG 600 Monolaurate [Armak]
Lipopeg 6-L [Lipo]
Mapeg 600ML [Mazer]
Nopalcol 6-L [Diamond Shamrock]
Pegosperse 600ML [Glyco]
PGE-600-ML [Hefti]

POE (12) monolaurate (cont'd.)

CATEGORY:
Wetting agent, emulsifier, detergent, antistat, chemical intermediate, dispersant, lubricant, softening agent, solubilizer, spreading agent, surfactant, thickening agent

APPLICATIONS:
Bath products: bath oils (Kessco PEG 600 Monolaurate; Lipopeg 6-L)
Cosmetic industry preparations: (Cithrol 6ML; Kessco PEG 600 Monolaurate; Lipopeg 6-L; Mapeg 600ML; Nopalcol 6-L); hair rinses (Kessco PEG 600 Monolaurate); perfumery (Kessco PEG 600 Monolaurate); shampoos (Kessco PEG 600 Monolaurate)
Farm products: (Kessco PEG 600 Monolaurate)
Food applications: (Kessco PEG 600 Monolaurate)
Industrial applications: adhesives (Nopalcol 6-L); dyes and pigments (Nopalcol 6-L); glass processing (Cithrol 6ML); metalworking (Cithrol 6ML; Mapeg 600ML); paper mfg. (Nopalcol 6-L); plastics (Kessco PEG 600 Monolaurate); textile/leather processing (Cithrol 6ML; Emerest 2661; Mapeg 600ML; Nopalcol 6-L)
Industrial cleaners: drycleaning compositions (Nopalcol 6-L)
Pharmaceutical applications: (Kessco PEG 600 Monolaurate; Mapeg 600ML)

PROPERTIES:
Form:
Liquid (Alkamuls 600ML; Emerest 2661; Kessco PEG 600 Monolaurate; Lipopeg 6-L; Mapeg 600ML; Pegosperse 600ML; PGE-600-ML)
Solid (CPH-43-N)
Color:
Light yellow (Kessco PEG 600 Monolaurate; Mapeg 600ML)
Yellow (Lipopeg 6-L; Pegosperse 600ML)
Gardner 2 (Emerest 2661)
Composition:
100% active (Alkamuls 600ML; Lipopeg 6-L)
100% conc. (CPH-43-N; Pegosperse 600ML; PGE-600-ML)
Solubility:
Sol. in carbon tetrachloride (Kessco PEG 600 Monolaurate)
Sol. in ethanol (Kessco PEG 600 Monolaurate; Pegosperse 600ML)
Sol. in ethyl acetate (Kessco PEG 600 Monolaurate; Pegosperse 600ML)
Sol. in isopropanol (Kessco PEG 600 Monolaurate; Mapeg 600ML)
Sol. methanol (Pegosperse 600ML)
Partly sol. in min. oil (Pegosperse 600ML)
Sol. hot in naphtha (Pegosperse 600ML)
Disp. in propylene glycol (Mapeg 600ML)
Sol. sodium sulfate (Kessco PEG 600 Monolaurate)
Sol. in toluol (Kessco PEG 600 Monolaurate; Mapeg 600ML); miscible (Pegosperse 600ML)
Miscible with veg. oil (Pegosperse 600ML)
Sol. in water (Emerest 2661; Kessco PEG 600 Monolaurate; Mapeg 600ML)

191

POE (12) monolaurate (cont'd.)

Sol. in xylene (Emerest 2661)
Ionic Nature: Nonionic
M.W.:
600 (Cithrol 6ML)
Sp.gr.:
1.01 (Pegosperse 600ML)
1.02 (Mapeg 600ML)
1.050 (Kessco PEG 600 Monolaurate)
Density:
8.6 lb/gal (Emerest 2661)
8.8 lb/gal (Kessco PEG 600 Monolaurate)
Visc.:
60 cst (100 F) (Emerest 2661)
F.P.:
23 C (Kessco PEG 600 Monolaurate)
Pour Pt.:
14 C (Emerest 2661)
Solidification Pt.:
16–21 C (Pegosperse 600ML)
Flash Pt.:
475 F (Kessco PEG 600 Monolaurate)
525 F (Emerest 2661)
Cloud Pt.:
63 C (Emerest 2661)
HLB:
14.6 (Alkamuls 600ML; CPH-43-N)
14.6 ± 0.5 (Pegosperse 600ML)
14.6 ± 1 (Lipopeg 6-L)
14.8 (Emerest 2661)
14.9 (Kessco PEG 600 Monolaurate; Mapeg 600ML)
15.0 (PGE-600-ML)
15.5 (Nopalcol 6-L)
15.8 (Cithrol 6ML)
Acid No.:
< 1 (Pegosperse 600ML)
5 max. (Kessco PEG 600 Monolaurate; Lipopeg 6-L; Mapeg 600ML)
Iodine No.:
< 4 (Pegosperse 600ML)
5 max. (Kessco PEG 600 Monolaurate; Mapeg 600ML)
Saponification No.:
64–74 (Cithrol 6ML; Kessco PEG 600 Monolaurate; Mapeg 600ML)
65–75 (Pegosperse 600ML)
65–76 (Lipopeg 6-L)

pH:
6–8 (5% aq. disp.) (Pegosperse 600ML)
STD. PKGS.:
450-lb steel drums (Kessco PEG 600 Monolaurate)

POE (4) monooleate

SYNONYMS:
PEG-4 oleate (CTFA)
PEG 200 monooleate
PEG 200 oleate
STRUCTURE:

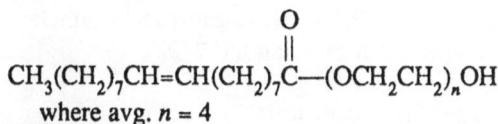

$$CH_3(CH_2)_7CH=CH(CH_2)_7C\overset{\overset{\displaystyle O}{\displaystyle \|}}{}-(OCH_2CH_2)_nOH$$

where avg. $n = 4$

CAS No.:
9004-96-0 (generic)
RD No.: 977060-09-5
TRADENAME EQUIVALENTS:
Alkamuls 200-MO [Alkaril]
Cithrol 2MO [Croda]
CPH-39-N [C.P. Hall]
Emerest 2624 [Emery]
Ethylan A2 [Lankro Chem. Ltd.]
Kessco PEG 200 Monooleate [Armak]
Mapeg 200MO [Mazer]
Radiasurf 7402 [Oleofina S.A.]
CATEGORY:
Emulsifier, o/w emulgent, surfactant, defoamer, lubricant, softener, antistat, wetting agent, spreading agent, penetrant, detergent, scouring aid, solubilizer, thickener, viscosity modifier, dispersant, textile auxiliary, opacifier, plasticizer, rust inhibitor, antifog aid
APPLICATIONS:
Bath products: bath oils (Kessco PEG 200 Monooleate)
Cosmetic industry preparations: (Cithrol 2MO; Kessco PEG 200 Monooleate; Mapeg 200MO; Radiasurf 7402); hair preparations (Kessco PEG 200 Monooleate); perfumery (Kessco PEG 200 Monooleate); shampoos (Kessco PEG 200 Monooleate)
Degreasers: (Alkamuls 200-MO; Emerest 2624)
Farm products: (Kessco PEG 200 Monooleate); insecticides/pesticides (Alkamuls 200-MO; Cithrol 2MO; Emerest 2624; Radiasurf 7402)

POE (4) monooleate *(cont'd.)*

Food applications: (Kessco PEG 200 Monooleate)

Industrial applications: (Kessco PEG 200 Monooleate); dyes and pigments (Cithrol 2MO; Radiasurf 7402); glass processing (Radiasurf 7402); lubricating/cutting oils (Alkamuls 200-MO; Cithrol 2MO; Emerest 2624; Mapeg 200MO; Radiasurf 7402); metalworking (Cithrol 2MO; Mapeg 200MO); paint mfg. (Radiasurf 7402); plastics (Kessco PEG 200 Monooleate; Radiasurf 7402); polishes and waxes (Radiasurf 7402); printing inks (Radiasurf 7402); textile/leather processing (Alkamuls 200-MO; Cithrol 2MO; Emerest 2624; Mapeg 200MO; Radiasurf 7402)

Industrial cleaners: metal processing surfactants (Emerest 2624); solvent cleaners (Alkamuls 200-MO)

Pharmaceutical applications: (Kessco PEG 200 Monooleate; Mapeg 200MO; Radiasurf 7402)

PROPERTIES:

Form:

Liquid (Alkamuls 200-MO; Cithrol 2MO; CPH-39-N; Emerest 2624; Ethylan A2; Kessco PEG 200 Monooleate; Mapeg 200MO; Radiasurf 7402)

Color:

Light amber (Ethylan A2; Kessco PEG 200 Monooleate)

Amber (Radiasurf 7402)

Yellow (Alkamuls 200-MO; Mapeg 200MO)

Gardner 3 (Emerest 2624)

Composition:

100% conc. (Cithrol 2MO; CPH-39-N; Mapeg 200MO)

100% active (Ethylan A2)

Solubility:

Sol. in acetone (Kessco PEG 200 Monooleate)

Disp. in aromatic solvent (@ 10%) (Alkamuls 200-MO)

Sol. in benzene (@ 10%) (Radiasurf 7402)

Disp. in butyl stearate (@ 5%) (Emerest 2624)

Sol. in carbon tetrachloride (Kessco PEG 200 Monooleate)

Sol. in ethyl acetate (Kessco PEG 200 Monooleate)

Sol. in glycerol trioleate (@ 5%) (Emerest 2624)

Sol. cloudy in hexane (@ 10%) (Radiasurf 7402)

Sol. in isopropanol (Kessco PEG 200 Monooleate; Mapeg 200MO); sol. (@ 10%) (Radiasurf 7402)

Sol. cloudy in min. oil (@ 10%) (Radiasurf 7402); disp. (@ 10%) (Alkamuls 200-MO); (@ 5%) (Emerest 2624)

Disp. in min. spirits (@ 10%) (Alkamuls 200-MO)

Disp. in perchloroethylene (@ 10%) (Alkamuls 200-MO)

Sol. in soybean oil (Mapeg 200MO)

Disp. in Stoddard solvent (@ 5%) (Emerest 2624)

Sol. in toluol (Kessco PEG 200 Monooleate; Mapeg 200MO)

Sol. in trichlorethylene (@ 10%) (Radiasurf 7402)

Sol. cloudy in veg. oil (@ 10%) (Radiasurf 7402)

Disp. in water (Kessco PEG 200 Monooleate; Mapeg 200MO); disp. (@ 10%) (Alkamuls 200-MO; Radiasurf 7402); (@ 5%) (Emerest 2624)

Sol. in xylene (@ 5%) (Emerest 2624)

Ionic Nature:

Nonionic (Cithrol 2MO; CPH-39-N; Ethylan A2; Kessco PEG 200 Monooleate; Mapeg 200MO; Radiasurf 7402)

M.W.:

200 (Cithrol 2MO)

512 avg. (Radiasurf 7402)

Sp.gr.:

0.962 (37.8 C) (Radiasurf 7402)

0.973 (Kessco PEG 200 Monooleate)

0.976 (Ethylan A2)

0.98 (Mapeg 200MO)

Density:

0.99 g/ml (Alkamuls 200-MO)

8.1 lb/gal (Emerest 2624; Kessco PEG 200 Monooleate)

Visc.:

34 cSt (100 F) (Emerest 2624)

34 cps (37.8 C) (Radiasurf 7402)

67 cs (20 C) (Ethylan A2)

F.P.:

< –15 C (Kessco PEG 200 Monooleate)

M.P.:

< –10 C (Mapeg 200MO)

Pour Pt.:

< –15 C (Emerest 2624)

Flash Pt.:

218 C (COC) (Radiasurf 7402)

395 F (COC) (Kessco PEG 200 Monooleate)

415 F (Emerest 2624)

Cloud Pt.:

–10 C (Radiasurf 7402)

< 25 C (Emerest 2624)

HLB:

6.2 (Cithrol 2MO)

7.0 (Ethylan A2)

7.6 (CPH-39-N)

8.0 (Alkamuls 200-MO; Kessco PEG 200 Monooleate; Mapeg 200MO)

8.3 (Emerest 2624)

8.4 (Radiasurf 7402)

POE (4) monooleate *(cont'd.)*

Acid No.:
   < 3 (Radiasurf 7402)
   5 max. (Kessco PEG 200 Monooleate; Mapeg 200MO)
Iodine No.:
   50–60 (Radiasurf 7402)
   56 max. (Kessco PEG 200 Monooleate; Mapeg 200MO)
Saponification No.:
   105–125 (Radiasurf 7402)
   110–112 (Cithrol 2MO)
   111–121 (Alkamuls 200-MO)
   115–124 (Kessco PEG 200 Monooleate)
   115–125 (Mapeg 200MO)
Ref. Index:
   1.4645 (Radiasurf 7402)
pH:
   5.0 (3% disp.) (Kessco PEG 200 Monooleate)
Surface Tension:
   32.00 dynes/cm (0.1% aq.) (Radiasurf 7402)
STD. PKGS.:
   190-kg net bung drums or bulk (Radiasurf 7402)

POE (8) monooleate

SYNONYMS:
   PEG-8 oleate (CTFA)
   PEG 400 monooleate
STRUCTURE:

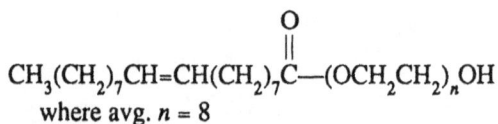

$$CH_3(CH_2)_7CH=CH(CH_2)_7\overset{\overset{\displaystyle O}{\|}}{C}-(OCH_2CH_2)_nOH$$

   where avg. $n = 8$
CAS No.
   9004-96-0 (generic)
TRADENAME EQUIVALENTS:
   Acconon 400 MO [Armstrong]
   Alkamuls 400 MO [Alkaril]
   Cithrol 4MO [Croda]
   Cithrol A [Croda]
   Collemul H4 [Allied Colloids Ltd.]
   Cyclochem PEG 400 MO [Cyclo]

TRADENAME EQUIVALENTS *(cont'd.):*
  Durpeg 400 MO [Durkee]
  Emerest 2646 [Emery]
  Empilan BQ100 [Albright & Wilson/Marchon]
  Emulsan O [Reilly-Whiteman]
  Kessco PEG 400 Monooleate [Armak]
  Lipopeg 4-O [Lipo]
  Lonzest PEG 4-O [Lonza]
  Mapeg 400 MO [Mazer]
  Nopalcol 4-O [Diamond Shamrock]
  Pegosperse 400-MO [Glyco]
  PGE-400-MO [Hefti Ltd.]
  Radiasurf 7403 [Oleofina]
  Scher PEG 400 Monooleate [Scher]
  Witconol H-31A [Witco/Organics]
CATEGORY:
  Emulsifier, surfactant, wetting agent, antifog aid, antistat, chemical intermediate,
    defoamer, detergent, dispersant, emollient, lubricant, o/w emulgent, opacifier,
    penetrant, plasticizer, rust inhibitor, scouring and detergent aid, softener, solubi-
    lizer, spreading agent, thickening agent, viscosity modifier
APPLICATIONS:
  Bath products: bath oils (Kessco PEG 400 Monooleate; PGE-400-MO; Witconol H-
    31A)
  Cleansers: hand cleanser (Empilan BQ 100; PGE-400-MO)
  Cosmetic industry preparations: (Acconon 400 MO; Alkamuls 400 MO; Cithrol 4MO;
    Collemul H4; Kessco PEG 400 Monooleate; Lonzest PEG-4-O; Mapeg 400 MO;
    Nopalcol 4-O; PGE-400-MO; Radiasurf 7403; Scher PEG 400 Monooleate; Wit-
    conol H-31A); hair rinses (Kessco PEG 400 Monooleate); perfumery (Kessco PEG
    400 Monooleate); shampoos (Collemul H4; Kessco PEG 400 Monooleate)
  Degreasers: (Emerest 2646; Lonzest PEG-4-O)
  Farm products: (Acconon 400 MO; Alkamuls 400 MO; Kessco PEG 400 Monooleate);
    animal feed (Durpeg 400 MO); insecticides/pesticides (Cithrol 4MO; Empilan BQ
    100; PGE-400-MO; Radiasurf 7403)
  Food applications: (Acconon 400 MO; Kessco PEG 400 Monooleate)
  Household detergents: laundry detergent (Empilan BQ 100; PGE-400-MO)
  Industrial applications: (PGE-400-MO; Witconol H-31A); adhesives (Empilan BQ
    100; PGE-400-MO); construction (Nopalcol 4-O; PGE-400-MO); dyes and pig-
    ments (Cithrol 4MO; Nopalcol 4-O; PGE-400-MO; Radiasurf 7403; Scher PEG
    400 Monooleate); glass processing (Radiasurf 7403); lubricating/cutting oils
    (Cithrol A; Emerest 2646; Empilan BQ 100; Nopalcol 4-O; PGE-400-MO; Scher
    PEG 400 Monooleate); metalworking (Cithrol 4MO; Mapeg 400MO; Scher PEG
    400 Monooleate); paint mfg. (Cithrol A; Nopalcol 4-O; Radiasurf 7403); paper mfg.
    (Cithrol A; Empilan BQ 100; Nopalcol 4-O; PGE-400-MO); plastics (Acconon 400

POE (8) monooleate (cont'd.)

MO; Emerest 2646; Empilan BQ 100; Kessco PEG 400 Monooleate; PGE-400-MO; Radiasurf 7403); polishes and waxes (Cithrol A; Nopalcol 4-O; Radiasurf 7403); printing inks (Radiasurf 7403); rubber (Cithrol A; Nopalcol 4-O); textile/leather processing (Cithrol 4MO; Cithrol A; Collemul H4; Emerest 2646; Mapeg 400MO; Nopalcol 4-O; PGE-400-MO; Radiasurf 7403; Scher PEG 400 Monooleate)

Industrial cleaners: drycleaning compositions (Lonzest PEG-4-O; Nopalcol 4-O; PGE-400-MO); metal processing surfactants (Cithrol A; Lonzest PEG-4-O; Nopalcol 4-O)

Pharmaceutical applications: (Kessco PEG 400 Monooleate; Lonzest PEG-4-O; Mapeg 400 MO; PGE-400-MO; Radiasurf 7403)

PROPERTIES:

Form:

Liquid (Acconon 400MO; Cithrol A; Collemul H4; Cyclochem PEG 400MO; Emerest 2646; Empilan BQ 100; Emulsan O; Kessco PEG 400 Monooleate; Lipopeg 4-O; Lonzest PEG-4-O; Mapeg 400MO; Nopalcol 4-O; Pegosperse 400MO; PGE-400-MO; Radiasurf 7403; Witconol H-31A)

Clear liquid (Alkamuls 400MO; Scher PEG 400 Monooleate)

Color:

Light amber (Kessco PEG 400 Monooleate)

Amber (Alkamuls 400MO; Cithrol A; Durpeg 400MO; Pegosperse 400MO; Radiasurf 7403)

Light straw (Cyclochem PEG 400MO)

Yellow (Lonzest PEG-4-O; Mapeg 400MO; Scher PEG 400 Monooleate)

Dark brown (Empilan BQ 100)

Gardner 2 (Emeret 2646)

Gardner 6 (Acconon 400MO)

Odor:

Slight agreeable (Cithrol A)

Mild, typical (Scher PEG 400 Monooleate)

Composition:

99% active (Acconon 400MO; Emulsan O)

100% active (Alkamuls 400MO; Durpeg 400MO; Empilan BQ 100)

100% conc. (Lonzest PEG-4-O; Mapeg 400MO; Pegosperse 400-MO; Witconol H-31A)

Solubility:

Sol. in acetate esters (Scher PEG 400 Monooleate)

Sol. in acetone (Kessco PEG 400 Monooleate; Pegosperse 400MO)

Sol. in aliphatic hydrocarbons (Scher PEG 400 Monooleate)

Sol. in most alcohols (Scher PEG 400 Monooleate)

Sol. in aromatic hydrocarbons (Scher PEG 400 Monooleate)

Sol. in benzene (Radiasurf 7403)

Disp. in butyl stearate (Emerest 2646)

Sol. in carbon tetrachloride (Kessco PEG 400 Monooleate)
Sol. in ethanol (Empilan BQ 100; Pegosperse 400MO)
Sol. in ethyl acetate (Kessco PEG 400 Monooleate; Pegosperse 400MO)
Disp. in glycerol trioleate (Emerest 2646)
Sol. in glycol ethers (Scher PEG 400 Monooleate)
Sol. in glycols (Scher PEG 400 Monooleate)
Sol. cloudy in hexane (Radiasurf 7403)
Sol. in isopropanol (Kessco PEG 400 Monooleate; Lonzest PEG-4-O; Mapeg 400MO; Radiasurf 7403)
Sol. cloudy in kerosene (Empilan BQ 100)
Sol. in ketones (Scher PEG 400 Monooleate)
Sol. in methanol (Pegosperse 400MO)
Sol. in methylated spirits (Cithrol A)
Sol. in min. oil (Lonzest PEG-4-O; Pegosperse 400MO; Radiasurf 7403); cloudy (Empilan BQ 100); partly sol. (Scher PEG 400 Monooleate)
Sol. in naphtha (Pegosperse 400MO); cloudy (Empilan BQ 100)
Sol. in oleic acid (Cithrol A; Empilan BQ 100)
Sol. in organic solvents (Acconon 400MO)
Sol. in paraffin (liquid) (Cithrol A)
Disp. in perchloroethylene (Emerest 2646); sol. cloudy (Empilan BQ 100)
Disp. in sodium sulfate (Kessco PEG 400 Monooleate)
Partly sol. in soybean oil (Mapeg 400MO)
Disp. in Stoddard solvent (Emerest 2646)
Sol. in toluol (Kessco PEG 400 Monooleate; Lonzest PEG-4-O; Mapeg 400MO; Pegosperse 400MO)
Sol. in trichlorethylene (Radiasurf 7403)
Sol. in veg. oil (Pegosperse 400MO; Radiasurf 7403); partly sol. (Scher PEG 400 Monooleate)
Disp. in water (Cithrol A; Emerest 2646; Empilan BQ 100; Kessco PEG 400 Monooleate; Lonzest PEG 4-O; Mapeg 400MO; Pegosperse 400 MO; Radiasurf 7403; Scher PEG 400 Monooleate)
Sol. in white spirit (Cithrol A); sol. cloudy (Empilan BQ 100)
Sol. in xylene (Empilan BQ 100)
Sol. in xylol (Cithrol A)
Ionic Nature: Nonionic
M.W.:
400 (Cithrol 4MO)
664 (Scher PEG 400 Monooleate)
728 (Radiasurf 7403)
Sp.gr.:
1.007 (37.8 C) (Radiasurf 7403)
1.01 (Acconon 400MO; Lonzest PEG-4-O; Mapeg 400MO; Pegosperse 400MO)
1.01–1.03 (Alkamuls 400MO)

POE (8) monooleate (cont'd.)

1.013 (Cithrol A; Kessco PEG 400 Monooleate)
1.016 (Scher PEG 400 Monooleate)
1.017 (20 C) (Empilan BQ100)

Density:
8.4 lb/gal (Acconon 400MO; Kessco PEG 400 Monooleate)
8.5 lb/gal (Emerest 2646)

Visc.:
130 cps (Empilan BQ 100)

F.P.:
> 10 C (Kessco PEG 400 Monooleate)

M.P.:
−1 C (Empilan BQ100)
0 C (Lonzest PEG-4-O)
< 10 C (Acconon 400MO; Mapeg 400MO)
10 C (Cyclochem PEG 400 MO)

Solidification Pt.:
< 40 F (Alkamuls 400MO)
< 0 C (Pegosperse 400MO)

Flash Pt.:
> 170 C (OC) (Scher PEG 400 Monooleate)
215 C (Cithrol A)
261 C (Radiasurf 7403)
510 F (Kessco PEG 400 Monooleate)

Fire Pt.:
585 F (Kessco PEG 400 Monooleate)

Cloud Pt.:
−1 C (Radiasurf 7403)
1 C (Scher PEG 400 Monooleate)
< 25 C (Emerest 2646)

HLB:
11.0 (Alkamuls 400MO)
11.0 ± 0.5 (Pegosperse 400MO)
11.1 (Durpeg 400MO)
11.4 (Cithrol 4MO; Kessco PEG 400 Monooleate; Lonzest PEG-4-O; Mapeg 400MO)
11.7 (Emerest 2646; Radiasurf 7403)
12.0 (Scher PEG 400 Monooleate)

Acid No.:
< 3 (Radiasurf 7403)
< 5 (Pegosperse 400MO)
5 max. (Cyclochem PEG 400MO; Kessco PEG 400 Monooleate; Mapeg 400MO; Scher PEG 400 Monooleate)
5–10 (Lonzest PEG-4-O)

Iodine No.:
31–40 (Lonzest PEG-4-O)
34–42 (Radiasurf 7403)
36–38 (Pegosperse 400MO)
39 max. (Kessco PEG 400 Monooleate; Scher PEG 400 Monooleate)

Saponification No.:
75–90 (Radiasurf 7403)
80–88 (Mapeg 400MO; Pegosperse 400MO)
80–89 (Kessco PEG 400 Monooleate)
80–90 (Scher PEG 400 Monooleate)
85 (Cyclochem PEG 400MO)
85–93 (Cithrol A)
86–96 (Lonzest PEG-4-O)

Ref. Index:
1.4646 (Scher PEG 400 Monooleate)
1.4672 (Radiasurf 7403)

pH:
4–6 (5% aq. disp.) (Pegosperse 400MO)
5.0 (3% disp.) (Kessco PEG 400 Monooleate)
6.5–7.5 (10% sol'n.) (Cithrol A)

Surface Tension:
34.00 dynes/cm (0.1% aq.) (Radiasurf 7403)

Biodegradable: (Acconon 400MO)

STD. PKGS.:
Mild steel drums (Cithrol A)
450-lb packages (Durpeg 400MO; Kessco PEG 400 Monooleate)
190-kg net bung drums or bulk (Radiasurf 7403)

POE (12) monooleate

SYNONYMS:
PEG-12 oleate (CTFA)
PEG 600 monooleate
PEG 600 oleate

STRUCTURE:

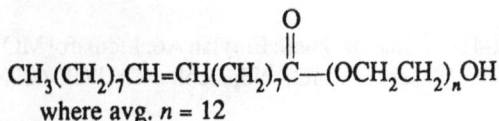

$$CH_3(CH_2)_7CH=CH(CH_2)_7C-(OCH_2CH_2)_nOH$$
where avg. $n = 12$

CAS No.:
9004-96-0 (generic)
RD No.: 977055-27-8

POE (12) monooleate (cont'd.)

TRADENAME EQUIVALENTS:
Alkamuls 600-MO [Alkaril]
Cithrol 6MO [Croda]
CPH-41-N [C.P. Hall]
Emerest 2660 [Emery]
Ethylan A6 [Lankro Chem. Ltd.]
Industrol MO-13 [BASF Wyandotte]
Kessco PEG 600 Monooleate [Armak]
Mapeg 600MO [Mazer]
Nopalcol 6-O [Diamond Shamrock]
PGE-600MO [Hefti Ltd.]
Radiasurf 7404 [Oleofina S.A.]

CATEGORY:
Wetting agent, emulsifier, o/w emulgent, surfactant, dispersant, penetrant, detergent, scouring aid, solubilizer, thickening agent, viscosity modifier, textile auxiliary, lubricant, emollient, softener, defoamer, stabilizer, mold release agent, opacifier, plasticizer, rust inhibitor, antifog aid, antistat

APPLICATIONS:
Bath products: bath oils (Kessco PEG 600 Monooleate)

Cosmetic industry preparations: (Cithrol 6MO; Kessco PEG 600 Monooleate; Mapeg 600MO; PGE-600MO; Radiasurf 7404); hair preparations (Kessco PEG 600 Monooleate); perfumery (Kessco PEG 600 Monooleate); shampoos (Kessco PEG 600 Monooleate)

Farm products: (Kessco PEG 600 Monooleate); herbicides (PGE-600MO); insecticides/pesticides (Cithrol 6MO; PGE-600MO; Radiasurf 7404)

Food applications: (Kessco PEG 600 Monooleate)

Industrial applications: (Kessco PEG 600 Monooleate; PGE-600MO); dyes and pigments (Ethylan A6; Radiasurf 7404); glass processing (Radiasurf 7404); lubricating/cutting oils (Cithrol 6MO; Mapeg 600MO; Radiasurf 7404); metalworking (Cithrol 6MO; Mapeg 600MO); paint mfg. (Radiasurf 7404); plastics (Ethylan A6; Kessco PEG 600 Monooleate; Radiasurf 7404); polishes and waxes (Radiasurf 7404); printing inks (Radiasurf 7404); textile/leather processing (Cithrol 6MO; Mapeg 600MO; PGE-600MO; Radiasurf 7404)

Pharmaceutical applications: (Kessco PEG 600 Monooleate; Mapeg 600MO; PGE-600MO; Radiasurf 7404)

PROPERTIES:
Form:
Liquid (Alkamuls 600-MO; CPH-41-N; Emerest 2660; Ethylan A6; Industrol MO-13; Kessco PEG 600 Monooleate; Mapeg 600MO; Nopalcol 6-O; PGE-600MO; Radiasurf 7404)

Color:
Light amber (Ethylan A6; Kessco PEG 600 Monooleate)
Amber (Radiasurf 7404)

Yellow (Mapeg 600MO)
Gardner 7 max. (Industrol MO-13)

Composition:
100% conc. (CPH-41-N; Emerest 2660; Mapeg 600MO; Nopalcol 6-O; PGE-600MO)
100% active (Alkamuls 600-MO; Ethylan A6; Industrol MO-13)

Solubility:
Sol. in acetone (Kessco PEG 600 Monooleate)
Sol. in benzene (@ 10%) (Radiasurf 7404)
Sol. in carbon tetrachloride (Kessco PEG 600 Monooleate)
Sol. in ethyl acetate (Kessco PEG 600 Monooleate)
Sol. cloudy in hexane (@ 10%) (Radiasurf 7404)
Sol. in isopropanol (Kessco PEG 600 Monooleate; Mapeg 600MO); sol. (@ 10%)
    (Radiasurf 7404)
Sol. cloudy in min. oil (@ 10%) (Radiasurf 7404)
Partly sol. in soybean oil (Mapeg 600MO)
Sol. in toluol (Kessco PEG 600 Monooleate; Mapeg 600MO)
Sol. in trichlorethylene (@ 10%) (Radiasurf 7404)
Sol. cloudy in veg. oil (@ 10%) (Radiasurf 7404)
Sol. in water (Industrol MO-13; Kessco PEG 600 Monooleate; Mapeg 600MO); (@
    10%) (Radiasurf 7404)

Ionic Nature:
Nonionic (CPH-41-N; Emerest 2660; Ethylan A6; Industrol MO-13; Kessco PEG 600
    Monooleate; Mapeg 600MO; PGE-600MO; Radiasurf 7404)

M.W.:
600 (Cithrol 6MO)
850 (Industrol MO-13)
920 avg. (Radiasurf 7404)

Sp.gr.:
1.030 (Mapeg 600MO); (37.8 C) (Radiasurf 7404)
1.037 (Ethylan A6; Kessco PEG 600 Monooleate)
1.04 (Industrol MO-13)

Density:
8.7 lb/gal (Kessco PEG 600 Monooleate)

Visc.:
66.30 cps (37.8 C) (Radiasurf 7404)
160 cs (20 C) (Ethylan A6)

F.P.:
23 C (Kessco PEG 600 Monooleate)

M.P.:
25 C (Mapeg 600MO)

Pour Pt.:
18 C (Industrol MO-13)

POE (12) monooleate *(cont'd.)*

Flash Pt.:
    254 C (COC) (Radiasurf 7404)
    525 F (COC) (Kessco PEG 600 Monooleate)
Fire Pt.:
    620 F (Kessco PEG 600 Monooleate)
Cloud Pt.:
    19 C (Radiasurf 7404)
HLB:
    12.3 (Ethylan A6)
    13.0 (Alkamuls 600-MO)
    13.1 (Cithrol 6MO)
    13.2 (Radiasurf 7404)
    13.5 (Kessco PEG 600 Monooleate; Mapeg 600MO; PGE-600MO)
    13.6 (CPH-41-N; Emerest 2660; Industrol MO-13)
    14.0 (Nopalcol 6-O)
Acid No.:
    2 max. (Industrol MO-13)
    < 3 (Radiasurf 7404)
    5 max. (Kessco PEG 600 Monooleate; Mapeg 600MO)
Iodine No.:
    27–37 (Radiasurf 7404)
    29 max. (Mapeg 600MO)
    30 max. (Kessco PEG 600 Monooleate)
Saponification No.:
    60–69 (Kessco PEG 600 Monooleate)
    60–70 (Mapeg 600MO)
    60–75 (Radiasurf 7404)
    65–75 (Cithrol 6MO)
pH:
    5.0 (3% disp.) (Kessco PEG 600 Monooleate)
    6.0–7.5 (5% aq.) (Industrol MO-13)
Surface Tension:
    37.00 dynes/cm (0.1% aq.) (Radiasurf 7404)
STD. PKGS.:
    190-kg net bung drums or bulk (Radiasurf 7404)
    55-gal (420 lb net) steel drums (Industrol MO-13)

POE (20) monooleate

SYNONYMS:
PEG-20 oleate (CTFA)
PEG 1000 monooleate

STRUCTURE:

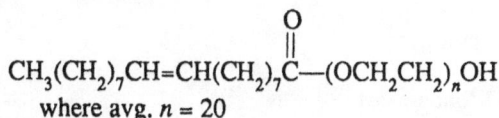

where avg. $n = 20$

CAS No.:
9004-96-0 (generic)
RD No.: 977063-50-5

TRADENAME EQUIVALENTS:
Chemax E-1000-MO [Chemax]
Kessco PEG 1000 Monooleate [Armak]

CATEGORY:
Surfactant, detergent, emulsifier, dispersant, thickener, solubilizer, lubricant

APPLICATIONS:
Bath products: bath oils (Kessco PEG 1000 Monooleate)
Cosmetic industry preparations: (Kessco PEG 1000 Monooleate); cream rinses (Kessco PEG 1000 Monooleate); hair preparations (Kessco PEG 1000 Monooleate); perfumery (Kessco PEG 1000 Monooleate); shampoos (Kessco PEG 1000 Monooleate)
Farm products: (Kessco PEG 1000 Monooleate)
Food applications: (Kessco PEG 1000 Monooleate)
Industrial applications: plastics (Kessco PEG 1000 Monooleate)
Pharmaceutical applications: (Kessco PEG 1000 Monooleate)

PROPERTIES:
Form:
Soft solid (Kessco PEG 1000 Monooleate)
Solid (Chemax E-1000-MO)
Color:
Cream (Kessco PEG 1000 Monooleate)
Composition:
100% conc. (Chemax E-1000-MO)
Solubility:
Sol. in acetone (Kessco PEG 1000 Monooleate)
Sol. in carbon tetrachloride (Kessco PEG 1000 Monooleate)
Sol. in ethyl acetate (Kessco PEG 1000 Monooleate)
Sol. in isopropanol (Kessco PEG 1000 Monooleate)
Sol. in Na_2SO_4 (5%) (Kessco PEG 1000 Monooleate)
Sol. hot in propylene glycol (Kessco PEG 1000 Monooleate)
Sol. in toluol (Kessco PEG 1000 Monooleate)
Sol. in water (Kessco PEG 1000 Monooleate)

POE (20) monooleate *(cont'd.)*

Ionic Nature:
   Nonionic (Chemax E-1000-MO; Kessco PEG 1000 Monooleate)
Sp.gr.:
   1.035 (65 C) (Kessco PEG 1000 Monooleate)
F.P.:
   39 C (Kessco PEG 1000 Monooleate)
Flash Pt.:
   515 F (COC) (Kessco PEG 1000 Monooleate)
Fire Pt.:
   595 F (Kessco PEG 1000 Monooleate)
HLB:
   15.4 (Kessco PEG 1000 Monooleate)
Acid No.:
   5.0 max. (Kessco PEG 1000 Monooleate)
Iodine No.:
   21.0 max. (Kessco PEG 1000 Monooleate)
Saponification No.:
   40–49 (Kessco PEG 1000 Monooleate)
pH:
   5.0 (3% disp.) (Kessco PEG 1000 Monooleate)

POE (4) monostearate

SYNONYMS:
   PEG-4 stearate (CTFA)
   PEG 200 monostearate
EMPIRICAL FORMULA:
   $C_{26}H_{52}O_6$
STRUCTURE:

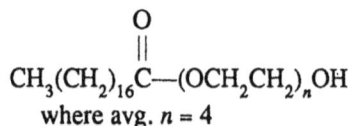

   where avg. $n = 4$
CAS No.:
   106-07-0; 9004-99-3 (generic)
TRADENAME EQUIVALENTS:
   Alkamuls 200-MS [Alkaril]
   Cithrol 2MS [Croda]
   Kessco PEG 200 Monostearate [Armak]
   Mapeg 200MS [Mazer]

POE (4) monostearate (cont'd.)

TRADENAME EQUIVALENTS *(cont'd.):*
Nikkol MYS-4 [Nikko]
Radiasurf 7412 [Oleofina S.A.]

CATEGORY:
Emulsifier, detergent, wetting agent, solubilizer, thickener, viscosity modifier, softener, lubricant, antifrothing agent, defoamer, foaming agent, opacifier, antistat, dispersant, plasticizer, rust inhibitor, antifog aid

APPLICATIONS:
Cosmetic industry preparations: (Cithrol 2MS; Kessco PEG 200 Monostearate; Mapeg 200MS; Nikkol MYS-4; Radiasurf 7412); hair preparations (Kessco PEG 200 Monostearate); perfumery (Cithrol 2MS; Kessco PEG 200 Monostearate); shampoos (Kessco PEG 200 Monostearate)

Farm products: (Kessco PEG 200 Monostearate); insecticides/pesticides (Cithrol 2MS; Radiasurf 7412)

Food applications: (Kessco PEG 200 Monostearate)

Industrial applications: (Kessco PEG 200 Monostearate); dyes and pigments (Radiasurf 7412); glass processing (Radiasurf 7412); lubricating/cutting oils (Radiasurf 7412); metalworking (Mapeg 200MS); paint mfg. (Radiasurf 7412); plastics (Kessco PEG 200 Monostearate; Radiasurf 7412); polishes and waxes (Radiasurf 7412); printing inks (Radiasurf 7412); silicone products (Radiasurf 7412); textile/leather processing (Alkamuls 200-MS; Mapeg 200MS; Radiasurf 7412)

Pharmaceutical applications: (Kessco PEG 200 Monostearate; Mapeg 200MS; Nikkol MYS-4; Radiasurf 7412)

PROPERTIES:
Form:
Paste (Radiasurf 7412)
Soft solid (Kessco PEG 200 Monostearate)
Solid (Alkamuls 200-MS; Cithrol 2MS; Mapeg 200MS; Nikkol MYS-4)

Color:
White (Alkamuls 200-MS; Mapeg 200MS; Radiasurf 7412)
White to cream (Kessco PEG 200 Monostearate)

Composition:
100% conc. (Cithrol 2MS; Mapeg 200MS; Nikkol MYS-4)

Solubility:
Sol. in acetone (Kessco PEG 200 Monostearate)
Disp. in aromatic solvent (@ 10%) (Alkamuls 200-MS)
Sol. in benzene (@ 10%) (Radiasurf 7412)
Sol. in carbon tetrachloride (Kessco PEG 200 Monostearate)
Sol. in ethyl acetate (Kessco PEG 200 Monostearate)
Sol. cloudy in hexane (@ 10%) (Radiasurf 7412)
Sol. in isopropanol (Kessco PEG 200 Monostearate; Mapeg 200MS); (@ 10%) (Radiasurf 7412)
Sol. in isopropyl myristate (Kessco PEG 200 Monostearate)

POE (4) monostearate (cont'd.)

Sol. cloudy in min. oil (@ 10%) (Radiasurf 7412); disp. (@ 10%) (Alkamuls 200-MS)

Disp. in min. spirits (@ 10%) (Alkamuls 200-MS)

Sol. in peanut oil (Kessco PEG 200 Monostearate)

Disp. in perchloroethylene (@ 10%) (Alkamuls 200-MS)

Sol. in soybean oil (Mapeg 200MS)

Sol. in toluol (Kessco PEG 200 Monostearate; Mapeg 200MS)

Sol. in trichlorethylene (@ 10%) (Radiasurf 7412)

Disp. in water (@ 10%) (Alkamuls 200-MS); disp. hot (Kessco PEG 200 Monostearate; Mapeg 200MS)

Ionic Nature:

Nonionic (Alkamuls 200-MS; Cithrol 2MS; Kessco PEG 200 Monostearate; Mapeg 200MS; Nikkol MYS-4; Radiasurf 7412)

M.W.:

200 (Cithrol 2MS)

522 avg. (Radiasurf 7412)

Sp.gr.:

0.913 (98.9 C) (Radiasurf 7412)

0.9360 (65 C) (Kessco PEG 200 Monostearate)

Visc.:

6.70 cps (98.9 C) (Radiasurf 7412)

M.P.:

31 C (Kessco PEG 200 Monostearate)

33 C (Mapeg 200MS)

36 C (Radiasurf 7412)

39–41 C (Cithrol 2MS)

Flash Pt.:

186 C (COC) (Radiasurf 7412)

410 F (COC) (Kessco PEG 200 Monostearate)

Fire Pt.:

450 F (Kessco PEG 200 Monostearate)

HLB:

6.3 (Cithrol 2MS)

6.5 (Nikkol MYS-4)

7.5 (Radiasurf 7412)

7.9 (Kessco PEG 200 Monostearate)

8.0 (Alkamuls 200-MS; Mapeg 200MS)

Acid No.:

< 3 (Radiasurf 7412)

5.0 max. (Kessco PEG 200 Monostearate; Mapeg 200MS)

Iodine No.:

0.5 max. (Kessco PEG 200 Monostearate)

1.0 max. (Mapeg 200MS; Radiasurf 7412)

Saponification No.:
110–120 (Cithrol 2MS)
120–129 (Kessco PEG 200 Monostearate)
120–130 (Alkamuls 200-MS; Mapeg 200MS)
120–135 (Radiasurf 7412)
pH:
5.0 (3% disp.) (Kessco PEG 200 Monostearate)
STD. PKGS.:
190-kg net bung drums or bulk (Radiasurf 7412)

POE (8) monostearate

SYNONYMS:
Macrogol Stearate 400
PEG-8 stearate (CTFA)
PEG 400 monostearate
POE 8 stearate
Polyoxyl 8 stearate
STRUCTURE:

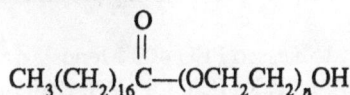

$$CH_3(CH_2)_{16}C\!-\!(OCH_2CH_2)_nOH$$
where avg. $n = 8$
CAS No.
9004-99-93 (generic); RD No.: 977055-39-2
TRADENAME EQUIVALENTS:
Alkamuls 400MS [Alkaril]
Cithrol 4MS [Croda]
Emerest 2640 [Emery]
Emerest 2711 [Emery]
Hodag 40-S [Hodag]
Kessco PEG 400 Monostearate [Armak]
Lipal 400S [PVO Int'l.]
Lipopeg 4-S [Lipo]
Mapeg 400MS [Mazer]
Myrj 45 [ICI]
Nonisol 300 [Ciba-Geigy]
Nopalcol 4-S [Diamond Shamrock]
Pegosperse 400MS [Glyco]
PGE-400-MS [Hefti]
Radiasurf 7413 [Oleofina]
Scher PEG 400 Monostearate [Scher]

POE (8) monostearate (cont'd.)

TRADENAME EQUIVALENTS *(cont'd.):*

Simulsol M45 [Seppic]

Varonic 400MS [Sherex]

Witconol H-35A [Witco/Organics]

CATEGORY:

Emulsifier, wetting agent, opacifier, thickening agent, antifog aid, antifrothing agent, antistat, chemical intermediate, defoamer, detergent, dispersant, foaming agent, lubricant, o/w emulgent, plasticizer, rust inhibitor, scouring and detergent aid, solubilizer, spreading agent, stabilizer, surfactant, viscosity modifier

APPLICATIONS:

Bath products: bath oils (Emerest 2711; Kessco PEG 400 Monostearate; Lipopeg 4-S)

Cleansers: body cleansers (Lipal 400S)

Cosmetic industry preparations: (Alkamuls 400MS; Cithrol 4MS; Emerest 2711; Kessco PEG 400 Monostearate; Lipopeg 4-S; Mapeg 400MS; Myrj 45; Pegosperse 400MS; PGE-400-MS; Radiasurf 7413; Simulsol M45; Varonic 400MS; Witconol H-35A); conditioners (Lipal 400S); creams and lotions (Scher PEG 400 Monostearate); hair rinses (Emerest 2711; Kessco PEG 400 Monostearate; Lipal 400S; Nonisol 300); makeup (Lipal 400S; Nonisol 300); perfumery (Cithrol 4MS; Emerest 2711; Kessco PEG 400 Monostearate); shampoos (Alkamuls 400MS; Emerest 2711; Kessco PEG 400 Monostearate; Lipal 400S); shaving preparations (Lipal 400S)

Farm products: (Alkamuls 400MS; Emerest 2711; Kessco PEG 400 Monostearate; Simulsol M45; Varonic 400MS); insecticides/pesticides (Cithrol 4MS; Radiasurf 7413)

Food applications: (Emerest 2711; Kessco PEG 400 Monostearate)

Industrial applications: dyes and pigments (Radiasurf 7413); glass processing (Radiasurf 7413); lubricating/cutting oils (Emerest 2640; Radiasurf 7413); metalworking (Mapeg 400MS); paint mfg. (Radiasurf 7413); paper mfg. (Emerest 2640); plastics (Emerest 2711; Kessco PEG 400 Monostearate; Radiasurf 7413); polishes and waxes (Radiasurf 7413); printing inks (Radiasurf 7413); textile/leather processing (Alkamuls 400MS; Emerest 2640; Mapeg 400MS; Radiasurf 7413)

Pharmaceutical applications: (Cithrol 4MS; Emerest 2711; Kessco PEG 400 Monostearate; Mapeg 400MS; Myrj 45; Nonisol 300; Pegosperse 400MS; PGE-400-MS; Radiasurf 7413; Witconol H-35A); antiseptic soaps (Lipal 400S)

PROPERTIES:

Form:

Paste (Alkamuls 400MS; Hodag 40S; Lipopeg 4-S; Nonisol 300; Nopalcol 4-S; Radiasurf 7413)

Soft waxy solid (Emerest 2640; Myrj 45)

Solid to liquid (Pegosperse 400MS)

Soft solid (Emerest 2711; Kessco PEG 400 Monostearate; Pegosperse 400MS)

Solid (Lipal 400S; Mapeg 400MS; PGE-400-MS; Scher PEG 400 Monostearate; Simulsol M45; Varonic 400MS)

Color:
White (Mapeg 400MS; Nonisol 300; Pegosperse 400MS; Radiasurf 7413)
White to cream (Kessco PEG 400 Monostearate; Scher PEG 400 Monostearate)
Cream (Lipopeg 4-S; Myrj 45)
Amber (Alkamuls 400MS)
Gardner 1 (Emerest 2640; Lipal 400S)

Odor:
Mild, typical (Scher PEG 400 Monostearate)

Composition:
100% active (Lipal 400S; Lipopeg 4-S; Varonic 400MS)
100% conc. (Mapeg 400MS; Myrj 45; Nonisol 300; Pegosperse 400MS)

Solubility:
Sol. in acetate esters (Scher PEG 400 Monostearate)
Sol. in acetone (Kessco PEG 400 Monostearate; Pegosperse 400MS)
Sol. in aliphatic hydrocarbons (Scher PEG 400 Monostearate)
Sol. in aromatic hydrocarbons (Scher PEG 400 Monostearate)
Sol. in most alcohols (Scher PEG 400 Monostearate)
Sol. in benzene (Radiasurf 7413)
Disp. in butyl stearate (Emerest 2640)
Sol. in carbon tetrachloride (Kessco PEG 400 Monostearate)
Sol. in ethanol (Myrj 45; Pegosperse 400MS)
Sol. in ethyl acetate (Kessco PEG 400 Monostearate; Pegosperse 400MS)
Disp. in glycerol trioleate (Emerest 2640)
Sol. in glycol ethers (Scher PEG 400 Monostearate)
Sol. in glycols (Scher PEG 400 Monostearate)
Sol. cloudy in hexane (Radiasurf 7413)
Sol. in isopropanol (Kessco PEG 400 Monostearate; Lipal 400S; Mapeg 400MS; Myrj 45)
Sol. in isopropyl myristate (Kessco PEG 400 Monostearate)
Sol. in ketones (Scher PEG 400 Monostearate)
Sol. in methanol (Myrj 45; Pegosperse 400MS)
Sol. in min. oil (Pegosperse 400MS; Radiasurf 7413); partly sol. (Scher PEG 400 Monostearate); disp (Emerest 2640)
Sol. hot in naphtha (Pegosperse 400MS)
Disp. in perchloroethylene (Emerest 2640)
Sol. in soybean oil (Mapeg 400MS)
Disp. in Stoddard solvent (Emerest 2640)
Sol. in toluol (Kessco PEG 400 Monostearate; Mapeg 400MS; Pegosperse 400MS)
Sol. in trichlorethylene (Radiasurf 7413)
Sol. in veg. oil (Pegosperse 400MS); partly sol. (Scher PEG 400 Monostearate)
Disp. in water (Emerest 2640; Kessco PEG 400 Monostearate; Lipal 400S; Mapeg 400MS; Myrj 45; Pegosperse 400MS)

Ionic Nature: Nonionic

POE (8) monostearate *(cont'd.)*

M.W.:
400 (Cithrol 4MS)
650 (Scher PEG 400 Monostearate)
722 (Radiasurf 7413)

Sp.gr.:
0.951 (98.9 C) (Radiasurf 7413)
0.9780 (65 C) (Kessco PEG 400 Monostearate)
1.0 (Myrj 45; Pegosperse 400MS); (30 C) (Scher PEG 400 Monostearate)

Visc.:
9.45 cps (98.9 C) (Radiasurf 7413)

M.P.:
28 C (Scher PEG 400 Monostearate)
30 C (Nonisol 300; Pegosperse 400MS)
31–34 C (Cithrol 4MS)
32 C (Emerest 2640; Kessco PEG 400 Monostearate)
33 C (Mapeg 400MS)
34 C (Radiasurf 7413)

Pour Pt.:
28 C (Myrj 45)

Flash Pt.:
> 170 C (OC) (Scher PEG 400 Monostearate)
248 C (Radiasurf 7413)
> 300 F (Myrj 45)
480 F (Kessco PEG 400 Monostearate)

Fire Pt.:
525 F (Kessco PEG 400 Monostearate)

Cloud Pt.:
< 25 C (Emerest 2640)

HLB:
11.0 (Cithrol 4MS)
11.1 (Myrj 45)
11.2 (Alkamuls 400MS)
11.2 ± 0.5 (Pegosperse 400MS)
11.2 ± 1 (Lipopeg 4-S)
11.3 ± 1 (Lipal 400S)
11.5 (Mapeg 400MS)
11.6 (Kessco PEG 400 Monostearate)
11.7 (Emerest 2640)
11.9 (Radiasurf 7413)
12.3 (Scher PEG 400 Monostearate)

Acid No.:
2 max. (Myrj 45)
< 3 (Pegosperse 400MS; Radiasurf 7413)

POE (8) monostearate (cont'd.)

5 max. (Kessco PEG 400 Monostearate; Lipal 400S; Lipopeg 4-S; Mapeg 400MS; Scher PEG 400 Monostearate)

Iodine No.:
0.5 max. (Kessco PEG 400 Monostearate)
< 1 (Pegosperse 400MS)
1 max. (Lipal 400S; Mapeg 400MS; Radiasurf 7413; Scher PEG 400 Monostearate)

Saponification No.:
75–90 (Radiasurf 7413)
80–90 (Lipopeg 4-S)
82–92 (Lipal 400S)
82–95 (Myrj 45)
83–92 (Kessco PEG 400 Monostearate)
83–93 (Scher PEG 400 Monostearate)
83–94 (Pegosperse 400MS)
84–93 (Mapeg 400MS)
95–105 (Cithrol 4MS)

Hydroxyl No.:
87–105 (Myrj 45)

Stability:
Stable over a wide pH range (Lipal 400S)

pH:
3.5–6.0 (5% aq. disp.) (Pegosperse 400MS)
5.0 (3% disp.) (Kessco PEG 400 Monostearate)

STD. PKGS.:
190-kg net bung drums or bulk (Radiasurf 7413)
450-lb steel drums (Kessco PEG 400 Monostearate)

POE (12) monostearate

SYNONYMS:
PEG-12 stearate (CTFA)
PEG 600 monostearate

STRUCTURE:

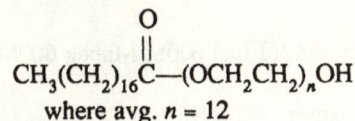

where avg. $n = 12$

CAS No.
9004-99-3 (generic)
RD No. 977055-40-5

POE (12) monostearate (cont'd.)

TRADENAME EQUIVALENTS:
Alkamuls 600MS [Alkaril]
Cithrol 6MS [Croda]
Emerest 2662 [Emery]
Kessco PEG 600 Monostearate [Armak]
Lipal 600S [PVO]
Mapeg 600MS [Mazer]
Nopalcol 6-S [Diamond Shamrock]
Pegosperse 600MS [Glyco]
Radiasurf 7414 [Oleofina]

CATEGORY:
Emulsifier, detergent, wetting agent, solubilizer, antifog aid, antifrothing agent, antistat, binding agent, chemical intermediate, defoamer, dispersant, foaming agent, lubricant, o/w emulgent, opacifier, plasticizer, rust inhibitor, scouring and detergent aid, surfactant, thickening agent, viscosity modifier

APPLICATIONS:
Bath products: bath oils (Kessco PEG 600 Monostearate)
Cosmetic industry preparations: (Cithrol 6MS; Emerest 2662; Kessco PEG 600 Monostearate; Lipal 600S; Mapeg 600MS; Nopalcol 6-S; Radiasurf 7414); hair rinses (Kessco PEG 600 Monostearate); perfumery (Cithrol 6MS; Kessco PEG 600 Monostearate); shampoos (Kessco PEG 600 Monostearate)
Farm products: (Kessco PEG 600 Monostearate); insecticides/pesticides (Cithrol 6MS; Radiasurf 7414)
Food applications: (Kessco PEG 600 Monostearate)
Industrial applications: adhesives (Nopalcol 6-S); dyes and pigments (Nopalcol 6-S; Radiasurf 7414); glass processing (Radiasurf 7414); lubricating/cutting oils (Radiasurf 7414); metalworking (Mapeg 600MS); paint mfg. (Radiasurf 7414); paper mfg. (Nopalcol 6-S); plastics (Kessco PEG 600 Monostearate; Radiasurf 7414); polishes and waxes (Radiasurf 7414); printing inks (Radiasurf 7414); textile/leather processing (Cithrol 6MS; Emerest 2662; Mapeg 600MS; Nopalcol 6-S; Radiasurf 7414)
Industrial cleaners: drycleaning compositions (Nopalcol 6-S)
Pharmaceutical applications: (Kessco PEG 600 Monostearate; Lipal 600S; Mapeg 600MS; Radiasurf 7414)

PROPERTIES:
Form:
Paste (Radiasurf 7414)
Solid (Alkamuls 600MS; Cithrol 6MS; Emerest 2662; Lipal 600S; Mapeg 600MS; Nopalcol 6-S)
Soft solid (Kessco PEG 600 Monostearate; Pegosperse 600MS)
Color:
White (Mapeg 600MS; Pegosperse 600MS; Radiasurf 7414)
White to cream (Kessco PEG 600 Monostearate)

Gardner 1 (Emerest 2662)

Composition:
100% active (Alkamuls 600MS; Lipal 600S)
100% conc. (Cithrol 6MS; Emerest 2662; Mapeg 600MS; Nopalcol 6-S; Pegosperse 600MS)

Solubility:
Sol. in acetone (Kessco PEG 600 Monostearate; Pegosperse 600MS)
Sol. in benzene (Radiasurf 7414)
Sol. in butyl stearate (Emerest 2662)
Sol. in carbon tetrachloride (Kessco PEG 600 Monostearate)
Sol. in ethanol (Pegosperse 600MS)
Sol. in ethyl acetate (Kessco PEG 600 Monostearate; Pegosperse 600MS)
Sol. cloudy in hexane (Radiasurf 7414)
Sol. in isopropanol (Kessco PEG 600 Monostearate; Mapeg 600MS; Radiasurf 7414)
Sol. in methanol (Pegosperse 600MS)
Sol. in min. oil (Radiasurf 7414); disp. hot (Mapeg 600MS); insol. (Emerest 2662)
Sol. in propylene glycol (Mapeg 600MS)
Sol. in soybean oil (Mapeg 600MS)
Sol. in Stoddard solvent (Emerest 2662)
Sol. in toluol (Kessco PEG 600 Monostearate; Mapeg 600MS; Pegosperse 600MS)
Sol. in trichlorethylene (Radiasurf 7414)
Sol. in veg. oil (Pegosperse 600MS; Radiasurf 7414)
Sol. in water (Kessco PEG 600 Monostearate; Mapeg 600MS); disp. (Emerest 2662; Pegosperse 600MS)
Sol. in xylene (Emerest 2662)

Ionic Nature: Nonionic
M.W.:
600 (Cithrol 6MS)
906 (Radiasurf 7414)

Sp.gr.:
0.981 (98.9 C) (Radiasurf 7414)
1.000 (65 C) (Kessco PEG 600 Monostearate)
1.01 (Pegosperse 600MS)

Density:
8.5 lb/gal (Emerest 2662)

Visc.:
12.60 cps (98.9 C) (Radiasurf 7414)

M.P.:
27–32 C (Pegosperse 600MS)
33–35 C (Cithrol 6MS)
36 C (Mapeg 600MS)
37 C (Kessco PEG 600 Monostearate)
38 C (Radiasurf 7414)

POE (12) monostearate *(cont'd.)*

40 C (Emerest 2662)
Flash Pt.:
  241 C (COC) (Radiasurf 7414)
  440 F (Emerest 2662)
  480 F (COC) (Kessco PEG 600 Monostearate)
Fire Pt.:
  550 F (Kessco PEG 600 Monostearate)
Cloud Pt.:
  55 C (Emerest 2662)
HLB:
  13.2 ± 0.5 (Pegosperse 600MS)
  13.3 (Radiasurf 7414)
  13.6 (Kessco PEG 600 Monostearate; Mapeg 600MS)
  13.8 (Emerest 2662)
  14.0 (Cithrol 6MS; Nopalcol 6-S)
Acid No.:
  < 3 (Radiasurf 7414)
  5 max. (Kessco PEG 600 Monostearate; Mapeg 600MS)
  < 8.5 (Pegosperse 600MS)
Iodine No.:
  0.25 max. (Kessco PEG 600 Monostearate)
  0.5 max. (Mapeg 600MS)
  < 1 (Pegosperse 600MS)
  1 max. (Radiasurf 7414)
Saponification No.:
  60–75 (Radiasurf 7414)
  61–70 (Kessco PEG 600 Monostearate)
  62–70 (Mapeg 600MS)
  68–76 (Cithrol 6MS)
  72–78 (Pegosperse 600MS)
pH:
  3–5 (5% aq. disp.) (Pegosperse 600MS)
  5.0 (3% disp.) (Kessco PEG 600 Monostearate)
STD. PKGS.:
  190-kg net bung drums or bulk (Radiasurf 7414)
  450-lb steel drums (Kessco PEG 600 Monostearate)

POE (20) monostearate

SYNONYMS:
PEG-20 stearate (CTFA)
PEG 1000 monostearate

STRUCTURE:

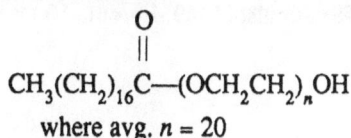

$$CH_3(CH_2)_{16}C—(OCH_2CH_2)_nOH$$

where avg. $n = 20$

CAS No.
9004-99-3 (generic)
RD No. 977055-41-6

TRADENAME EQUIVALENTS:
Cerasynt 840 [Van Dyk]
Chemax E-1000MS [Chemax]
Cithrol 10MS [Croda]
Emerest 2713 [Emery]
Kessco PEG 1000 Monostearate [Armak]
Lipopeg 10-S [Lipo]
Mapeg 1000MS [Mazer]
Pegosperse 1000MS [Glyco]
PGE-1000-MS [Hefti Ltd.]
Simulsol M49 [Seppic]
Varonic 1000MS [Sherex]

CATEGORY:
Emulsifier, detergent, wetting agent, solubilizer, antifrothing agent, chemical intermediate, dispersant, foaming agent, humectant, surfactant, thickening agent, softener, lubricant, stabilizer

APPLICATIONS:
Bath products: bath oils (Kessco PEG 1000 Monostearate; Lipopeg 10-S; Simulsol M49)
Cosmetic industry preparations: (Cerasynt 840; Chemax E-1000MS; Cithrol 10MS; Kessco PEG 1000 Monostearate; Mapeg 1000MS; PGE-1000-MS; Simulsol M49); creams and lotions (Cerasynt 840; Emerest 2713; Lipopeg 10-S); hair rinses (Kessco PEG 1000 Monostearate); perfumery (Cithrol 10MS; Kessco PEG 1000 Monostearate); shampoos (Kessco PEG 1000 Monostearate)
Farm products: (Kessco PEG 1000 Monostearate); insecticides/pesticides (Cithrol 10MS)
Food applications: (Kessco PEG 1000 Monostearate)
Industrial applications: metalworking (Mapeg 1000MS); plastics (Kessco PEG 1000 Monostearate); textile/leather processing (Chemax E-1000MS; Mapeg 1000MS)

217

POE (20) monostearate (cont'd.)

Pharmaceutical applications: (Kessco PEG 1000 Monostearate; Mapeg 1000MS)

PROPERTIES:

Form:

Solid (Cerasynt 840; Chemax E-1000MS; Emerest 2713; Lipopeg 10-S; Mapeg 1000MS; Pegosperse 1000MS; PGE-1000-MS; Simulsol M49; Varonic 1000MS)

Flake (Mapeg 1000MS)

Wax (Kessco PEG 1000 Monostearate)

Color:

White (Mapeg 1000MS)

Cream (Kessco PEG 1000 Monostearate; Pegosperse 1000MS)

Composition:

100% conc. (Cerasynt 840; Cithrol 10MS; Chemax E-1000MS; Lipopeg 10-S; Mapeg 1000MS; Pegosperse 1000MS; PGE-1000-MS; Simulsol M49)

Solubility:

Sol. in acetone (Kessco PEG 1000 Monostearate; Pegosperse 1000MS)

Sol. in carbon tetrachloride (Kessco PEG 1000 Monostearate)

Sol. in ethanol (Pegosperse 1000MS)

Sol. in ethyl acetate (Kessco PEG 1000 Monostearate; Pegosperse 1000MS)

Sol. in isopropanol (Kessco PEG 1000 Monostearate; Mapeg 1000MS)

Sol. in methanol (Pegosperse 1000MS)

Partly sol. in min. oil (Pegosperse 1000MS)

Sol. hot in naphtha (Pegosperse 1000MS)

Sol. in propylene glycol (Mapeg 1000MS)

Sol. in sodium sulfate (5%) (Kessco PEG 1000 Monostearate)

Sol. in toluol (Kessco PEG 1000 Monostearate; Mapeg 1000MS; Pegosperse 1000MS)

Sol. hot in veg. oil (Pegosperse 1000MS)

Sol. in water (Chemax E-1000MS; Kessco PEG 1000 Monostearate; Mapeg 1000MS); disp. hot (Pegosperse 1000MS)

Ionic Nature: Nonionic

M.W.:

1000 (Cithrol 10MS)

Sp.gr.:

1.02 (Pegosperse 1000MS)

1.030 (65 C) (Kessco PEG 1000 Monostearate)

M.P.:

34–40 C (Cithrol 10MS)

37–43 C (Pegosperse 1000MS)

41 C (Kessco PEG 1000 Monostearate)

42 C (Mapeg 1000MS)

Flash Pt.:

475 F (COC) (Kessco PEG 1000 Monostearate)

Fire Pt.:

495 F (Kessco PEG 1000 Monostearate)

HLB:

15.0 (Simulsol M49)
15.2 (Lipopeg 10-S)
15.2 ± 0.5 (Pegosperse 1000MS)
15.6 (Chemax E-1000MS; Emerest 2713; Kessco PEG 1000 Monostearate)
15.7 (Mapeg 1000MS)
16.0 (Cithrol 10MS)

Acid No.:

< 3 (Pegosperse 1000MS)
5.0 max. (Kessco PEG 1000 Monostearate; Mapeg 1000MS)

Iodine No.:

0.25 max. (Kessco PEG 1000 Monostearate)
0.5 max. (Mapeg 1000MS)
< 1 (Pegosperse 1000MS)

Saponification No.:

36–50 (Cithrol 10MS)
40–48 (Kessco PEG 1000 Monostearate)
41–49 (Mapeg 1000MS)
45–55 (Pegosperse 1000MS)

pH:

3–5 (5% aq. disp.) (Pegosperse 1000MS)
5.0 (3% disp.) (Kessco PEG 1000 Monostearate)

STD. PKGS.:

450 lb steel drums (Kessco PEG 1000 Monostearate)

POE (40) monostearate

SYNONYMS:

PEG-40 stearate (CTFA)
PEG 2000 monostearate
Polyoxyl 40 stearate
Stearethate 40

STRUCTURE:

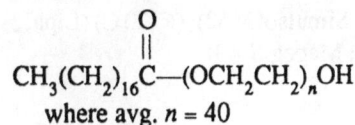

$$CH_3(CH_2)_{16}\overset{\overset{\textstyle O}{\textstyle \|}}{C}-(OCH_2CH_2)_nOH$$

where avg. $n = 40$

CAS No.:

9004-99-3 (generic)
RD No.: 977009-12-3

POE (40) monostearate (cont'd.)

TRADENAME EQUIVALENTS:
Atlas G-2198 [ICI United States]
Emerest 2715 [Emery]
Hetoxamate SA-40 (DF) [Heterene] (1,4 dioxane-free)
Industrol MS-40 [BASF Wyandotte]
Lipal 39S [PVO Int'l.]
Lipopeg 39-S [Lipo]
Mapeg S-40 [Mazer]
Myrj 52, 52C, 52S [ICI United States]
Nikkol MYS-40 [Nikko]
Pegosperse 1750MS [Glyco]
RS-55-40 [Hefti Ltd.]
Simulsol M52 [Seppic]

CATEGORY:
Emulsifier, surfactant, detergent, solubilizer, wetting agent, dispersant, spreading agent, lubricant, stabilizer, emollient, moisturizer, thickener, conditioner, foam modifier, plasticizer

APPLICATIONS:
Bath products: bath oils (Lipopeg 39-S)
Cleansers: hand cleanser (Hetoxamate SA-40 (DF)); germicidal skin cleanser (Lipal 39S)
Cosmetic industry preparations: (Emerest 2715; Mapeg S-40; Myrj 52, 52C, 52S; Nikkol MYS-40; Pegosperse 1750MS; RS-55-40); conditioners (Pegosperse 1750MS); creams and lotions (Hetoxamate SA-40 (DF); Lipal 39S; Lipopeg 39-S); hair preparations (Lipal 39S; Pegosperse 1750MS); makeup (Lipal 39S; Pegosperse 1750MS); shampoos (Lipal 39S; Pegosperse 1750MS); shaving preparations (Lipal 39S); toiletries (Pegosperse 1750MS)
Household detergents: (Pegosperse 1750MS)
Industrial applications: lubricating/cutting oils (Mapeg S-40); metalworking (Mapeg S-40); textile/leather processing (Hetoxamate SA-40 (DF); Mapeg S-40); waxes and oils (Hetoxamate SA-40 (DF))
Pharmaceutical applications: (Emerest 2715; Mapeg S-40; Myrj 52, 52C, 52S; Nikkol MYS-40; Pegosperse 1750MS; RS-55-40)

PROPERTIES:
Form:
Solid (Mapeg S-40; Nikkol MYS-40; RS-55-40; Simulsol M52); (@ 20 C) (Lipal 39S)
Flake (Emerest 2715; Hetoxamate SA-40 (DF); Mapeg S-40)
Pieces (Myrj 52C)
Solid wax (Lipopeg 39-S)
Wax (Industrol MS-40)
Waxy solid (Atlas G-2198; Myrj 52)
Waxy granular solid (Myrj 52S)

Color:
    White (Lipopeg 39-S; Mapeg S-40; Myrj 52S)
    Ivory (Myrj 52)
    Light tan (Atlas G-2198)
    Gardner 1 max. (Industrol MS-40)
    Gardner 2 max. (Hetoxamate SA-40 (DF))
    Gardner 4 (Lipal 39S)
Composition:
    100% active (Industrol MS-40; Lipal 39S; Lipopeg 39-S; Myrj 52, 52S)
    100% conc. (Emerest 2715; Mapeg S-40; Myrj 52C; Nikkol MYS-40; RS-55-40; Simulsol M52)
Solubility:
    Sol. in acetone (Atlas G-2198; Myrj 52, 52S)
    Sol. in lower alcohols (Atlas G-2198)
    Sol. in 5% aluminum chloride aq. sol'n. (Myrj 52S)
    Sol. in aniline (Atlas G-2198; Myrj 52S)
    Sol. in carbon tetrachloride (Atlas G-2198; Myrj 52S)
    Sol. in Cellosolve (Atlas G-2198; Myrj 52S)
    Sol. in dioxane (Atlas G-2198; Myrj 52S)
    Sol. in ethanol (Myrj 52, 52S)
    Sol. in ether (Atlas G-2198; Myrj 52, 52S)
    Sol. in ethyl acetate (Atlas G-2198; Myrj 52S)
    Sol. in isopropanol (Hetoxamate SA-40 (DF); Lipal 39S; Mapeg S-40)
    Sol. in methanol (Myrj 52, 52S)
    Insol. in min. oil (Hetoxamate SA-40 (DF))
    Sol. in many organic solvents (Atlas G-2198)
    Disp. in peanut oil (Lipal 39S)
    Sol. in propylene glycol (Mapeg S-40); disp. (Lipal 39S)
    Sol. in 5% sodium sulfate aq. sol'n. (Myrj 52S)
    Sol. in 5% sulfuric acid aq. sol'n. (Myrj 52S)
    Sol. in toluol (Mapeg S-40; Myrj 52S)
    Sol. in water (Atlas G-2198; Hetoxamate SA-40 (DF); Industrol MS-40; Lipal 39S; Mapeg S-40; Myrj 52, 52S)

Ionic Nature:
    Nonionic (Atlas G-2198; Industrol MS-40; Lipal 39S; Lipopeg 39-S; Mapeg S-40; Myrj 52, 52S; Pegosperse 1750-MS)
M.W.:
    2030 (Industrol MS-40)
Sp.gr.:
    1.1 (Myrj 52, 52S)
M.P.:
    48 C (Industrol MS-40; Mapeg S-40)

POE (40) monostearate *(cont'd.)*

Pour Pt.:
38 C (Atlas G-2198; Myrj 52, 52S)
Flash Pt.:
> 300 F (Atlas G-2198; Myrj 52, 52S)
Fire Pt.:
> 300 F (Atlas G-2198; Myrj 52)
HLB:
16.9 (Atlas G-2198; Emerest 2715; Industrol MS-40; Myrj 52, 52C, 52S; Simulsol M52)
16.9 ± 1 (Lipal 39S; Lipopeg 39-S)
17.0 (RS-55-40)
17.2 (Hetoxamate SA-40 (DF))
17.4 (Mapeg S-40)
17.5 (Nikkol MYS-40)
Acid No.:
1.0 max. (Lipopeg 39-S; Mapeg S-40; Myrj 52)
1.5 max. (Myrj 52S)
2 (Lipal 39S)
2 max. (Industrol MS-40)
3.0 max. (Hetoxamate SA-40 (DF))
Iodine No.:
0.5 max. (Mapeg S-40)
2 (Lipal 39S)
Saponification No.:
23–35 (Lipal 39S; Lipopeg 39-S; Mapeg S-40; Myrj 52, 52S)
24–34 (Hetoxamate SA-40 (DF))
Hydroxyl No.:
25–37 (Myrj 52S)
27–40 (Myrj 52)
33–43 (Hetoxamate SA-40 (DF))
Stability:
Stable over a wide pH range
pH:
5.5–7.0 (5% aq.) (Industrol MS-40)
STD. PKGS.:
30-gal fiber leverpak (Myrj 52)
55-gal (450 lb net) steel drums (Industrol MS-40)

SYNONYMS:
  PEG-75 stearate (CTFA)
  PEG 4000 monostearate
STRUCTURE:

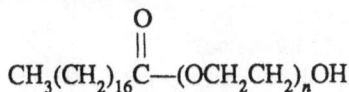

  where avg. $n = 75$
CAS No.
  9004-99-3 (generic)
  RD No. 977055-46-1
TRADENAME EQUIVALENTS:
  Cithrol 40MS [Croda]
  Kessco PEG 4000 Monostearate [Armak]
  Mapeg 4000MS [Mazer]
  Pegosperse 4000MS [Glyco]
CATEGORY:
  Emulsifier, detergent, wetting agent, antifrothing agent, foaming agent, solubilizer, surfactant, thickening agent
APPLICATIONS:
  Bath products: bath oils (Kessco PEG 4000 Monostearate; Mapeg 4000MS)
  Cosmetic industry preparations: (Cithrol 40MS; Kessco PEG 4000 Monostearate; Mapeg 4000MS); hair rinses (Kessco PEG 4000 Monostearate; Mapeg 4000MS); perfumery (Cithrol 40MS; Kessco PEG 4000 Monostearate; Mapeg 4000MS); shampoos (Kessco PEG 4000 Monostearate; Mapeg 4000MS)
  Farm products: (Kessco PEG 4000 Monostearate; Mapeg 4000MS); insecticides/pesticides (Cithrol 40MS)
  Food applications: (Kessco PEG 4000 Monostearate; Mapeg 4000MS)
  Industrial applications: plastics (Kessco PEG 4000 Monostearate; Mapeg 4000MS)
  Pharmaceutical applications: (Kessco PEG 4000 Monostearate; Mapeg 4000MS)
PROPERTIES:
Form:
  Liquid (Mapeg 4000MS)
  Wax (Kessco PEG 4000 Monostearate)
  Waxy solid (Pegosperse 4000MS)
Color:
  Cream (Kessco PEG 4000 Monostearate; Pegosperse 4000MS
Composition:
  100% conc. (Cithrol 40MS; Mapeg 4000MS; Pegosperse 4000MS)
Solubility:
  Sol. in acetone (Kessco PEG 4000 Monostearate) sol. hot (Pegosperse 4000MS)
  Sol. in carbon tetrachloride (Kessco PEG 4000 Monostearate)
  Sol. hot in ethanol (Pegosperse 4000MS)

POE (75) monostearate (cont'd.)

Sol. in ethyl acetate (Kessco PEG 4000 Monostearate); sol. hot (Pegosperse 4000MS)
Sol. in isopropanol (Kessco PEG 4000 Monostearate)
Sol. hot in methanol (Pegosperse 4000MS)
Partly sol. hot in min. oil (Pegosperse 4000MS)
Disp. in naphtha (Pegosperse 4000MS)
Sol. in sodium sulfate (5%) (Kessco PEG 4000 Monostearate)
Sol. in toluol (Kessco PEG 4000 Monostearate); sol. hot (Pegosperse 4000MS)
Sol. hot in veg. oil (Pegosperse 4000MS)
Sol. in water (Kessco PEG 4000 Monostearate); sol. hot (Pegosperse 4000MS)

Ionic Nature: Nonionic

M.W.:
4000 (Cithrol 40MS)

Sp.gr.:
1.075 (65 C) (Kessco PEG 4000 Monostearate)
1.10 (Pegosperse 4000MS)

M.P.:
54–61 C (Pegosperse 4000MS)
56 C (Kessco PEG 4000 Monostearate)

Flash Pt.:
465 F (Kessco PEG 4000 Monostearate)

Fire Pt.:
520 F (Kessco PEG 4000 Monostearate)

HLB:
18.0 ± 0.5 (Pegosperse 4000MS)
18.6 (Kessco PEG 4000 Monostearate)
18.7 (Mapeg 4000MS)
18.8 (Cithrol 40MS)

Acid No.:
5.0 max. (Kessco PEG 4000 Monostearate)
< 6 (Pegosperse 4000MS)

Iodine No.:
0.1 max. (Kessco PEG 4000 Monostearate)
< 1 (Pegosperse 4000MS)

Saponification No.:
10–18 (Kessco PEG 4000 Monostearate)
12–17 (Cithrol 40MS)
17–22 (Pegosperse 4000MS)

pH:
3–5 (5% aq. disp.) (Pegosperse 4000MS)
5.0 (3% disp.) (Kessco PEG 4000 Monostearate)

STD. PKGS.:
450 lb steel drums (Kessco PEG 4000 Monostearate)

POE (1500) monostearate

SYNONYMS:
PEG-6-32 stearate (CTFA)
PEG-1500 monostearate
CAS No.:
9004-99-3 (generic)
TRADENAME EQUIVALENTS:
Cithrol 15MS [Croda]
Hodag 150-S [Hodag]
Lipopeg 15-S [Lipo]
Mapeg 1500 MS [Mazer]
Pegosperse 1500 MS [Glyco]
Radiasurf 7417 [Oleofina]
Tefose 1500 [Gattefosse]
CATEGORY:
Emulsifier, detergent, wetting agent, solubilizer, antifog aid, antifrothing agent, antistat, defoamer, dispersant, foaming agent, lubricant, opacifier, o/w emulgent, plasticizer, rust inhibitor, scouring and detergent aid, thickening agent, viscosity modifier
APPLICATIONS:
Bath products: bath oils (Lipopeg 15-S)
Cosmetic industry preparations: (Cithrol 15 MS; Hodag 150-S; Mapeg 1500 MS; Radiasurf 7417); cosmetic base (Tefose 1500); perfumery (Cithrol 15 MS; Hodag 150-S; Mapeg 1500 MS)
Farm products: insecticides/pesticides (Cithrol 15 MS; Hodag 150-S; Mapeg 1500 MS; Radiasurf 7417)
Industrial applications: dyes and pigments (Radiasurf 7417); glass processing (Radiasurf 7417); lubricating/cutting oils (Radiasurf 7417); paint mfg. (Radiasurf 7417); plastics (Radiasurf 7417); polishes and waxes (Radiasurf 7417); printing inks (Radiasurf 7417); textile/leather processing (Radiasurf 7417)
Pharmaceutical applications: (Radiasurf 7417)
PROPERTIES:
Form:
Solid (Hodag 150-S; Lipopeg 15-S; Mapeg 1500 MS; Radiasurf 7417; Tefose 1500)
Waxy solid (Pegosperse 1500 MS
Color:
White (Radiasurf 7417)
Cream (Pegosperse 1500 MS)
Composition:
100% conc. (Pegosperse 1500 MS)
Solubility:
Sol. in benzene (Radiasurf 7417)
Sol. in ethanol (Pegosperse 1500 MS)
Sol. in ethyl acetate (Pegosperse 1500 MS)

Sol. hot in methanol (Pegosperse 1500 MS)
Sol. hot in naphtha (Pegosperse 1500 MS)
Sol. in toluol (Pegosperse 1500 MS)
Sol. in trichlorethylene (Radiasurf 7417)
Sol. in veg. oil (Pegosperse 1500 MS)
Miscible with water (Pegosperse 1500 MS)

Ionic Nature: Nonionic

M.W.:
1500 (Cithrol 15MS)
1812 (Radiasurf 7417)

Sp.gr.:
1.023 (98.9 C) (Radiasurf 7417)
1.05 (Pegosperse 1500 MS)

Visc.:
28.40 cps (98.9 C) (Radiasurf 7417)

M.P.:
27–31 C (Pegosperse 1500 MS)
47 C (Radiasurf 7417)

Flash Pt.:
248 C (COC) (Radiasurf 7417)

HLB:
13.8 ± 0.5 (Pegosperse 1500 MS)
16.5 avg. (Radiasurf 7417)
17.0 (Cithrol 15MS)

Acid No.:
< 3 (Radiasurf 7417)
< 3.5 (Pegosperse 1500 MS)

Iodine No.:
< 1 (Pegosperse 1500 MS; Radiasurf 7417)

Saponification No.:
30–40 (Radiasurf 7417)
30–55 (Cithrol 15MS)
57–67 (Pegosperse 1500MS)

pH:
3–5 (5% aq. disp.) (Pegosperse 1500 MS)

Surface Tension:
46–50 dynes/cm (0.1% aq.) (Radiasurf 7417)

STD. PKGS.:
25-kg net multiply paper bags or bulk (Radiasurf 7417)

SYNONYMS:
  PEG-8 tallate (CTFA)
  PEG 400 monotallate
STRUCTURE:

O
‖
RC—(OCH$_2$CH$_2$)$_n$OH

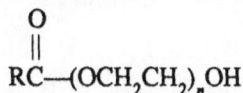

  where   RCO⁻ represents tall oil acid radical and
        avg. $n = 8$
CAS No.
  61791-00-2 (generic)
  RD No.: 977055-49-4
TRADENAME EQUIVALENTS:
  Industrol 400MOT [BASF Wyandotte]
  Mapeg 400MOT [Mazer]
  Pegosperse 400MOT [Glyco]
  Witconol H-31 [Witco/Organics]
CATEGORY:
  Surfactant, emulsifier, o/w emulsifier, defoamer, dispersant
APPLICATIONS:
  Industrial applications: metalworking (Witconol H-31)
PROPERTIES:
Form:
  Liquid (Industrol 400MOT; Mapeg 400MOT; Pegosperse 400MOT; Witconol H-31)
Color:
  Yellow (Pegosperse 400MOT)
  Gardner 10 max. (Industrol 400MOT)
Composition:
  100% conc. (Pegosperse 400MOT)
Solubility:
  Sol. in acetone (Pegosperse 400MOT)
  Sol. in ethanol (Pegosperse 400MOT)
  Sol. in ethyl acetate (Pegosperse 400MOT)
  Sol. in methanol (Pegosperse 400MOT)
  Partly sol. hot in min. oil (Pegosperse 400MOT)
  Sol. hot in naphtha (Pegosperse 400MOT)
  Sol. in oil (Witconol H-31)
  Sol. in toluol (Pegosperse 400MOT)
  Partly sol. hot in veg. oil (Pegosperse 400MOT)
  Disp. in water (Industrol 400MOT)
Ionic Nature: Nonionic
Sp.gr.:
  1.02 (Pegosperse 400MOT)

227

POE (8) monotallate *(cont'd.)*

    1.04 (Industrol 400MOT)
Visc.:
    400 cps (Industrol 400MOT)
Pour Pt.:
    −5 C (Industrol 400MOT)
Solidification Pt.:
    < 0 C (Pegosperse 400MOT)
Cloud Pt.:
    < 25 C (1% aq.) (Industrol 400MOT)
HLB:
    11.0 (Industrol 400MOT)
    11.0 ± 0.5 (Pegosperse 400MOT)
Acid No.:
    1 max. (Industrol 400MOT)
    < 5–8 (Pegosperse 400MOT)
Iodine No.:
    54–59 (Pegosperse 400MOT)
Saponification No.:
    83–89 (Pegosperse 400MOT)
pH:
    3.5–5.0 (5% aq. disp.) (Pegosperse 400MOT)
    6.0–7.5 (5% aq. disp.) (Industrol 400MOT)
STD. PKGS.:
    55 gal (450 lb net) steel drums (Industrol 400MOT)

POE (10) monotallate

SYNONYMS:
    PEG-10 tallate (CTFA)
    PEG 500 monotallate
STRUCTURE:

$$RC\!-\!(OCH_2CH_2)_n OH$$
$$\overset{O}{\overset{\|}{}}$$

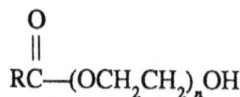

    where   R represents the tall oil fatty radical and
               avg. $n = 10$
CAS No.:
    61791-00-2 (generic)
TRADENAME EQUIVALENTS:
    Chemax TO-10 [Chemax]
    Polyfac MT-610 [Westvaco]

CATEGORY:
Emulsifier, coemulsifier, solubilizer
APPLICATIONS:
Degreasers: (Chemax TO-10)
Industrial applications: lubricating/cutting oils (Chemax TO-10; Polyfac MT-610); textile/leather processing (Polyfac MT-610)
PROPERTIES:
Form:
Liquid (Chemax TO-10; Polyfac MT-610)
Color:
Amber (Polyfac MT-610)
Solubility:
Sol. in isopropanol (@ 10%) (Polyfac MT-610)
Insol. in min. oil (@ 10%) (Polyfac MT-610)
Disp. in water (@ 10%) (Polyfac MT-610)
Sol. in xylene (@ 10%) (Polyfac MT-610)
Density:
8.7 lb/gal (Polyfac MT-610)
Visc.:
300 cps (Polyfac MT-610)
Pour Pt.:
41 F (Polyfac MT-610)
Cloud Pt.:
77 F (Polyfac MT-610)
HLB:
11.8 (Chemax TO-10; Polyfac MT-610)
Saponification No.:
70–80 (Polyfac MT-610)
Stability:
Relatively stable in alkaline and acidic media; unaffected by hard water or the presence of metal ions (Polyfac MT-610)
TOXICITY/HANDLING:
Nonhazardous DOT classification (Polyfac MT-610)
STD. PKGS.:
55-gal drums, tank trucks, rail cars (Polyfac MT-610)

POE (2) oleyl ether

SYNONYMS:
Oleth-2 (CTFA)
Oleyl polyglycol ether (2 EO)

POE (2) oleyl ether *(cont'd.)*

PEG-2 oleyl ether
PEG 100 oleyl ether
POE (2) oleyl alcohol

STRUCTURE:

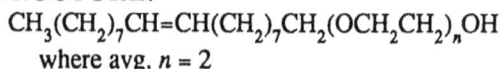

$CH_3(CH_2)_7CH=CH(CH_2)_7CH_2(OCH_2CH_2)_nOH$
   where avg. $n = 2$

CAS No.:

9004-98-2 (generic); 25190-05-0 (generic)
RD No.: 977057-49-0

TRADENAME EQUIVALENTS:

Alkasurf OA-2 [Alkaril]
Ameroxol OE-2 [Amerchol]
Brij 92, 93 [ICI United States]
Ethosperse OA-2 [Glyco]
Hetoxol OL2 [Heterene]
Lipocol O-2 [Lipo]
Macol OA-2 [Mazer]
OL-55-F-2 [Hefti Ltd.]
Serdox NOL2 [Servo B.V.]
Simulsol 92 [Seppic]

CATEGORY:

Emulsifier, coemulsifier, solubilizer, dispersant, spreading agent, emollient, surfactant, detergent, thickener, stabilizer, defoamer, wetting agent, conditioning agent, leveling agent, intermediate

APPLICATIONS:

Bath products: bath oils (Ameroxol OE-2; Ethosperse OA-2; Hetoxol OL2)
Cosmetic industry preparations: (Alkasurf OA-2; Macol OA-2; OL-55-F-2); conditioners (Ethosperse OA-2); creams and lotions (Ameroxol OE-2; Ethosperse OA-2; Hetoxol OL2; Lipocol O-2); hair preparations (Ethosperse OA-2); shampoos (Hetoxol OL2)
Household products: (Hetoxol OL2; Macol OA-2)
Industrial applications: (Macol OA-2); dyes and pigments (Hetoxol OL2; Lipocol O-2); lubricating/cutting oils (Macol OA-2); metalworking (Macol OA-2); silicone products (Hetoxol OL2); textile/leather processing (Hetoxol OL2; Macol OA-2); waxes (Hetoxol OL2)
Pharmaceutical applications: (OL-55-F-2); antiperspirant/deodorant (Ethosperse OA-2; Lipocol O-2); depilatories (Lipocol O-2)

PROPERTIES:
Form:

Liquid (Alkasurf OA-2; Brij 92, 93; Ethosperse OA-2; Hetoxol OL2; Lipocol O-2; Macol OA-2; OL-55-F-2; Serdox NOL2; Simulsol 92)
Clear liquid (Ameroxol OE-2)

Color:
Colorless (Ethosperse OA-2; Macol OA-2)
Pale straw (Ameroxol OE-2)
Yellow (Brij 92, 93; Lipocol O-2)
Odor:
Bland (Ameroxol OE-2)
Composition:
100% active (Lipocol O-2)
100% conc. (Macol OA-2; OL-55-F-2; Serdox NOL2; Simulsol 92)
Solubility:
Sol. in alcohols (Brij 92, 93)
Sol. in aromatic solvent (@ 10%) (Alkasurf OA-2)
Sol. in cottonseed oil (Brij 92, 93)
Sol. in ethanol (Ethosperse OA-2); sol. in anhyd. ethanol (Ameroxol OE-2)
Sol. in isopropanol (Hetoxol OL2; Macol OA-2)
Sol. in isopropyl esters (Ameroxol OE-2)
Sol. in min. oil (Ameroxol OE-2; Brij 92, 93; Ethosperse OA-2; Hetoxol OL2; Macol
    OA-2); (@ 10%) (Alkasurf OA-2)
Disp. in min. spirits (@ 10%) (Alkasurf OA-2)
Sol. in oil (Alkasurf OA-2)
Sol. in perchloroethylene (@ 10%) (Alkasurf OA-2)
Sol. in propylene glycol (Brij 92, 93)
Disp. in veg. oil (Ethosperse OA-2)
Disp. hot in water (Ethosperse OA-2); insol. (@ 10%) (Alkasurf OA-2)
Ionic Nature:
Nonionic (Ameroxol OE-2; Brij 92, 93; Ethosperse OA-2; Lipocol O-2; Hetoxol OL2;
    Lipocol O-2; Macol OA-2; OL-55-F-2; Serdox NOL2; Simulsol 92)
Density:
0.90 g/ml (Alkasurf OA-2)
Visc.:
30 cps (Brij 92, 93; Ethosperse OA-2)
Pour Pt.:
10 C (Brij 92, 93)
Flash Pt.:
> 300 F (Brij 92, 93)
Fire Pt.:
> 300 F (Brij 92, 93)
Cloud Pt.:
47–50 C (10% in 25% butyl Carbitol) (Alkasurf OA-2)
HLB:
4.0 ± 1 (Ethosperse OA-2)
4.9 (Alkasurf OA-2; Brij 92, 93; Hetoxol OL2; Macol OA-2; Simulsol 92)
4.9 ± 1 (Lipocol O-2)

POE (2) oleyl ether (cont'd.)

5.0 (OL-55-F-2; Serdox NOL2)
Acid No.:
0.5 max. (Ethosperse OA-2)
1.0 max. (Lipocol O-2)
Hydroxyl No.:
152–162 (Hetoxol OL2; Macol OA-2)
160–180 (Lipocol O-2)
165–180 (Ethosperse OA-2)
Stability:
Chemically stable; resistant to acid hydrolysis and alkaline saponification (Ethosperse OA-2)
Stable to hydrolysis by strong acids and alkalis (Macol OA-2)
Biodegradable: (Serdox NOL2)

POE (4) oleyl ether

SYNONYMS:
Oleth-4 (CTFA)
PEG-4 oleyl ether
PEG 200 oleyl ether
STRUCTURE:

$CH_3(CH_2)_7CH=CH(CH_2)_7CH_2(OCH_2CH_2)_nOH$
where avg. $n = 4$
CAS No.:
9004-98-2 (generic); 25190-05-0 (generic)
RD No.: 977065-24-9
TRADENAME EQUIVALENTS:
Chemal OA-4 [Chemax]
Hetoxol OL-4 [Heterene]
Macol OA-4 [Mazer]
CATEGORY:
Emulsifier, detergent, wetting agent, intermediate, dispersant, leveling agent, solubilizer, coupling agent, detergent, stabilizer, anticoagulant
APPLICATIONS:
Bath products: bath oils (Hetoxol OL-4)
Cosmetic industry preparations: (Macol OA-4); creams and lotions (Hetoxol OL-4); shampoos (Hetoxol OL-4); topical preparations (Chemal OA-4)
Food applications: citrus wax coatings (Chemal OA-4)
Household detergents: (Hetoxol OL-4; Macol OA-4)
Industrial applications: (Macol OA-4); dyes and pigments (Hetoxol OL-4); metal-

working lubricants (Macol OA-4); plastics (Chemal OA-4); textile/leather process-
ing (Hetoxol OL-4; Macol OA-4)
Industrial cleaners: textile cleaning (Hetoxol OL-4)
PROPERTIES:
Form:
Liquid (Chemal OA-4; Hetoxol OL-4; Macol OA-4)
Color:
Colorless (Macol OA-4)
Composition:
100% conc. (Macol OA-4)
Solubility:
Sol. in isopropanol (Hetoxol OL-4; Macol OA-4)
Sol. in min. oil (Hetoxol OL-4; Macol OA-4)
Ionic Nature: Nonionic
HLB:
8.0 (Hetoxol OL-4; Macol OA-4)
Hydroxyl No.:
120–135 (Chemal OA-4; Hetoxol OL-4)
120–136 (Macol OA-4)

POE (10) oleyl ether

SYNONYMS:
Oleth-10 (CTFA)
PEG-10 oleyl ether
PEG 500 oleyl ether
POE (10) oleyl alcohol
STRUCTURE:
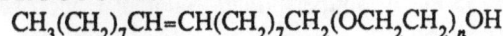
$CH_3(CH_2)_7CH{=}CH(CH_2)_7CH_2(OCH_2CH_2)_nOH$
where avg. $n = 10$
CAS No.:
9004-98-2 (generic); 25190-05-0 (generic)
RD No.: 977057-52-5
TRADENAME EQUIVALENTS:
Alkasurf OA-10 [Alkaril]
Ameroxol OE-10 [Amerchol]
Brij 96 [ICI United States]
Brij 97 [ICI United States]
Chemal OA-10 [Chemax]
Empilan KL10 [Albright & Wilson/Marchon]

POE (10) oleyl ether (cont'd.)

TRADENAME EQUIVALENTS *(cont'd.):*
  Eumulgin O10 [Henkel]
  Hetoxol OL-10 [Heterene]
  Lipocol O-10 [Lipo]
  Macol OA-10 [Mazer]
  Nikkol BO-10TX [Nikko]
  OL-55-F-10 [Hefti Ltd.]
  Simulsol 96 [Seppic]
  Standamul O-10 [Henkel]
  Volpo 10 [Croda]
  Volpo O10 [Croda Chem. Ltd.]
Distilled:
  Volpo N10 [Croda Chem. Ltd.]

CATEGORY:
  Emulsifier, detergent, surfactant, wetting agent, solubilizer, coupling agent, dispersant, stabilizer, anticoagulant, defoamer, conditioning agent, emollient, lubricant, superfatting agent

APPLICATIONS:
  Bath products: bubble bath (Volpo 10); bath oils (Volpo 10)
  Cosmetic industry preparations: (Empilan KL10; Hetoxol OL-10; Macol OA-10; OL-55-F-10; Volpo 10, N10, O10); astringent creams and lotions (Volpo 10); cold waves (Volpo 10); creams and lotions (Eumulgin O10; Lipocol O-10); hair straighteners (Volpo 10); makeup (Volpo 10); perfumery (Ameroxol OE-10; OL-55-F-10); shampoos (Volpo 10); topical preparations (Alkasurf OA-10; Chemal OA-10)
  Household detergents: (Macol OA-10)
  Industrial applications: (Macol OA-10); dyes and pigments (Lipocol O-10); lubricating/cutting oils (OL-55-F-10); metalworking lubricants (Macol OA-10); plastics (Chemal OA-10); textile/leather processing (Macol OA-10; OL-55-F-10)
  Industrial cleaners: (Volpo N10, O10); textile cleaning (Empilan KL10)
  Pharmaceutical applications: (OL-55-F-10); antiperspirant/deodorant (Lipocol O-10); depilatories (Lipocol O-10; Volpo 10)

PROPERTIES:
Form:
  Liquid (Brij 97; Lipocol O-10; Nikkol BO-10TX)
  Liquid with some solids (Brij 96)
  Liquid/paste (OL-55-F-10)
  Gel (Simulsol 96)
  Paste (Alkasurf OA-10; Empilan KL10; Eumulgin O10; Volpo N10, O10)
  Semisolid (Ameroxol OE-10; Hetoxol OL-10; Volpo 10)
  Solid (Macol OA-10)
  Soft waxy solid (Standamul O-10)

POE (10) oleyl ether (cont'd.)

Color:
White (Ameroxol OE-10; Volpo 10)
Cream (Empilan KL10; Macol OA-10)
Pale yellow (Brij 97)
Yellow (Brij 96; Lipocol O-10)
Gardner 3 max. (Hetoxol OL-10)

Odor:
Bland (Ameroxol OE-10)

Composition:
97% conc. (Volpo N10, O10)
100% conc. (Eumulgin O10; Macol OA-10; Nikkol BO-10TX; OL-55-F-10; Simulsol 96; Standamul O-10)
100% active (Empilan KL10; Lipocol O-10)

Solubility:
Sol. in alcohols (Brij 96, 97; Standamul O-10; Volpo 10)
Sol. in aromatic solvent (@ 10%) (Alkasurf OA-10); sol. in most aromatic solvents (Volpo 10)
Sol. in most chlorinated solvents (Volpo 10)
Sol. in ethanol (Ameroxol OE-10)
Sol. in glycols (Ameroxol OE-10; Volpo 10)
Sol. in hydroalcoholics (Ameroxol OE-10)
Sol. in hydrocarbons (Standamul O-10)
Sol. in isopropanol (Hetoxol OL-10; Macol OA-10)
Sol. in ketones (Volpo 10)
Disp. in min. oil (Macol OA-10); disp. (@ 10%) (Alkasurf OA-10); insol. (Hetoxol OL-10)
Disp. in min. spirits (@ 10%) (Alkasurf OA-10)
Sol. in most organic solvents (Standamul O-10)
Disp. in perchloroethylene (@ 10%) (Alkasurf OA-10)
Sol. in water (Ameroxol OE-10; Brij 96, 97; Hetoxol OL-10; Macol OA-10; Volpo 10); sol. in cold water (Empilan KL10); sol. (@ 10%) (Alkasurf OA-10)

Ionic Nature:
Nonionic (Ameroxol OE-10; Brij 96; Eumulgin O10; Lipocol O-10; Macol OA-10; Nikkol BO-10TX; OL-55-F-10; Simulsol 96; Standamul O-10; Volpo N10, O10)

Density:
0.9 g/cm^3 (Empilan KL10)

Visc.:
100 cps (Brij 96, 97)

M.P.:
16 C (Macol OA-10)
40 C (Empilan KL10)

Pour Pt.:
16 C (Brij 96, 97)

POE (10) oleyl ether *(cont'd.)*

Flash Pt.:
> 300 F (Brij 96, 97)

Fire Pt.:
> 300 F (Brij 96, 97)

Cloud Pt.:
47–55 C (Ameroxol OE-10)
53–60 C (1% DW) (Alkasurf OA-10)

HLB:
12.4 (Alkasurf OA-10; Brij 96, 97; Hetoxol OL-10; Macol OA-10; Simulsol 96)
12.4 ± 1 (Lipocol O-10)
13.5 (OL-55-F-10)
14.5 (Nikkol BO-10TX)

Acid No.:
2 max. (Lipocol O-10; Volpo 10)

Iodine No.:
31–37 (Wijs) (Volpo 10)

Hydroxyl No.:
74–84 (Lipocol O-10)
77–83 (Hetoxol OL-10)
79–91 (Volpo 10)
80–95 (Macol OA-10)

Stability:
Stable to acids and alkalis (Empilan KL10; Lipocol O-10)
Stable to hydrolysis by strong acids and alkalis (Macol OA-10)
Stable to many alkalis, acids under extreme pH conditions (Standamul O-10)
Stable in sol'ns. of metallic ions and to many acids and alkalis (Volpo 10)

pH:
5–7 (3% aq. sol'n.) (Volpo 10)
8–10 (Empilan KL10)

TOXICITY/HANDLING:
Risk of skin and eye irritation occurs at conc. in excess of 3% (Volpo 10)

STD. PKGS.:
55-gal closed-head steel drum (Brij 97)

POE (20) oleyl ether

SYNONYMS:
Oleth-20 (CTFA)
PEG-20 oleyl ether
PEG 1000 oleyl ether

STRUCTURE:
$CH_3(CH_2)_7CH=CH(CH_2)_7CH_2(OCH_2CH_2)_nOH$
where avg. $n = 20$

CAS No.:
9004-98-2 (generic); 25190-05-0 (generic)
RD No.: 977057-53-6

TRADENAME EQUIVALENTS:
Ahco 3998 [ICI Americas]
Ameroxol OE-20 [Amerchol]
Brij 98, 99 [ICI United States]
Chemal OA-20, OA-20/70 [Chemax]
Empilan KL20 [Albright & Wilson/Marchon]
Emulphor ON-870 [GAF]
Hetoxol OA-20 Special, OL-20 [Heterene]
Hostacerin O-20 [American Hoechst]
Lipal 20 OA [PVO Int'l.]
Lipocol O-20 [Lipo]
Macol OA-20 [Mazer]
Nikkol BO-20TX [Nikko]
OL-55-F-20 [Hefti Ltd.]
Simulsol 98 [Seppic]
Standamul O20 [Henkel]
Volpo 20 [Croda Inc.]
Volpo O20 [Croda Chem. Ltd.]
Distilled:
Volpo N20 [Croda Chem. Ltd.]

CATEGORY:
Emulsifier, detergent, wetting agent, solubilizer, coupling agent, spreading agent, dispersant, surfactant, stabilizer, anticoagulant, emollient, lubricant, superfatting agent

APPLICATIONS:
Bath products: bubble bath (Volpo 20); bath oils (Ameroxol OE-20)
Cleansers: germicidal skin cleanser (Lipal 20 OA)
Cosmetic industry preparations: (Empilan KL20; Hetoxol OA-20 Special, OL-20; Macol OA-20; OL-55-F-20; Volpo 20, N20); astringent creams and lotions (Volpo 20); cold waves (Volpo 20); creams and lotions (Ameroxol OE-20; Emulphor ON-870; Lipal 20 OA); hair preparations (Lipal 20 OA); hair straighteners (Volpo 20); makeup (Lipal 20 OA); perfumery (Emulphor ON-870); shampoos (Volpo 20); shaving preparations (Lipal 20 OA); topical preparations (Chemal OA-20, OA-20/70)
Degreasers: (Emulphor ON-870)
Food applications: citrus wax coatings (Chemal OA-20, OA-20/70)
Household detergents: (Macol OA-20)

POE (20) oleyl ether (cont'd.)

Industrial applications: (Macol OA-20; Volpo N20); adhesives (Emulphor ON-870); dyes and pigments (Emulphor ON-870); metalworking lubricants (Macol OA-20); paint mfg. (Emulphor ON-870); plastics (Chemal OA-20, OA-20/70); textile/leather processing (Empilan KL20; Emulphor ON-870; Macol OA-20; OL-55-F-20)

Industrial cleaners: metal processing surfactants (Emulphor ON-870); textile cleaning (Empilan KL20)

Pharmaceutical applications: (Emulphor ON-870; OL-55-F-20); antiperspirant/deodorant (Lipal 20 OA); depilatories (Volpo 20); drugs (Emulphor ON-870); ointments (Lipal 20 OA)

PROPERTIES:

Form:
Liquid (Ahco 3998)
Semisolid (Chemal OA-20, OA-20/70)
Soft solid (Empilan KL20; Volpo O20)
Solid (Brij 99; Hetoxol OA-20 Special, OL-20; Lipal 20 OA; Lipocol O-20; Macol OA-20; Nikkol BO-20TX; Simulsol 98; Volpo 20, N20)
Soft wax (OL-55-F-20)
Solid wax (Emulphor ON-870)
Wax (Hostacerin O-20)
Soft waxy solid (Brij 98)
Waxy solid (Ameroxol OE-20; Standamul O20)

Color:
White (Ameroxol OE-20; Emulphor ON-870; Hetoxol OA-20 Special; Volpo 20)
Cream (Brij 98, 99; Empilan KL20; Macol OA-20)
Pale straw (Volpo O20)
Gardner 2 max. (Hetoxol OL-20)
Gardner 3 (Lipal 20 OA)

Odor:
Bland (Ameroxol OE-20)

Composition:
97% conc. (Volpo N20, O20)
100% active (Empilan KL20; Emulphor ON-870; Lipal 20 OA)
100% conc. (Ahco 3998; Hostacerin O-20; Lipocol O-20; Macol OA-20; Nikkol BO-20TX; OL-55-F-20; Simulsol 98; Volpo 20)

Solubility:
Sol. in alcohols (Brij 98, 99; Standamul O20; Volpo 20)
Partly sol. in arachis oil (Volpo O20)
Sol. in aromatic solvents (Volpo 20)
Sol. in butyl Cellosolve (Emulphor ON-870)
Partly sol. in butyl stearate (Volpo O20)
Sol. in chlorinated solvents (Volpo 20)
Sol. in ethanol (Ameroxol OE-20; Emulphor ON-870; Volpo O20)

Sol. in ethylene glycol (Emulphor ON-870)
Sol. in glycols (Ameroxol OE-20; Volpo 20)
Sol. in hydroalcoholics (Ameroxol OE-20)
Sol. in hydrocarbons (Standamul O20)
Sol. in isopropanol (Hetoxol OA-20 Special, OL-20; Lipal 20 OA; Macol OA-20)
Sol. in ketones (Volpo 20)
Partly sol. in oleic acid (Volpo O20)
Sol. in most organic solvents (Standamul O20)
Sol. in propylene glycol (Brij 98, 99)
Sol. in trichlorethylene (Volpo O20)
Sol. in water (Ameroxol OE-20; Brij 98, 99; Emulphor ON-870; Hetoxol OA-20 Special, OL-20; Lipal 20 OA; Macol OA-20; Volpo 20, O20); sol. in cold water (Empilan KL20)
Sol. in xylene (Emulphor ON-870)

Ionic Nature:
Nonionic (Ahco 3998; Ameroxol OE-20; Brij 98, 99; Emulphor ON-870; Lipal 20 OA; Macol OA-20; Standamul O20; Volpo 20, N20, O20)

Sp.gr.:
1.04 (Emulphor ON-870)

Density:
0.9 g/cm^3 (Empilan KL20)

M.P.:
30 C (Macol OA-20)
42 C (Empilan KL20)

Pour Pt.:
30 C (Brij 98)
33 C (Brij 99)
46 C (Emulphor ON-870)

Solidification Pt.:
43 C (Emulphor ON-870)

Flash Pt.:
> 93 C (PMCC) (Emulphor ON-870)
> 300 F (Brij 98, 99)

Fire Pt.:
> 300 F (Brij 98, 99)

Cloud Pt.:
67–72 C (1% sol'n.) (Lipal 20 OA)
78 C (brine) (Volpo O20)
87–93 C (Ameroxol OE-20)
> 100 C (Emulphor ON-870)

HLB:
15.3 (Ahco 3998; Brij 98, 99; Hetoxol OA-20 Special, OL-20; Lipocol O-20; Macol OA-20; Simulsol 98)

POE (20) oleyl ether *(cont'd.)*

    15.4 (Volpo 20)
    15.4 ± 1 (Lipal 20 OA)
    15.5 (OL-55-F-20; Volpo O20)
    17.0 (Nikkol BO-20TX)
Acid No.:
    1.0 max. (Volpo O20)
    2.0 max. (Hetoxol OA-20 Special, OL-20; Volpo 20)
Iodine No.:
    15–20 (Volpo O20)
    18–25 (Hetoxol OA-20 Special; Volpo 20)
Hydroxyl No.:
    47–62 (Lipal 20 OA)
    50–58 (Hetoxol OA-20 Special, OL-20; Volpo 20)
    50–60 (Volpo O20)
    50–62 (Chemal OA-20, OA-20/70)
    50–65 (Macol OA-20)
Stability:
    Stable to acids and alkalis (Empilan KL20)
    Stable to hydrolysis by strong acids and alkalis (Macol OA-20)
    Stable over wide pH range (Lipal 20 OA)
    Stable in sol'ns. of metallic ions and to many acids and alkalis (Volpo 20)
    Stable to alkalis and strong mineral acids (Volpo O20)
    Stable to many alkalis and acids under extreme pH conditions (Standamul O20)
pH:
    5.0–7.0 (3% aq. sol'n.) (Volpo 20)
    6.0–7.5 (3%) (Volpo O20)
    8.0–10.0 (Empilan KL20)
Surface Tension:
    37 dynes/cm (0.1% conc.) (Emulphor ON-870)
    41 dynes/cm (0.1% aq.) (Volpo O20)
Biodegradable: (Volpo O20)
TOXICITY/HANDLING:
    Probable eye irritant (Emulphor ON-870)
    Risk of skin and eye irritation occurs in conc. in excess of 3% (Volpo 20)

POE (4) oleyl ether phosphate

SYNONYMS:
    Oleth-4 phosphate (CTFA)
    PEG-4 oleyl ether phosphate
    PEG 200 oleyl ether phosphate

POE (4) oleyl ether phosphate *(cont'd.)*

CAS No.:
39464-69-2 (generic)
RD No.: 977065-25-0
TRADENAME EQUIVALENTS:
Chemfac PB-184 [Chemax]
CATEGORY:
Detergent, wetting agent, emulsifier, antistatic agent, coupling agent
APPLICATIONS:
Household detergents: (Chemfac PB-184)
PROPERTIES:
Form:
Liquid (Chemfac PB-184)
Composition:
99% active min. (Chemfac PB-184)
Solubility:
Excellent solubility in alkali sol'ns. and other electrolytes (Chemfac PB-184)
Ionic Nature: Anionic

POE (25) propylene glycol monostearate

SYNONYMS:
PEG-25 propylene glycol stearate (CTFA)
STRUCTURE:

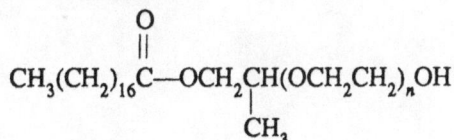

where avg. $n = 25$

CAS No.:
RD No.: 977055-29-0
TRADENAME EQUIVALENTS:
Atlas G-2162 [ICI United States]
CATEGORY:
Surfactant
PROPERTIES:
Form:
Semisolid (Atlas G-2162)
Color:
Cream (Atlas G-2162)

POE (25) propylene glycol monostearate *(cont'd.)*

Composition:
  100% active (Atlas G-2162)
Solubility:
  Sol. in ethanol (Atlas G-2162)
  Sol. in methanol (Atlas G-2162)
  Sol. in water (Atlas G-2162)
Ionic Nature: Nonionic
Flash Pt.:
  > 300 F (Atlas G-2162)
Fire Pt.:
  > 300 F (Atlas G-2162)
HLB:
  16 (Atlas G-2162)

POE (4) sorbitan monolaurate

SYNONYMS:
  Polysorbate 21 (CTFA)
STRUCTURE:

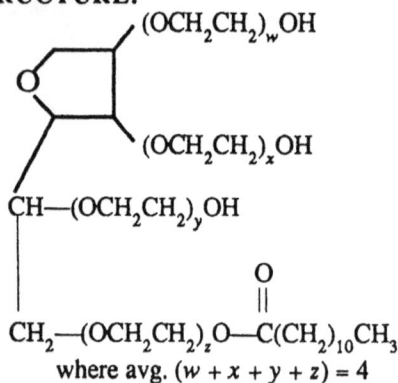

where avg. $(w + x + y + z) = 4$

CAS No.:
  9005-64-5 (generic)
  RD No.: 977053-95-4
TRADENAME EQUIVALENTS:
  Ahco 7596D [ICI United States]
  Alkamuls PSML-4 [Alkaril]
  Crillet 11 [Croda Chem.]
  Emsorb 6916 [Emery]
  Hetsorb L-4 [Heterene]

POE (4) sorbitan monolaurate (cont'd.)

TRADENAME EQUIVALENTS *(cont'd.):*
ML-55-F-4 [Hefti Ltd.]
Tween 21 [ICI United States]
CATEGORY:
Emulsifier, coemulsifier, solubilizer, coupling agent, dispersant, antistat, lubricant, leveling agent
APPLICATIONS:
Cosmetic industry preparations: (Crillet 11); creams and lotions (Crillet 11; ML-55-F-4); perfumery (Crillet 11)
Farm products: animal feed (Crillet 11)
Industrial applications: dyes and pigments (Ahco 7596D; ML-55-F-4); hydraulic fluids (ML-55-F-4); plastics (Alkamuls PSML-4; Emsorb 6916); polishes and waxes (ML-55-F-4); polymers/polymerization (Alkamuls PSML-4); textile/leather processing (ML-55-F-4; Tween 21)
Pharmaceutical applications: (Crillet 11); ointments (ML-55-F-4)
PROPERTIES:
Form:
Liquid (Ahco 7596D; Alkamuls PSML-4; Emsorb 6916; Hetsorb L-4; ML-55-F-4)
Clear liquid (Crillet 11)
Oily liquid (Tween 21)
Color:
Yellow (Alkamuls PSML-4; Crillet 11; Tween 21)
Gardner 7 (Emsorb 6916)
Gardner 8 max. (Hetsorb L-4)
Composition:
100% active (Tween 21)
100% conc. (Ahco 7596D; ML-55-F-4)
Solubility:
Sol. in aniline (Tween 21)
Disp. in aromatic solvent (@ 10%) (Alkamuls PSML-4)
Disp. in butyl stearate (Emsorb 6916)
Sol. in carbon tetrachloride (Tween 21)
Sol. in Cellosolve (Tween 21)
Sol. in corn oil (Tween 21)
Sol. in cottonseed oil at high concs. (Tween 21)
Sol. in diethyl ether at high concs. (Tween 21)
Sol. in dioxane (Tween 21)
Sol. in ethanol (Crillet 11; Tween 21)
Sol. in ethyl acetate at high concs. (Tween 21)
Sol. in ethylene glycol at high concs. (Tween 21)
Sol. in isopropanol (Hetsorb L-4)
Partly sol. in isopropyl myristate (Crillet 11)
Sol. in kerosene (Crillet 11)

POE (4) sorbitan monolaurate *(cont'd.)*

Sol. in methanol (Tween 21)
Sol. in min. oil at high concs. (Tween 21); disp. in min. oil (Hetsorb L-4); disp. (@ 10%)
   (Alkamuls PSML-4)
Sol. hazy in min. spirits (Tween 21); disp. (@ 10%) (Alkamuls PSML-4)
Sol. in oleic acid (Crillet 11)
Sol. in oleyl alcohol (Crillet 11)
Partly sol. in olive oil (Crillet 11)
Disp. in perchloroethylene (Emsorb 6916)
Disp. in Stoddard solvent (Emsorb 6916)
Sol. hazy in toluol (Tween 21)
Partly sol. in trichlorethylene (Crillet 11)
Partly sol. in water (Crillet 11); disp. (Emsorb 6916; Hetsorb L-4; Tween 21); gels at
   high concs. (Tween 21); disp. (@ 10%) (Alkamuls PSML-4)
Partly sol. in xylene (Crillet 11)
Ionic Nature:
Nonionic (Ahco 7596D; ML-55-F-4; Tween 21)
Sp.gr.:
1.1 (Tween 21)
Density:
1.0 g/ml (Alkamuls PSML-4)
9.0 lb/gal (Emsorb 6916)
Visc.:
500 cps (Tween 21)
700 cs (Emsorb 6916)
Flash Pt.:
> 300 F (Tween 21)
Fire Pt.:
> 300 F (Tween 21)
Cloud Pt.:
< 25 C (Emsorb 6916)
HLB:
12.0 (ML-55-F-4)
12.1 (Emsorb 6916)
13.3 (Ahco 7596D; Alkamuls PSML-4; Crillet 11; Tween 21)
Acid No.:
3.0 max. (Hetsorb L-4; Tween 21)
Saponification No.:
100–115 (Alkamuls PSML-4; Crillet 11; Hetsorb L-4; Tween 21)
Hydroxyl No.:
215–255 (Alkamuls PSML-4)
225–255 (Hetsorb L-4; Tween 21)
Surface Tension:
34.7 dynes/cm (0.1%) (Crillet 11)

SYNONYMS:
Polysorbate 20 (CTFA)
Sorbimacrogol laurate 300
Sorbitan, monododecanoate, poly (oxy-1-2-ethanediyl) derivatives
Sorbitan POE (20) monolaurate

STRUCTURE:

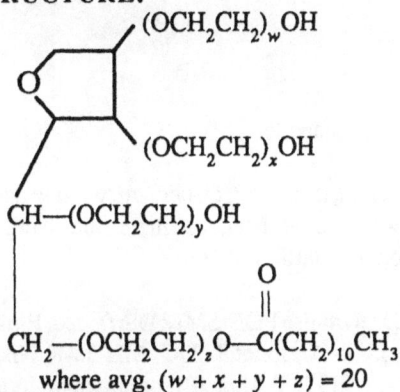

$(OCH_2CH_2)_w OH$

$(OCH_2CH_2)_x OH$

$CH—(OCH_2CH_2)_y OH$

$$CH_2—(OCH_2CH_2)_z O—\overset{\overset{\displaystyle O}{\|}}{C}(CH_2)_{10}CH_3$$

where avg. $(w + x + y + z) = 20$

CAS No.:
9005-64-5 (generic)

TRADENAME EQUIVALENTS:
Accosperse 20 [Armstrong]
Ahco 7596T [ICI United States]
Alkamuls PSML-20 [Alkaril]
Armotan PML20 [Akzo Chemie]
Crillet 1 [Croda]
Drewmulse POE-SML [PVO Int'l.]
Durfax 20 [Durkee Foods]
Emasol L-120 [Kao Corp.]
Emsorb 6915 [Emery]
Glycosperse L-20 [Glyco]
Glycosperse L-20X [Glyco] (anhyd.)
Hetsorb L-20 [Heterene]
Hodag PSML-20 [Hodag]
Industrol L-20-S [BASF Wyandotte]
Ionet T-20C [Sanyo]
Liposorb L-20 [Lipo]
Lonzest SML-20 [Lonza]
ML-55-F [Hefti Ltd.]
Montanox 20 [Seppic]
Nikkol TL-10 [Nikko]
Nissan Nonion LT-221 [Nippon Oil & Fats]

POE (20) sorbitan monolaurate (cont'd.)

TRADENAME EQUIVALENTS *(cont'd.):*
Radiamuls 137 [Oleofina S.A.]
Radiamuls SORB 2137 [Oleofina S.A.]
Radiasurf 7137 [Oleofina S.A.]
Sorbax PML-20 [Chemax]
Sorbon T-20 [Toho Chem. Industry Co.]
Sorgen TW20 [Dai-ichi Kogyo Seiyaku Co.]
T-Maz 20 [Mazer]
Tween 20 [ICI United States]
Tween 20SD [Atlas Chem. Ind. N.V.] (specially deodorized)

CATEGORY:
Emulsifier, solubilizer, leveling agent, surfactant, detergent, dispersant, wetting agent, detergent, viscosity control agent, bodying aid, antistat, stabilizer, lubricant, descouring aid, anticorrosive agent, antifog aid, antifoaming agent

APPLICATIONS:
Cosmetic industry preparations: (Crillet 1; Drewmulse POE-SML; Durfax 20; Emasol L-120; Glycosperse L-20, L-20X; Hetsorb L-20; Ionet T-20C; ML-55-F; Nikkol TL-10; Nissan Nonion LT-221; Radiasurf 7137; T-Maz 20); creams and lotions (Crillet 1; Drewmulse POE-SML; Ionet T-20C); perfumery (Crillet 1; Drewmulse POE-SML; ML-55-F); shampoos (Drewmulse POE-SML; Durfax 20); shaving preparations (Drewmulse POE-SML); toiletries (Durfax 20)
Farm products: animal feed (Crillet 1); insecticides/pesticides (Ionet T-20C; Radiasurf 7137)
Food applications: (Nissan Nonion LT-221; Sorgen TW20; T-Maz 20); food emulsifying (Glycosperse L-20, L-20X; Radiamuls 137, SORB 2137)
Household detergents: (Durfax 20; Radiasurf 7137)
Industrial applications: construction (Radiasurf 7137); dyes and pigments (Ahco 7596T; Ionet T-20C; Radiasurf 7137); lubricating/cutting oils (Hetsorb L-20; Radiasurf 7137); metalworking (Ionet T-20C; T-Maz 20); paper mfg. (Radiasurf 7137); plastics (Radiasurf 7137); polymers/polymerization (Emasol L-120); textile/leather processing (Durfax 20; Ionet T-20C; Radiasurf 7137; T-Maz 20; Tween 20, 20SD); waxes and oils (Hetsorb L-20)
Industrial cleaners: drycleaning compositions (Radiasurf 7137); metal processing surfactants (Radiasurf 7137)
Pharmaceutical applications: (Crillet 1; Drewmulse POE-SML; Emasol L-120; Glycosperse L-20, L-20X; ML-55-F; Nikkol TL-10; Nissan Nonion LT-221; Radiasurf 7137; T-Maz 20); disposable diapers (Durfax 20); germicides (Drewmulse POE-SML); ointments (Drewmulse POE-SML); vitamins (ML-55-F; Radiamuls SORB 2137)

PROPERTIES:
Form:
Liquid (Accosperse 20; Ahco 7596T; Alkamuls PSML-20; Armotan PML20; Drewmulse POE-SML; Emasol L-120; Glycosperse L-20, L-20X; Hetsorb L-20;

POE (20) sorbitan monolaurate *(cont'd.)*

Hodag PSML-20; Industrol L-20-S; Ionet T-20C; Lonzest SML-20; ML-55-F; Montanox 20; Nikkol TL-10; Radiamuls 137, SORB 2137; Radiasurf 7137; Sorbax PML-20; Sorbon T-20; Sorgen TW20; T-Maz 20; Tween 20, 20SD)
Clear liquid (Crillet 1)
Oily liquid (Nissan Nonion LT-221)
Solid (Emsorb 6915)
Color:
  Amber (Radiasurf 7137)
  Pale yellow (Tween 20)
  Yellow (Alkamuls PSML-20; Crillet 1; Glycosperse L-20, L-20X; Ionet T-20C; T-Maz 20)
  Gardner 6 max. (Nissan Nonion LT-221)
Odor:
  Low (Crillet 1)
Composition:
  97% active (Alkamuls PSML-20)
  97% active min. (T-Maz 20)
  100% active (Ionet T-20C; Tween 20, 20SD)
  100% conc. (Accosperse 20; Ahco 7596T; Armotan PML20; Emasol L-120; Emsorb 6915; Glycosperse L-20, L-20X; Hodag PSML-20; Ionet T-20C; Lonzest SML-20; ML-55-F; Montanox 20; Nikkol TL-10; Nissan Nonion LT-221; Sorbax PML-20; Sorbon T-20; Sorgen TW20)
Solubility:
  Sol. in acetone (Glycosperse L-20, L-20X; Nissan Nonion LT-221; T-Maz 20)
  Sol. in benzene (@ 10%) (Radiasurf 7137)
  Sol. in cottonseed oil (Tween 20)
  Sol. in ethanol (Crillet 1; Glycosperse L-20, L-20X; Nissan Nonion LT-221; T-Maz 20; Tween 20)
  Sol. in ethyl acetate (Glycosperse L-20, L-20X)
  Sol. in ethyl ether (Nissan Nonion LT-221)
  Sol. in ethylene glycol (Nissan Nonion LT-221; Tween 20)
  Sol. in isopropanol (Hetsorb L-20; Tween 20); (@ 10%) (Radiasurf 7137)
  Sol. in methanol (Glycosperse L-20, L-20X; Nissan Nonion LT-221; Tween 20)
  Sol. cloudy in min. oil (@ 10%) (Radiasurf 7137)
  Sol. in oleic acid (Crillet 1)
  Sol. in oleyl alcohol (Crillet 1)
  Sol. in propylene glycol (Tween 20)
  Sol. in tetrachloromethane (Nissan Nonion LT-221)
  Misc. with toluol in certain proportions (Glycosperse L-20, L-20X; T-Maz 20)
  Sol. in trichlorethylene (@ 10%) (Radiasurf 7137)
  Sol. cloudy in veg. oils (@ 10%) (Radiasurf 7137); misc. in certain proportions (Glycosperse L-20, L-20X; T-Maz 20)
  Sol. in water (Crillet 1; Glycosperse L-20, L-20X; Hetsorb L-20; Industrol L-20-S;

POE (20) sorbitan monolaurate (cont'd.)

Nissan Nonion LT-221; Sorgen TW20; T-Maz 20; Tween 20); sol. (@ 10%) (Radia-surf 7137)

Sol. in xylene (Nissan Nonion LT-221)

Ionic Nature:

Nonionic (Accosperse 20; Ahco 7596T; Armotan PML20; Drewmulse POE-SML; Durfax 20; Glycosperse L-20, L-20X; Hodag PSML-20; Industrol L-20-S; Ionet T-20C; Lonzest SML-20; ML-55-F; Montanox 20; Nikkol TL-10; Nissan Nonion LT-221; Radiamuls 137; Radiasurf 7137; Sorbax PML-20; Sorbon T-20; Sorgen TW20; T-Maz 20; Tween 20)

M.W.:

1340 avg. (Radiasurf 7137)

Sp.gr.:

1.095 (37.8 C) (Radiasurf 7137)

1.1 (Glycosperse L-20, L-20X; T-Maz 20; Tween 20)

1.13 (Industrol L-20-S)

Visc.:

172.3 cps (37.8 C) (Radiasurf 7137)

400 cps (Glycosperse L-20, L-20X; Industrol L-20-S; T-Maz 20; Tween 20)

Solidification Pt.:

0 C max. (Nissan Nonion LT-221)

Flash Pt.:

248 C (COC) (Radiasurf 7137)

> 300 F (Tween 20)

Fire Pt.:

> 300 F (Tween 20)

Cloud Pt.:

−10 C (Radiasurf 7137)

HLB:

15.7 (Emsorb 6915)

16.0 (Radiamuls 137; Radiasurf 7137)

16.4 (Radiamuls SORB 2137)

16.5 (Durfax 20; ML-55-F)

16.7 (Ahco 7596T; Alkamuls PSML-20; Crillet 1; Drewmulse POE-SML; Emasol L-120; Glycosperse L-20, L-20X; Hetsorb L-20; Hodag PSML-20; Industrol L-20-S; Ionet T-20C; Montanox 20; Nissan Nonion LT-221; Sorbax PML-20; Sorbon T-20; Sorgen TW20; T-Maz 20; Tween 20)

16.7 ± 1 (Ionet T-20C)

16.9 (Nikkol TL-10)

Acid No.:

2.0 max. (Glycosperse L-20, L-20X; Industrol L-20-S; Radiamuls 137; Radiasurf 7137; T-Maz 20; Tween 20)

Iodine No.:

5.0 max. (Radiamuls 137; Radiasurf 7137)

POE (20) sorbitan monolaurate *(cont'd.)*

Saponification No.:
   39–52 (Durfax 20; Glycosperse L-20, L-20X)
   40–50 (Hetsorb L-20; Industrol L-20-S; Ionet T-20C; T-Maz 20; Tween 20)
   40–51 (Crillet 1; Drewmulse POE-SML)
   40–55 (Radiasurf 7137)
Hydroxyl No.:
   95–108 (Hetsorb L-20)
   96–108 (Industrol L-20-S; Ionet T-20C; T-Maz 20; Tween 20)
   96–114 (Durfax 20; Glycosperse L-20)
   100–117 (Glycosperse L-20X)
Ref. Index:
   1.4712 (Radiasurf 7137)
Surface Tension:
   38.5 dynes/cm (0.1% sol'n.) (Crillet 1)
Biodegradable: (Accosperse 20)
STD. PKGS.:
   190-kg net bung drums (Radiamuls 137, SORB 2137)
   190-kg net bung drums or bulk (Radiasurf 7137)
   55-gal (470 lb net) steel drum (Industrol L-20-S)

POE (5) sorbitan monooleate

SYNONYMS:
   Polysorbate 81 (CTFA)
STRUCTURE:

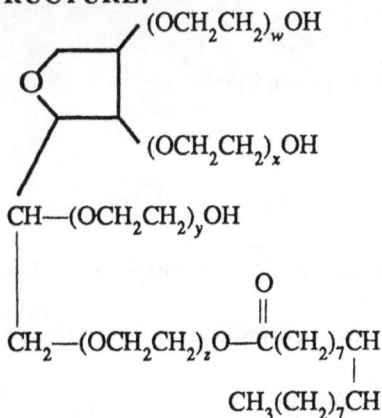

where avg. $(w + x + y + z) = 5$

POE (5) sorbitan monooleate (cont'd.)

CAS No.:
9005-65-6 (generic)
RD No.: 977053-93-2
TRADENAME EQUIVALENTS:
Alkamuls PSMO-5 [Alkaril]
Glycosperse O-5 [Glyco]
Liposorb O-5 [Lipo]
Lonzest SMO [Lonza]
Montanox 81 [Seppic]
Sorbax PMO-5 [Chemax]
T-Maz 81 [Mazer]
Tween 81 [ICI United States]
CATEGORY:
Emulsifier, surfactant, solubilizer, lubricant, antistat, corrosion inhibitor
APPLICATIONS:
Cosmetic industry preparations: (Alkamuls PSMO-5; Glycosperse O-5)
Farm products: herbicides (Alkamuls PSMO-5); insecticides/pesticides (Alkamuls PSMO-5)
Food applications: (Alkamuls PSMO-5; Glycosperse O-5; Lonzest SMO)
Industrial applications: paint mfg. (Alkamuls PSMO-5); textile/leather processing (Lonzest SMO; Tween 81)
Pharmaceutical applications: (Glycosperse O-5)
PROPERTIES:
Form:
Liquid (Alkamuls PSMO-5; Liposorb O-5; Lonzest SMO; Montanox 81; Sorbax PMO-5; T-Maz 81)
Liquid, gels on standing (Glycosperse O-5)
Oily liquid, may gel on standing (Tween 81)
Color:
Amber (Alkamuls PSMO-5; Glycosperse O-5; Tween 81)
Odor:
Typical (Alkamuls PSMO-5)
Composition:
98.5% active (Alkamuls PSMO-5)
100% active (Tween 81)
100% conc. (Glycosperse O-5; Liposorb O-5; Lonzest SMO; Montanox 81; T-Maz 81)
Solubility:
Sol. in acetone at certain concs. (Tween 81); misc. in certain proportions (Glycosperse O-5)
Disp. in 5% sol'n. of aluminum chloride (Tween 81)
Sol. in aniline (Tween 81)
Sol. in carbon tetrachloride at certain concs. (Tween 81)
Sol. in Cellosolve (Tween 81)

Sol. in corn oil (Tween 81)
Sol. in cottonseed oil at certain concs. (Tween 81)
Sol. in dioxane (Tween 81)
Sol. in ethanol (Glycosperse O-5; Tween 81)
Disp. in ether (Tween 81)
Sol. in ethyl acetate (Glycosperse O-5; Tween 81)
Disp. in ethylene glycol (Tween 81)
Sol. in methanol (Glycosperse O-5; Tween 81)
Sol. in min. oil (Glycosperse O-5; Tween 81)
Sol. in min. spirits at certain concs. (Tween 81)
Misc. with naphtha in certain proportions (Glycosperse O-5)
Disp. in 5% sol'n. of sodium hydroxide (Tween 81)
Disp. in 5% sol'n. of sodium sulfate (Tween 81)
Disp. in 5% sol'n. of sulfuric acid (Tween 81)
Sol. in toluol at certain concs. (Tween 81); misc. in certain proportions (Glycosperse O-5)
Disp. in water (Glycosperse O-5; Tween 81); insol. (Alkamuls PSMO-5)

Ionic Nature:
Nonionic (Alkamuls PSMO-5; Glycosperse O-5; Liposorb O-5; Lonzest SMO; Montanox 81; T-Maz 81; Tween 81)

Sp.gr.:
1.0 (Glycosperse O-5; Tween 81)

Visc.:
450 cps (Glycosperse O-5; Tween 81)

Flash Pt.:
> 300 F (Tween 81)

Fire Pt.:
> 300 F (Tween 81)

HLB:
10.0 (Alkamuls PSMO-5; Glycosperse O-5; Liposorb O-5; Montanox 81; Sorbax PMO-5; T-Maz 81; Tween 81)

Acid No.:
2.0 max. (Glycosperse O-5; Tween 81)

Saponification No.:
95–105 (Glycosperse O-5)
96–104 (Tween 81)

Hydroxyl No.:
134–150 (Tween 81)
136–152 (Glycosperse O-5)

POE (20) sorbitan monooleate

SYNONYMS:

Polysorbate 80 (CTFA)

Sorbitan, mono-9-octadecenoate, poly (oxy-1,2-ethanediyl) derivatives

Sorbimacrogol oleate 300

STRUCTURE:

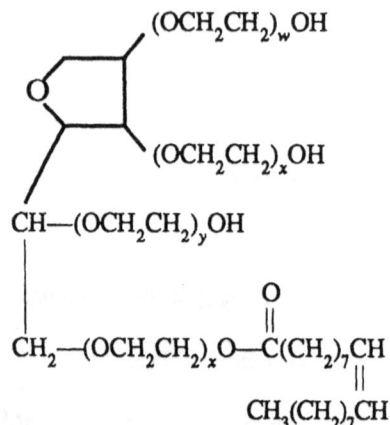

where avg. $(w + x + y + z) = 20$

CAS No.:

9005-65-6 (generic)

TRADENAME EQUIVALENTS:

Accosperse 80 [Armstrong]

Ahco DFO-150 [ICI United States]

Alkamuls PSMO-20 [Alkaril]

Amerchol Polysorbate 80 [Amerchol]

Armotan PMO-20 [Akzo]

Crillet 4 [Croda]

Drewmulse POE-SMO [PVO Int'l.]

Drewpone 80 [PVO Int'l.]

Durfax 80, 80K [Durkee]

Emasol O-120 [Kao]

Emsorb 6900 [Emery]

Glycosperse O-20 [Glyco]

Glycosperse O-20 Veg. [Glyco] (vegetable grade)

Glycosperse O-20X [Glyco] (anhyd.)

Hetsorb O-20 [Heterene]

Industrol O-20-S [BASF Wyandotte]

Liposorb O-20 [Lipo]

Lonzest SMO-20 [Lonza]

Montanox 80 [Seppic]

Nikkol TO-10 [Nikko]

POE (20) sorbitan monooleate (cont'd.)

TRADENAME EQUIVALENTS *(cont'd.):*
Sorbax PMO-20 [Chemax]
T-Maz 80 [Mazer]
Tween 80 [ICI United States] (USP grade)

CATEGORY:
Wetting agent, emulsifier, detergent, surfactant, lubricant, surfactant, antistat, solubilizer, stabilizer, dispersant, viscosity modifier, antifog agent

APPLICATIONS:
Cleansers: hand cleanser (Crillet 4)

Cosmetic industry preparations: (Alkamuls PSMO-20; Amerchol Polysorbate 80; Crillet 4; Drewmulse POE-SMO; Emsorb 6900; Glycosperse O-20, O-20 VEG, O-20X; Hetsorb O-20; Nikkol TO-10; T-Maz 80); creams and lotions (Drewmulse POE-SMO); hair preparations (Durfax 80); personal care products (Lonzest SMO-20); perfumery (Drewmulse POE-SMO; Emsorb 6900); shampoos (Drewmulse POE-SMO; Durfax 80); shaving preparations (Drewmulse POE-SMO; Durfax 80); toiletries (Amerchol Polysorbate 80)

Farm products: animal feed (Crillet 4); herbicides (Alkamuls PSMO-20; Crillet 4); insecticides/pesticides (Alkamuls PSMO-20; Crillet 4; Emsorb 6900); veterinary products (Crillet 4)

Food applications: (Alkamuls PSMO-20; Lonzest SMO-20; T-Maz 80); flavors (Drewmulse POE-SMO; Drewpone 80); food emulsifying (Drewpone 80; Durfax 80, 80K; Glycosperse O-20, O-20 VEG, O-20X; Nikkol TO-10)

Household products: (Emsorb 6900)

Industrial applications: (Lonzest SMO-20); coatings (Emsorb 6900); dyes and pigments (Emsorb 6900); lubricating/cutting oils (Emsorb 6900; Hetsorb O-20); metalworking (T-Maz 80); paint mfg. (Alkamuls PSMO-20); petroleum industry (Crillet 4); plastics (Durfax 80); polishes (Crillet 4; Durfax 80); silicone products (Crillet 4); textile/leather processing (Ahco DFO-150; Alkamuls PSMO-20; Crillet 4; Emsorb 6900; T-Maz 80; Tween 80); waxes (Crillet 4; Emsorb 6900; Hetsorb O-20)

Industrial cleaners: marine tank cleaners (Crillet 4)

Pharmaceutical applications: (Amerchol Polysorbate 80; Drewmulse POE-SMO; Glycosperse O-20, O-20 VEG, O-20X; Nikkol TO-10; Crillet 4); drug industry (T-Maz 80); germicides (Drewmulse POE-SMO); ointments (Drewmulse POE-SMO); vitamins (Drewmulse POE-SMO; Drewpone 80)

PROPERTIES:
Form:
Liquid (Accosperse 80; Ahco DFO-150; Alkamuls PSMO-20; Amerchol Polysorbate 80; Armotan PMO 20; Drewmulse POE-SMO; Drewpone 80; Durfax 80, 80K; Emasol O-120; Emsorb 6900; Glycosperse O-20, O-20 VEG, O-20X; Industrol O-20-S; Liposorb O-20; Lonzest SMO-20; Montanox 80; Nikkol TO-10; Sorbax PMO-20; T-Maz 80; Tween 80)
Clear liquid (Crillet 4)

POE (20) sorbitan monooleate *(cont'd.)*

Color:
Pale yellow (Tween 80)
Yellow (Alkamuls PSMO-20; Drewpone 80; Durfax 80K; Glycosperse O-20, O-20 VEG, O-20X; Liposorb O-20; T-Maz 80)
Yellow amber (Crillet 4)
Amber (Accosperse 80)
Gardner 5 (Emsorb 6900)

Odor:
Faint characteristic fatty (Crillet 4)
Typical (Alkamuls PSMO-20)

Composition:
65–69.5% ETO (Drewpone 80)
97% active (Alkamuls PSMO-20)
97% active min. (T-Maz 80)
98% active (Accosperse 80)
100% active (Amerchol Polysorbate 80; Durfax 80K; Liposorb O-20; Tween 80)
100% conc. (Ahco DFO-150; Armotan PMO 20; Emasol O-120; Emsorb 6900; Glycosperse O-20, O-20 VEG, O-20X; Lonzest SMO-20; Montanox 80; Nikkol TO-10)

Solubility:
Partly sol. in acetone (Glycosperse O-20, O-20 VEG, O-20X); poorly sol. (T-Maz 80)
Sol. in alcohol (Amerchol Polysorbate 80)
Sol. in cottonseed oil at low levels (Durfax 80K)
Sol. in ethanol (Crillet 4; Glycosperse O-20, O-20 VEG, O-20X; T-Maz 80; Tween 80)
Sol. in ethyl acetate (Glycosperse O-20, O-20 VEG, O-20X)
Disp. in glycerol trioleate (Emsorb 6900)
Sol. in isopropanol (Tween 80)
Partly sol. in isopropyl myristate (Crillet 4)
Sol. in methanol (Glycosperse O-20, O-20 VEG, O-20X)
Sol. in oleic acid (Crillet 4)
Sol. in oleyl alcohol (Crillet 4)
Sol. in toluol (Glycosperse O-20, O-20 VEG, O-20X; T-Maz 80)
Sol. in veg. oil (Glycosperse O-20, O-20 VEG, O-20X; T-Maz 80)
Sol. in water (Alkamuls PSMO-20; Amerchol Polysorbate 80; Crillet 4; Durfax 80, 80K; Emsorb 6900; Glycosperse O-20, O-20 VEG, O-20X; Industrol O-20-S; T-Maz 80; Tween 80)
Partly sol. in xylene (Crillet 4)

Ionic Nature:
Nonionic (Ahco DFO-150; Alkamuls PSMO-20; Armotan PMO 20; Drewmulse POE-SMO; Durfax 80, 80K; Emasol O-120; Emsorb 6900; Glycosperse O-20, O-20 VEG, O-20X; Industrol O-20-S; Liposorb O-20; Lonzest SMO-20; Montanox 80; Nikkol TO-10; T-Maz 80)

POE (20) sorbitan monooleate *(cont'd.)*

Sp.gr.:
1.0 (Glycosperse O-20, O-20 VEG, O-20X; T-Maz 80)
1.08 (Industrol O-20-S; Tween 80)
1.1 (Accosperse 80)

Density:
9.0 lb/gal (Emsorb 6900)

Visc.:
400 cps (Emsorb 6900; Glycosperse O-20, O-20 VEG; Industrol O-20-S; T-Maz 80)
400–450 (Accosperse 80)
425 cps (Tween 80)
600 cps (Glycosperse O-20X)

Flash Pt.:
> 300 F (Tween 80)

Fire Pt.:
> 300 F (Tween 80)

Cloud Pt.:
> 100 C (Emsorb 6900)

HLB:
15.0 (Ahco DFO-150; Amerchol Polysorbate 80; Crillet 4; Drewmulse POE-SMO; Drewpone 80; Emasol O-120; Glycosperse O-20; Industrol O-20-S; Montanox 80; Nikkol TO-10; Sorbax PMO-20; T-Maz 80; Tween 80)
15.0 ± 1 (Liposorb O-20)
15.1 (Emsorb 6900)
15.9 (Durfax 80, 80K)

Acid No.:
1.0 max. (Industrol O-20-S)
2.0 max. (Accosperse 80; Glycosperse O-20, O-20 VEG, O-20X; T-Maz 80; Tween 80)

Saponification No.:
44–56 (Glycosperse O-20, O-20 VEG, O-20X)
45–55 (Crillet 4; Drewmulse POE-SMO; Drewpone 80; Durfax 80; Industrol O-20-S; Liposorb O-20; T-Maz 80; Tween 80)

Hydroxyl No.:
64–81 (Glycosperse O-20, O-20 VEG, O-20X)
65–80 (Durfax 80; Industrol O-20-S; Liposorb O-20; T-Maz 80; Tween 80)
65–81 (Accosperse 80)

Surface Tension:
42.5 dynes/cm (0.1%) (Crillet 4)

TOXICITY/HANDLING:
Low toxicity (Lonzest SMO-20)

POE (20) sorbitan monooleate *(cont'd.)*

STD. PKGS.:
  55-gal lined bung drums (Accosperse 80)
  55-gal (450 lb net) steel drums (Industrol O-20-S)
  470 lb packaging, bulk (Durfax 80K)

POE (20) sorbitan monopalmitate

SYNONYMS:
  Polysorbate 40 (CTFA)
  Sorbimacrogol palmitate 300
  Sorbitan, monohexadecanoate, poly (oxy-1,2-ethanediyl) derivatives
  Sorbitan POE (20) monopalmitate

STRUCTURE:

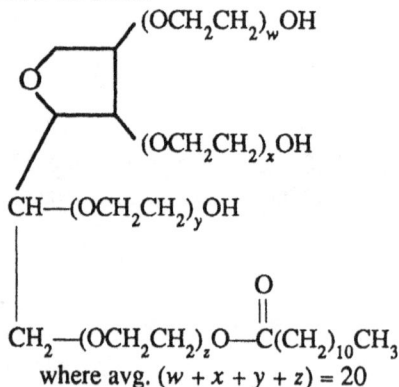

where avg. $(w + x + y + z) = 20$

CAS No.:
  9005-66-7 (generic)

TRADENAME EQUIVALENTS:
  Ahco DFP-156 [ICI United States]
  Crillet 2 [Croda]
  Emasol P-120 [Kao Corp.]
  Emsorb 6910 [Emery]
  Glycosperse P-20 [Glyco]
  Hetsorb P-20 [Heterene]
  Hodag PSMP-20 [Hodag]
  Liposorb P-20 [Lipo]
  Lonzest SMP-20 [Lonza]
  Montanox 40 [Seppic]
  Sorbax PMP-20 [Chemax]
  T-Maz 40 [Mazer]
  Tween 40 [ICI United States]

POE (20) sorbitan monopalmitate (cont'd.)

CATEGORY:
Emulsifier, solubilizer, stabilizer, dispersant, penetrant, leveling agent, lubricant, corrosion inhibitor, antistat, detergent, wetting agent, viscosity modifier

APPLICATIONS:
Cosmetic industry preparations: (Crillet 2; Emasol P-120; Emsorb 6910; Glycosperse P-20; Hetsorb P-20; T-Maz 40); perfumery (Sorbax PMP-20)

Farm products: (Crillet 2); herbicides (Crillet 2); insecticides/pesticides (Crillet 2)

Food applications: (Lonzest SMP-20; T-Maz 40); flavors (Sorbax PMP-20); food emulsifying (Glycosperse P-20)

Industrial applications: construction (Emsorb 6910); lubricating/cutting oils (Hetsorb P-20); metalworking (T-Maz 40); paper mfg. (Emsorb 6910); textile/leather processing (Ahco DFP-156; Crillet 2; Emsorb 6910; Lonzest SMP-20; T-Maz 40; Tween 40); waxes (Crillet 2; Emsorb 6910; Hetsorb P-20)

Pharmaceutical applications: (Crillet 2; Emasol P-120; Glycosperse P-20; T-Maz 40)

PROPERTIES:

Form:
Liquid (Ahco DFP-156; Emasol P-120; Glycosperse P-20; Hetsorb P-20; Hodag PSMP-20; Liposorb P-20; Sorbax PMP-20; T-Maz 40)

Liquid (may gel on standing) (Tween 40)

Pasty liquid (Crillet 2)

Viscous liquid (Emsorb 6910)

Gel (Montanox 40)

Solid (Lonzest SMP-20)

Color:
Yellow amber (Crillet 2)

Pale yellow (Tween 40)

Yellow (Glycosperse P-20; T-Maz 40)

Gardner 6 (Emsorb 6910)

Composition:
97% active min. (T-Maz 40)

100% active (Liposorb P-20; Tween 40)

100% conc. (Ahco DFP-156; Emasol P-120; Emsorb 6910; Glycosperse P-20; Hodag PSMP-20; Lonzest SMP-20; Montanox 40)

Solubility:
Sol. in acetone (Glycosperse P-20; T-Maz 40)

Sol. in cottonseed oil (Tween 40)

Sol. in ethanol (Crillet 2; Glycosperse P-20; T-Maz 40; Tween 40)

Sol. in ethyl acetate (Glycosperse P-20)

Sol. in ethylene glycol (Tween 40)

Sol. in glycerol trioleate (Emsorb 6910)

Sol. in isopropanol (Hetsorb P-20; Tween 40)

Sol. in methanol (Glycosperse P-20; Tween 40)

Misc. with min. oil in certain proportions (Hetsorb P-20)

POE (20) sorbitan monopalmitate *(cont'd.)*

    Misc. with naphtha in certain proportions (Glycosperse P-20; T-Maz 40)
    Sol. in oleic acid (Crillet 2)
    Sol. in oleyl alcohol (Crillet 2)
    Disp. in perchloroethylene (Emsorb 6910)
    Misc. with toluol in certain proportions (Glycosperse P-20; T-Maz 40)
    Misc. with veg. oil in certain proportions (Glycosperse P-20; T-Maz 40)
    Sol. in water (Crillet 2; Emsorb 6910; Glycosperse P-20; Hetsorb P-20; T-Maz 40; Tween 40)

Ionic Nature:
    Nonionic (Emsorb 6910; Glycosperse P-20; Lonzest SMP-20; T-Maz 40; Tween 40)

Sp.gr.:
    1.0 (Glycosperse P-20; T-Maz 40)
    1.08 (Tween 40)

Density:
    9.2 lb/gal (Emsorb 6910)

Visc.:
    550 cps (Glycosperse P-20)
    600 cps (T-Maz 40)
    600 cs (Emsorb 6910)

Flash Pt.:
    > 300 F (Tween 40)

Fire Pt.:
    > 300 F (Tween 40)

Cloud Pt.:
    > 100 C (Emsorb 6910)

HLB:
    15.6 (Ahco DFP-156; Crillet 2; Emasol P-120; Glycosperse P-20; Hetsorb P-20; Hodag PSMP-20; Montanox 40; Sorbax PMP-20; T-Maz 40; Tween 40)
    15.6 ± 1 (Liposorb P-20)
    15.8 (Emsorb 6910)

Acid No.:
    2.0 max. (Glycosperse P-20; T-Maz 40; Tween 40)

Saponification No.:
    40–53 (Glycosperse P-20; Liposorb P-20)
    41–52 (Tween 40)
    43–49 (Crillet 2; Hetsorb P-20; T-Maz 40)

Hydroxyl No.:
    89–105 (Hetsorb P-20; T-Maz 40)
    90–105 (Tween 40)
    90–107 (Liposorb P-20)
    91–107 (Glycosperse P-20)

Surface Tension:
    41.5 dynes/cm (0.1%) (Crillet 2)

POE (4) sorbitan monostearate

SYNONYMS:
PEG 4 sorbitan monostearate
Polysorbate 61 (CTFA)

STRUCTURE:

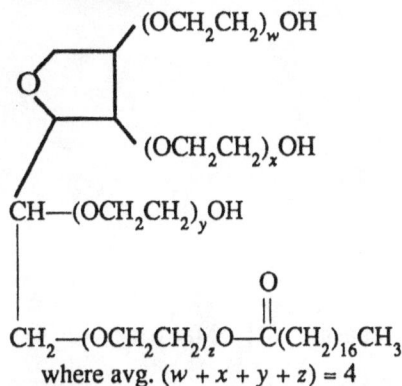

$(OCH_2CH_2)_w OH$

$(OCH_2CH_2)_x OH$

$CH-(OCH_2CH_2)_y OH$

$$CH_2-(OCH_2CH_2)_z O-\overset{\displaystyle O}{\overset{\|}{C}}(CH_2)_{16}CH_3$$

where avg. $(w + x + y + z) = 4$

CAS No.:
9005-67-8 (generic)
RD No.: 977053-94-3

TRADENAME EQUIVALENTS:
Ahco DFS-96 [ICI United States]
Alkamuls PSMS-4 [Alkaril]
Crillet 31 [Croda]
Emsorb 6906 [Emery]
Hetsorb S-4 [Heterene]
Montanox 61 [Seppic]
T-Maz 61 [Mazer]
Tween 61 [ICI United States]

CATEGORY:
Emulsifier, surfactant, solubilizer, dispersant, penetrant, leveling agent, lubricant, antistat, wetting agent, softener, stabilizer, bodying agent, opacifier, defoamer, corrosion inhibitor

APPLICATIONS:
Cosmetic industry preparations: (Alkamuls PSMS-4; Crillet 31); perfumery (Tween 61)
Farm products: (Crillet 31); herbicides (Alkamuls PSMS-4; Crillet 31); insecticides/pesticides (Alkamuls PSMS-4; Crillet 31)
Food applications: (Alkamuls PSMS-4); flavors (Tween 61)
Industrial applications: dyes and pigments (Crillet 31); paint mfg. (Alkamuls PSMS-4); polishes and waxes (Crillet 31); textile/leather processing (Ahco DFS-96; Alkamuls PSMS-4; Crillet 31)
Pharmaceutical applications: (Crillet 31); vitamins (Tween 61)

POE (4) sorbitan monostearate (cont'd.)

PROPERTIES:
Form:
 Solid (Ahco DFS-96; Alkamuls PSMS-4; Emsorb 6906; Montanox 61; T-Maz 61;
  Tween 61)
 Semisolid (Hetsorb S-4)
 Waxy solid (Crillet 31)
Color:
 Cream pale yellow (Crillet 31)
 Tan (Alkamuls PSMS-4)
 Gardner 8 max. (Hetsorb S-4)
Odor:
 Typical (Alkamuls PSMS-4)
Composition:
 97% active (Alkamuls PSMS-4)
 100% conc. (Ahco DFS-96; Emsorb 6906; Montanox 61; T-Maz 61; Tween 61)
Solubility:
 Sol. in isopropanol (Hetsorb S-4)
 Disp. in min. oil (Hetsorb S-4)
 Sol. in oleic acid (Crillet 31)
 Sol. in oleyl alcohol (Crillet 31)
 Disp. in water (Hetsorb S-4); insol. (Alkamuls PSMS-4)
Ionic Nature:
 Nonionic (Ahco DFS-96; Alkamuls PSMS-4; Emsorb 6906; Montanox 61; T-Maz 61;
  Tween 61)
HLB:
 9.6 (Ahco DFS-96; Alkamuls PSMS-4; Crillet 31; Montanox 61; T-Maz 61; Tween 61)
 15.2 (Emsorb 6906)
Acid No.:
 2.0 max. (Hetsorb S-4)
Saponification No.:
 95–115 (Hetsorb S-4)
 98–113 (Crillet 31)
Hydroxyl No.:
 170–200 (Hetsorb S-4)
Surface Tension:
 41.5 dynes/cm (0.1%) (Crillet 31)

POE (20) sorbitan monostearate

SYNONYMS:

Polysorbate 60 (CTFA)

Sorbimacrogol stearate 300

Sorbitan, monooctadecanoate, poly (oxy-1,2-ethanediyl) derivatives

Sorbitan POE (20) monostearate

STRUCTURE:

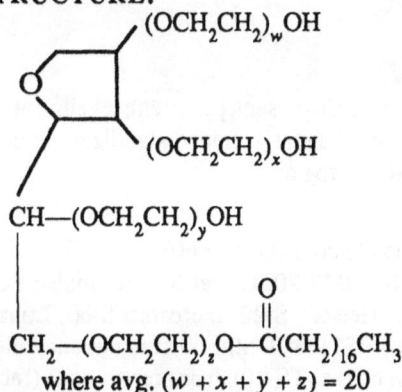

$$(OCH_2CH_2)_w OH$$

$$(OCH_2CH_2)_x OH$$

$$CH-(OCH_2CH_2)_y OH$$

$$CH_2-(OCH_2CH_2)_z O-\overset{\overset{\displaystyle O}{\|}}{C}(CH_2)_{16}CH_3$$

where avg. $(w + x + y + z) = 20$

CAS No.:

9005-67-8 (generic)

TRADENAME EQUIVALENTS:

Accosperse 60 [Armstrong]

Ahco DFS-149 [ICI United States]

Alkamuls PSMS-20 [Alkaril]

Capmul POE-S [Capital City]

Crillet 3 [Croda]

Drewmulse POE-SMS [PVO Int'l.]

Drewpone 60 [PVO Int'l.]

Durfax 60, 60K [Durkee Foods]

Emasol S-120 [Kao Corp.]

Emsorb 6905 [Emery]

Glycosperse S-20 [Glyco]

Hetsorb S-20 [Heterene]

Hodag PSMS-20 [Hodag]

Incrosorb S-60 [Croda Surfactants]

Industrol S-20-S [BASF Wyandotte]

Liposorb S-20 [Lipo]

Lonzest SMS-20 [Lonza]

MS-55-F [Hefti Ltd.]

Montanox 60 [Seppic]

Nikkol TS-10, TS-10 (FF) [Nikko]

Nissan Nonion ST-221 [Nippon Oil & Fats]

261

POE (20) sorbitan monostearate (cont'd.)

TRADENAME EQUIVALENTS *(cont'd.):*
Radiamuls 147 [Oleofina S.A.]
Radiamuls SORB 2147 [Oleofina S.A.]
Radiasurf 7147 [Oleofina S.A.]
Sorbax PMS-20 [Chemax]
T-Maz 60 [Mazer]
Tween 60 [ICI United States]

CATEGORY:
Emulsifier, solubilizer, wetting agent, surfactant, dispersant, penetrant, leveling agent, lubricant, antistat, detergent, viscosity control agent, softener, stabilizer, descouring aid, anticorrosive agent, bodying aid, antifog aid

APPLICATIONS:
Cleansers: (Incrosorb S-60); waterless hand cleanser (Durfax 60)

Cosmetic industry preparations: (Alkamuls PSMS-20; Crillet 3; Drewmulse POE-SMS; Emsorb 6905; Glycosperse S-20; Hetsorb S-20; Incrosorb S-60; Lonzest SMS-20; MS-55-F; Nikkol TS-10; Radiasurf 7147; T-Maz 60); creams and lotions (Drewmulse POE-SMS; Incrosorb S-60; Nikkol TS-10); hair preparations (Incrosorb S-60; Nikkol TS-10); makeup (Incrosorb S-60); perfumery (Drewmulse POE-SMS; Incrosorb S-60; Sorbax PMS-20); shampoos (Drewmulse POE-SMS; Durfax 60); shaving preparations (Drewmulse POE-SMS; Durfax 60; Incrosorb S-60; Nikkol TS-10); skin preparations (Incrosorb S-60); toiletries (Incrosorb S-60)

Farm products: (Crillet 3); herbicides (Alkamuls PSMS-20; Crillet 3); insecticides/ pesticides (Alkamuls PSMS-20; Crillet 3; Radiasurf 7147)

Food applications: (Alkamuls PSMS-20; Capmul POE-S; Hodag PSMS-20; Lonzest SMS-20; T-Maz 60); flavors (Drewmulse POE-SMS; Sorbax PMS-20); food emulsifying (Drewpone 60; Durfax 60, 60K; Glycosperse S-20; Radiamuls 147, SORB 2147)

Household products: (Emsorb 6905)

Industrial applications: (Hodag PSMS-20; Lonzest SMS-20; MS-55-F); construction (Emsorb 6905; Radiasurf 7147); dyes and pigments (Radiasurf 7147); lubricating/ cutting oils (Hetsorb S-20; Radiasurf 7147); metalworking (T-Maz 60); paint mfg. (Alkamuls PSMS-20; Nikkol TS-10); paper mfg. (Emsorb 6905; Radiasurf 7147); plastics (Radiasurf 7147); polishes and waxes (Crillet 3; Emsorb 6905; Hetsorb S-20; Nikkol TS-10); polymers/polymerization (MS-55-F); textile/leather processing (Ahco DFS-149; Crillet 3; Emsorb 6905; Radiasurf 7147; T-Maz 60; Tween 60)

Industrial cleaners: (Radiasurf 7147); drycleaning compositions (Radiasurf 7147); metal processing surfactants (Radiasurf 7147)

Pharmaceutical applications: (Drewmulse POE-SMS; Glycosperse S-20; MS-55-F; Nikkol TS-10; Radiasurf 7147; T-Maz 60); antiperspirant/deodorant (Nikkol TS-10); germicides (Drewmulse POE-SMS); lip pomade (Nikkol TS-10); ointments (Drewmulse POE-SMS; Nikkol TS-10); sunscreens (Nikkol TS-10); vitamins (Drewmulse POE-SMS)

POE (20) sorbitan monostearate (cont'd.)

PROPERTIES:
Form:
    Liquid (Accosperse 60; Capmul POE-S; Glycosperse S-20; Hetsorb S-20; Hodag
       PSMS-20; Industrol S-20-S; Lonzest SMS-20; Nikkol TS-10 (FF); Nissan Nonion
       ST-221; Radiamuls 147, SORB 2147; Radiasurf 7147; Sorbax PMS-20; T-Maz 60)
    Liquid to soft solid (Drewpone 60)
    Liquid gel (Incrosorb S-60)
    Liquid (may gel on standing) (Tween 60)
    Liquid gels to soft solid on cooling (Crillet 3)
    Liquid to paste (MS-55-F)
    Viscous liquid (Durfax 60K; Nikkol TS-10)
    Viscous liquid, gel on standing (Alkamuls PSMS-20)
    Gel (Accosperse 60; Montanox 60)
    Paste (Durfax 60; Liposorb S-20)
    Solid (Ahco DFS-149; Drewmulse POE-SMSEmasol S-120)
    Waxy semisolid (Emsorb 6905)
Color:
    Amber (Accosperse 60; Radiasurf 7147)
    Pale yellow (Incrosorb S-60; Tween 60)
    Pale yellow to yellow (Nikkol TS-10)
    Yellow (Alkamuls PSMS-20; Crillet 3; Drewpone 60; Durfax 60K; Glycosperse S-20;
      Liposorb S-20; T-Maz 60)
    Gardner 3 (Emsorb 6905)
Odor:
    Typical (Alkamuls PSMS-20)
Composition:
    65–69.5% ETO (Drewpone 60)
    97% active (Alkamuls PSMS-20)
    97% active min. (T-Maz 60)
    98% active (Accosperse 60)
    100% active (Durfax 60K; Liposorb S-20; Tween 60)
    100% conc. (Ahco DFS-149; Capmul POE-S; Emasol S-120; Emsorb 6905; Glyco-
      sperse S-20; Hodag PSMS-20; Lonzest SMS-20; Montanox 60; MS-55-F; Nikkol
      TS-10, TS-10(FF); Nissan Nonion ST-221)
Solubility:
    Misc. with acetone at certain proportions (Glycosperse S-20; T-Maz 60)
    Sol. cloudy in benzene (@ 10%) (Radiasurf 7147)
    Sol. in cottonseed oil (at low levels) (Durfax 60K)
    Sol. in ethanol (Crillet 3; Nikkol TS-10; Tween 60); misc. at certain proportions
      (Glycosperse S-20; T-Maz 60)
    Sol. in ethyl acetate (Glycosperse S-20; Nikkol TS-10)
    Sol. in isopropanol (Tween 60); sol. cloudy (@ 10%) (Radiasurf 7147); misc. (Hetsorb
      S-20)

POE (20) sorbitan monostearate *(cont'd.)*

Partly sol. in isopropyl myristate (Crillet 3)
Misc. with methanol in certain proportions (Glycosperse S-20)
Sol. in min. oil (@ 10%) (Radiasurf 7147)
Misc. with naphtha at certain proportions (Glycosperse S-20; T-Maz 60)
Sol. in oleic acid (Crillet 3)
Sol. in oleyl alcohol (Crillet 3)
Disp. in perchloroethylene (Emsorb 6905)
Sol. in Stoddard solvent (Emsorb 6905)
Sol. in toluene (Nikkol TS-10)
Sol. in toluol (Glycosperse S-20; T-Maz 60)
Sol. in trichlorethylene (@ 10%) (Radiasurf 7147); partly sol. in (Crillet 3)
Sol. cloudy in veg. oil (@ 10%) (Radiasurf 7147)
Sol. in water (Durfax 60, 60K; Emsorb 6905; Glycosperse S-20; Hetsorb S-20;
    Industrol S-20-S; Nikkol TS-10; T-Maz 60; Tween 60); sol. cloudy (@ 10%)
    (Radiasurf 7147); partly sol. (Crillet 3)
Partly sol. in xylene (Crillet 3)

Ionic Nature:
Nonionic (Alkamuls PSMS-20; Drewmulse POE-SMS; Durfax 60, 60K; Emsorb
    6905; Glycosperse S-20; Hodag PSMS-20; Incrosorb S-60; Industrol S-20-S;
    Liposorb S-20; MS-55-F; Nissan Nonion ST-221; Radiasurf 7147; T-Maz 60;
    Tween 60)

Sp.gr.:
1.068 (37.8 C) (Radiasurf 7147)
1.1 (Accosperse 60; Glycosperse S-20; T-Maz 60; Tween 60)
1.13 (Industrol S-20-S)

Density:
8.9 lb/gal (Emsorb 6905)

Visc.:
202.6 cps (37.8 C) (Radiasurf 7147)
500 cs (Emsorb 6905)
550 cps (T-Maz 60)
550–600 cps (Accosperse 60)
600 cps (Industrol S-20-S; Tween 60)

Pour Pt.:
23–25 C (T-Maz 60)

Flash Pt.:
260 C (COC) (Radiasurf 7147)
> 300 F (Tween 60)

Fire Pt.:
> 300 F (Tween 60)

Cloud Pt.:
21.5 C (Radiasurf 7147)
> 100 C (Emsorb 6905)

POE (20) sorbitan monostearate (cont'd.)

HLB:

14.9 (Accosperse 60; Ahco DFS-149; Capmul POE-S; Crillet 3; Drewmulse POE-SMS; Drewpone 60; Durfax 60, 60K; Emasol S-120; Glycosperse S-20; Hetsorb S-20; Industrol S-20-S; Montanox 60; Nikkol TS-10(FF); Nissan Nonion ST-221; Sorbax PMS-20; T-Maz 60; Tween 60)

14.9 ± 1 (Liposorb S-20)

15.0 (MS-55-F; Radiamuls 147; Radiasurf 7147)

15.1 (Radiamuls SORB 2147)

15.2 (Emsorb 6905)

15.6 (Hodag PSMS-20)

Acid No.:

1.5 max. (Nikkol TS-10)

2.0 max. (Accosperse 60; Glycosperse S-20; Incrosorb S-60; Industrol S-20-S; Radiamuls 147; Radiasurf 7147; T-Maz 60; Tween 60)

Iodine No.:

1.0 max. (Radiamuls 147; Radiasurf 7147)

Saponification No.:

43–49 (Nikkol TS-10)

44–56 (Glycosperse S-20; Radiasurf 7147)

45–55 (Accosperse 60; Crillet 3; Drewmulse POE-SMS; Drewpone 60; Durfax 60; Hetsorb S-20; Incrosorb S-60; Industrol S-20-S; Liposorb S-20; T-Maz 60; Tween 60)

Hydroxyl No.:

80–96 (Accosperse 60; Glycosperse S-20; Hetsorb S-20)

81–96 (Durfax 60; Incrosorb S-60; Industrol S-20-S; Liposorb S-20; T-Maz 60; Tween 60)

84–100 (Nikkol TS-10)

pH:

5.7–7.7 (5%) (Nikkol TS-10)

Surface Tension:

42.5 dynes/cm (0.1%) (Crillet 3)

TOXICITY/HANDLING:

Avoid prolonged contact with skin and eyes (Incrosorb S-60)

STORAGE/HANDLING:

Store in a cool, dry place (Incrosorb S-60)

STD. PKGS.:

15-kg net petroleum can (Nikkol TS-10)

190-kg net bung drums (Radiamuls 147, SORB 2147)

190-kg net bung drums or bulk (Radiasurf 7147)

470-lb packaging, bulk (Durfax 60K)

55-gal (425 lb net) lined drums (Incrosorb S-60)

55-gal (450 lb net) steel drums (Industrol S-20-S)

POE (20) sorbitan trioleate

SYNONYMS:
Polysorbate 85 (CTFA)
Sorbimacrogol trioleate 300
Sorbitan, tri-9-octadecenoate, poly (oxy-1,2-ethanediyl) derivatives

STRUCTURE:

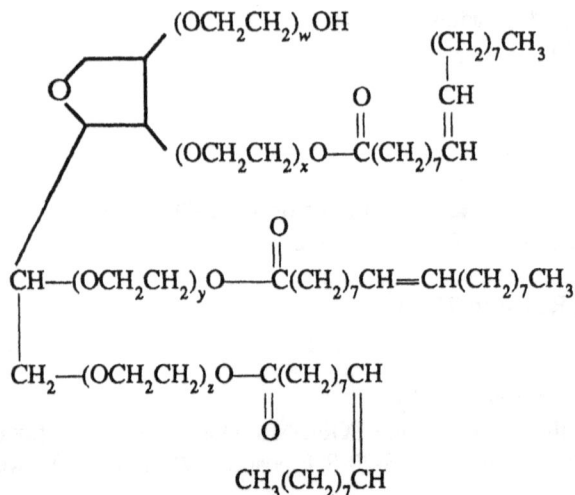

where avg. $(w + x + y + z) = 20$

CAS No.:
9005-70-3 (generic)

TRADENAME EQUIVALENTS:
Ahco DFO-110 [ICI United States]
Alkamuls PSTO-20 [Alkaril]
Crillet 45 [Croda]
Emasol O-320 [Kao Corp.]
Emsorb 6903 [Emery]
Glycosperse TO-20 [Glyco]
Hetsorb TO-20 [Heterene]
Hodag PSTO-20 [Hodag]
Liposorb TO-20 [Lipo]
Lonzest STO-20 [Lonza]
Montanox 85 [Seppic]
Nikkol TO-30 [Nikko]
Sorbax PTO-20 [Chemax]
T-Maz 85 [Mazer]
TO-55-F [Hefti Ltd.]
Tween 85 [ICI United States]

POE (20) sorbitan trioleate (cont'd.)

CATEGORY:

Emulsifier, solubilizer, stabilizer, dispersant, lubricant, detergent, antistat, wetting agent, viscosity modifier

APPLICATIONS:

Automobile products: lubricant additive (Emsorb 6903)

Cleansers: hand cleanser (Crillet 45)

Cosmetic industry preparations: (Alkamuls PSTO-20; Crillet 45; Glycosperse TO-20; Lonzest STO-20; Nikkol TO-30; T-Maz 85; TO-55-F); perfumery (Sorbax PTO-20)

Farm products: animal feed (Crillet 45); herbicides (Alkamuls PSTO-20; Crillet 45); insecticides/pesticides (Alkamuls PSTO-20; Crillet 45); veterinary products (Crillet 45)

Food applications: (Alkamuls PSTO-20; Hodag PSTO-20; Lonzest STO-20; T-Maz 85); flavors (Sorbax PTO-20); food emulsifying (Glycosperse TO-20)

Industrial applications: (Hodag PSTO-20; Lonzest STO-20; TO-55-F); dyes and pigments (TO-55-F); glass processing (Emsorb 6903); lubricating/cutting oils (Emsorb 6903); metalworking (Emsorb 6903; T-Maz 85); paint mfg. (Alkamuls PSTO-20); petroleum industry (Crillet 45); polishes and waxes (Crillet 45; Emsorb 6903; TO-55-F); textile/leather processing (Ahco DFO-110; Crillet 45; Emsorb 6903; T-Maz 85; Tween 85); wood preservation (TO-55-F)

Industrial cleaners: marine tank cleaners (Crillet 45)

Pharmaceutical applications: (Crillet 45; Glycosperse TO-20; Lonzest STO-20; Nikkol TO-30; T-Maz 85; TO-55-F)

PROPERTIES:

Form:

Liquid (Ahco DFO-110; Alkamuls PSTO-20; Emasol O-320; Emsorb 6903; Hetsorb TO-20; Hodag PSTO-20; Liposorb TO-20; Lonzest STO-20; Montanox 85; Nikkol TO-30; T-Maz 85; TO-55-F)

Clear viscous liquid (Crillet 45)

Liquid-gel (Sorbax PTO-20)

Liquid (may gel) (Tween 85)

Liquid (gels on standing) (Glycosperse TO-20)

Color:

Amber (Alkamuls PSTO-20; Crillet 45; Tween 85)

Yellow (Glycosperse TO-20; Liposorb TO-20; T-Maz 85)

Gardner 6 (Emsorb 6903)

Gardner 8 max. (Hetsorb TO-20)

Odor:

Typical (Alkamuls PSTO-20)

Composition:

95% active (Emsorb 6903)

95% active min. (T-Maz 85)

95% conc. (Glycosperse TO-20)

POE (20) sorbitan trioleate (cont'd.)

97% active (Alkamuls PSTO-20)
100% active (Liposorb TO-20; Tween 85)
100% conc. (Ahco DFO-110; Emasol O-320; Hodag PSTO-20; Lonzest STO-20; Montanox 85; Nikkol TO-30; TO-55-F)

Solubility:
Sol. in acetone (Tween 85); misc. with acetone in certain proportions (Glycosperse TO-20); misc. hot in certain proportions (T-Maz 85)
Sol. in lower alcohols (Tween 85)
Sol. in aromatic solvents (Tween 85)
Sol. in butyl stearate (Crillet 45; Emsorb 6903)
Sol. in carbon tetrachloride (Tween 85)
Sol. in Cellosolve (Tween 85)
Sol. in dioxane (Tween 85)
Sol. in ethanol (Crillet 45; Glycosperse TO-20; T-Maz 85)
Sol. in ethyl acetate (Glycosperse TO-20; Tween 85)
Sol. in ethylene glycol (Tween 85)
Sol. in glycerol trioleate (Emsorb 6903)
Sol. in isopropanol (Hetsorb TO-20)
Sol. in isopropyl myristate (Crillet 45)
Sol. in kerosene (Crillet 45)
Sol. in methanol (Glycosperse TO-20)
Sol. in min. oil (Hetsorb TO-20; Tween 85); partly sol. (Glycosperse TO-20); disp. (Emsorb 6903);
Sol. in min. spirits (Tween 85)
Misc. with naphtha in certain proportions (Glycosperse TO-20); misc. hot in certain proportions (T-Maz 85)
Sol. in oleic acid (Crillet 45)
Sol. in oleyl alcohol (Crillet 45)
Disp. in perchloroethylene (Emsorb 6903)
Sol. in Stoddard solvent (Emsorb 6903)
Misc. with toluol in certain proportions (Glycosperse TO-20); misc. hot in certain proportions (T-Maz 85)
Sol. in trichlorethylene (Crillet 45)
Sol. in most veg. oils (Tween 85); misc. with veg. oil in certain proportions (Glycosperse TO-20); misc. hot in certain proportions (T-Maz 85)
Disp. in water (Emsorb 6903; Glycosperse TO-20; Hetsorb TO-20; T-Maz 85; Tween 85); insol. (Alkamuls PSTO-20)

Ionic Nature:
Nonionic (Alkamuls PSTO-20; Emsorb 6903; Glycosperse TO-20; Hodag PSTO-20; Liposorb TO-20; Nikkol TO-30; T-Maz 85; TO-55-F; Tween 85)

Sp.gr.:
1.0 (Glycosperse TO-20; T-Maz 85; Tween 85)

POE (20) sorbitan trioleate (cont'd.)

Density:
8.6 lb/gal (Emsorb 6903)

Visc.:
300 cs (Emsorb 6903)
300 cps (Glycosperse TO-20; T-Maz 85)

Flash Pt.:
> 300 F (Tween 85)

Fire Pt.:
> 300 F (Tween 85)

Cloud Pt.:
< 25 C (Emsorb 6903)

HLB:
10.9 (Emsorb 6903)
11.0 (Ahco DFO-110; Alkamuls PSTO-20; Crillet 45; Emasol O-320; Glycosperse TO-20; Hodag PSTO-20; Nikkol TO-30; Sorbax PTO-20; T-Maz 85; TO-55-F; Tween 85)
11.0 ± 1 (Liposorb TO-20)

Acid No.:
2.0 max. (Glycosperse TO-20; Hetsorb TO-20; T-Maz 85; Tween 85)

Saponification No.:
80–95 (Tween 85)
82–95 (Crillet 45; Glycosperse TO-20; Hetsorb TO-20; Liposorb TO-20)
83–93 (T-Maz 85)

Hydroxyl No.:
39–52 (Glycosperse TO-20; Hetsorb TO-20; Liposorb TO-20; T-Maz 85; Tween 85)

Surface Tension:
41 dynes/cm (0.1%) (Crillet 45)

POE (20) sorbitan tristearate

SYNONYMS:
Polysorbate 65 (CTFA)
Sorbimacrogol tristearate 300
Sorbitan, trioctadecanoate, poly (oxy-1,2-ethanediyl) derivatives
Sorbitan POE (20) tristearate

POE (20) sorbitan tristearate (cont'd.)

STRUCTURE:

$(OCH_2CH_2)_wOH$

$(OCH_2CH_2)_xO\!-\!\overset{\overset{\displaystyle O}{\|}}{C}(CH_2)_{16}CH_3$

$CH\!-\!(OCH_2CH_2)_yO\!-\!\overset{\overset{\displaystyle O}{\|}}{C}(CH_2)_{16}CH_3$

$CH_2\!-\!(OCH_2CH_2)_zO\!-\!\underset{\underset{\displaystyle O}{\|}}{C}(CH_2)_{16}CH_3$

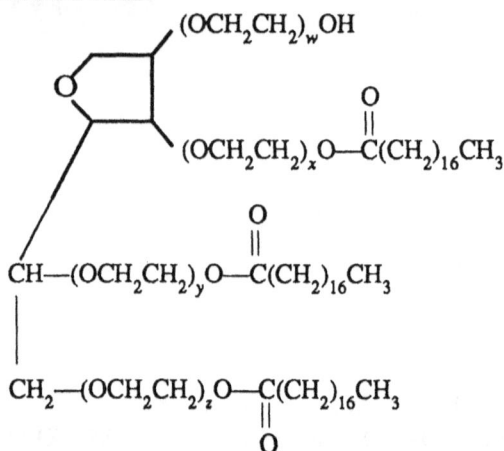

where avg. $(w + x + y + z) = 20$

CAS No.:
9005-71-4 (generic)

TRADENAME EQUIVALENTS:
Ahco 7166T [ICI United States]
Alkamuls PSTS-20 [Alkaril]
Crillet 35 [Croda]
Drewmulse POE-STS [PVO Int'l.]
Drewpone 65 [PVO Int'l.]
Durfax 65, 65K [Durkee Foods]
Emasol S-320 [Kao Corp.]
Emsorb 6907 [Emery]
Glycosperse TS-20 [Glyco]
Hetsorb TS-20 [Heterene]
Hodag PSTS-20 [Hodag]
Industrol STS-20-S [BASF Wyandotte]
Liposorb TS-20 [Lipo]
Lonzest STS-20 [Lonza]
Montanox 65 [Seppic]
Nikkol TS-30 [Nikko]
Sorbax PTS-20 [Chemax]
T-Maz 65 [Mazer]
TS-55-F [Hefti Ltd.]
Tween 65 [ICI United States]

CATEGORY:
Emulsifier, solubilizer, stabilizer, wetting agent, dispersant, penetrant, leveling agent, lubricant, softener, antistat, surfactant, detergent, viscosity control agent

270

POE (20) sorbitan tristearate (cont'd.)

APPLICATIONS:

Cosmetic industry preparations: (Alkamuls PSTS-20; Crillet 35; Drewmulse POE-STS; Emasol S-320; Glycosperse TS-20; Lonzest STS-20; T-Maz 65; TS-55-F); creams and lotions (Drewmulse POE-STS); perfumery (Drewmulse POE-STS; Sorbax PTS-20); shampoos (Drewmulse POE-STS); shaving preparations (Drewmulse POE-STS; Durfax 65)

Farm products: (Crillet 35); herbicides (Alkamuls PSTS-20; Crillet 35); insecticides/pesticides (Alkamuls PSTS-20; Crillet 35; Durfax 65)

Food applications: (Alkamuls PSTS-20; Emasol S-320; Glycosperse TS-20; Hodag PSTS-20; Lonzest STS-20; T-Maz 65); flavors (Drewmulse POE-STS; Sorbax PTS-20); food emulsifying (Drewpone 65; Durfax 65, 65K)

Industrial applications: (Emasol S-320; Hodag PSTS-20; Lonzest STS-20); dyes and pigments (Crillet 35); metalworking (T-Maz 65); paint mfg. (Alkamuls PSTS-20; TS-55-F); polishes and waxes (Crillet 35; TS-55-F); textile/leather processing (Crillet 35; Emsorb 6907; T-Maz 65; Tween 65)

Pharmaceutical applications: (Crillet 35; Drewmulse POE-STS; Glycosperse TS-20; T-Maz 65; TS-55-F); germicides (Drewmulse POE-STS); ointments (Drewmulse POE-STS); vitamins (Drewmulse POE-STS)

PROPERTIES:

Form:

Liquid (Ahco 7166T; Nikkol TS-30)

Soft solid (Drewpone 65)

Solid (Alkamuls PSTS-20; Drewmulse POE-STS; Durfax 65, 65K; Emasol S-320; Hetsorb TS-20; Hodag PSTS-20; Industrol STS-20-S; Lonzest STS-20; Montanox 65; Sorbax PTS-20; T-Maz 65; TS-55-F)

Solid wax (Liposorb TS-20)

Waxy solid (Crillet 35; Emsorb 6907; Glycosperse TS-20; Tween 65)

Color:

Cream/buff (Crillet 35)

Amber (Drewpone 65)

Tan (Alkamuls PSTS-20; Durfax 65K; Glycosperse TS-20; Industrol STS-20-S; Liposorb TS-20; T-Maz 65; Tween 65)

Gardner 5 (Emsorb 6907)

Gardner 7 max. (Hetsorb TS-20)

Odor:

Typical (Alkamuls PSTS-20)

Composition:

46–50% ETO (Drewpone 65)

97% active (Alkamuls PSTS-20; Durfax 65K)

97% active min. (T-Maz 65)

100% active (Liposorb TS-20; Tween 65)

100% conc. (Ahco 7166T; Emasol S-320; Emsorb 6907; Glycosperse TS-20; Hodag PSTS-20; Lonzest STS-20; Montanox 65; Nikkol TS-30; TS-55-F)

271

POE (20) sorbitan tristearate *(cont'd.)*

Solubility:
Sol. in acetone (Glycosperse TS-20; T-Maz 65)
Sol. in butyl stearate (Emsorb 6907)
Sol. in cottonseed oil at low levels (Durfax 65K)
Sol. in ethanol (Crillet 35; Glycosperse TS-20; T-Maz 65; Tween 65)
Sol. in ethyl acetate (Glycosperse TS-20)
Sol. in glycerol trioleate (Emsorb 6907)
Sol. in isopropanol (Hetsorb TS-20; Tween 65)
Partly sol. in isopropyl myristate (Crillet 35)
Partly sol. in kerosene (Crillet 35)
Sol. in methanol (Glycosperse TS-20)
Sol. in min. oil (Emsorb 6907; Glycosperse TS-20; Hetsorb TS-20; Tween 65); sol. hot
   (T-Maz 65); partly sol. in (Crillet 35)
Sol. in naphtha (Glycosperse TS-20; T-Maz 65)
Sol. in oleic acid (Crillet 35)
Sol. in oleyl alcohol (Crillet 35)
Partly sol. in olive oil (Crillet 35)
Disp. in perchloroethylene (Emsorb 6907)
Sol. in Stoddard solvent (Emsorb 6907)
Disp. in toluol (Glycosperse TS-20; T-Maz 65)
Sol. in trichlorethylene (Crillet 35)
Sol. in veg. oil (Glycosperse TS-20; Tween 65); sol. hot (T-Maz 65)
Partly sol. in water (Crillet 35); disp. (Durfax 65, 65K; Emsorb 6907; Glycosperse TS-
   20; Hetsorb TS-20; Industrol STS-20-S; T-Maz 65); insol. (Alkamuls PSTS-20)
Partly sol. in xylene (Crillet 35)
Ionic Nature:
Nonionic (Alkamuls PSTS-20; Drewmulse POE-STS; Durfax 65, 65K, 65K; Emsorb
   6907; Glycosperse TS-20; Hodag PSTS-20; Industrol STS-20-S; T-Maz 65; TS-55-
   F; Tween 65)
Sp.gr.:
1.05 (Glycosperse TS-20; Industrol STS-20-S; Tween 65)
1.1 (T-Maz 65)
M.P.:
30–32 C (T-Maz 65)
34 C (Emsorb 6907)
Pour Pt.:
92 F (Tween 65)
Flash Pt.:
> 300 F (Tween 65)
Fire Pt.:
> 300 F (Tween 65)
Cloud Pt.:
< 25 C (Emsorb 6907)

POE (20) sorbitan tristearate *(cont'd.)*

HLB:
10.5 (Ahco 7166T; Alkamuls PSTS-20; Crillet 35; Drewmulse POE-STS; Drewpone 65; Durfax 65; Emasol S-320; Glycosperse TS-20; Hodag PSTS-20; Industrol STS-20-S; Montanox 65; Sorbax PTS-20; T-Maz 65; TS-55-F; Tween 65)
10.5 ± 1 (Liposorb TS-20)
11.0 (Emsorb 6907; Nikkol TS-30)

Acid No.:
2.0 max. (Glycosperse TS-20; Hetsorb TS-20; Industrol STS-20-S; T-Maz 65; Tween 65)

Saponification No.:
88–98 (Crillet 35; Drewmulse POE-STS; Drewpone 65; Durfax 65; Glycosperse TS-20; Hetsorb TS-20; Industrol STS-20-S; Liposorb TS-20; T-Maz 65; Tween 65)

Hydroxyl No.:
44–60 (Durfax 65; Glycosperse TS-20; Hetsorb TS-20; Industrol STS-20-S; Liposorb TS-20; T-Maz 65; Tween 65)

Surface Tension:
42.5 dynes/cm (0.1%) (Crillet 35)

STD. PKGS.:
470-lb packaging, bulk (Durfax 65K)
55-gal (450 lb net) steel drums (Industrol STS-20-S)

POE (6) sorbitol beeswax

SYNONYMS:
PEG-6 sorbitan beeswax (CTFA)
PEG 300 sorbitan beeswax

CAS No.:
8051-15-8

TRADENAME EQUIVALENTS:
Atlas G-1702 [ICI United States]
Nikkol GBW-25 [Nikko]

CATEGORY:
Surfactant, emulsifier

APPLICATIONS:
Cosmetic industry preparations: (Nikkol GBW-25)

PROPERTIES:
Form:
Solid (Nikkol GBW-25)
Waxy solid (Atlas G-1702)

Color:
Tan (Atlas G-1702)

POE (6) sorbitol beeswax *(cont'd.)*

Composition:
  100% conc. (Nikkol GBW-25)
Solubility:
  Slightly sol. in veg. oils (Atlas G-1702)
Ionic Nature:
  Nonionic (Atlas G-1702; Nikkol GBW-25)
Flash Pt.:
  > 300 F (Atlas G-1702)
Fire Pt.:
  > 300 F (Atlas G-1702)
HLB:
  5.0 (Atlas G-1702)
  7.5 (Nikkol GBW-25)

POE (20) sorbitol beeswax

SYNONYMS:
  PEG-20 sorbitan beeswax (CTFA)
  PEG 1000 sorbitan beeswax
CAS No.:
  RD No.: 977055-32-5
TRADENAME EQUIVALENTS:
  Atlas G-1726 [ICI United States]
  Nikkol GBW-125 [Nikko]
CATEGORY:
  Surfactant, emulsifier
APPLICATIONS:
  Cosmetic industry preparations: (Nikkol GBW-125)
PROPERTIES:
Form:
  Solid (Nikkol GBW-125)
  Waxy solid (Atlas G-1726)
Color:
  Tan (Atlas G-1726)
Composition:
  100% conc. (Nikkol GBW-125)
Solubility:
  Slightly sol. in veg. oils (Atlas G-1726)
Ionic Nature:
  Nonionic (Atlas G-1726; Nikkol GBW-125)

POE (20) sorbitol beeswax *(cont'd.)*

Flash Pt.:
> 300 F (Atlas G-1726)
Fire Pt.:
> 300 F (Atlas G-1726)
HLB:
5.0 (Atlas G-1726)
9.5 (Nikkol GBW-125)

POE (5) soya amine

SYNONYMS:
PEG-5 soyamine (CTFA)
PEG (5) soya amine
STRUCTURE:

$$R-N \begin{cases} (CH_2CH_2O)_xH \\ (CH_2CH_2O)_yH \end{cases}$$

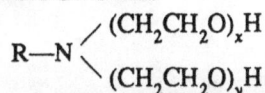

where   R represents the soya radical and
avg. $(x + y) = 5$
CAS No.:
61791-24-0 (generic)
RD No.: 977055-36-9
TRADENAME EQUIVALENTS:
Accomeen S5 [Armstrong]
Ethomeen S/15 [Armak]
Hetoxamine S5 [Heterene]
Mazeen S5 [Mazer]
Varonic L205 [Sherex]
CATEGORY:
Emulsifier, surfactant, antistat, dispersant, desizing agent, softener, rewetting agent,
lubricant
APPLICATIONS:
Cosmetic industry preparations: (Mazeen S5)
Farm products: agricultural oils/sprays (Hetoxamine S5); herbicides (Mazeen S5);
insecticides/pesticides (Mazeen S5)
Industrial applications: dyes and pigments (Varonic L205); latex (Varonic L205);
lubricating/cutting oils (Mazeen S5); printing inks (Mazeen S5); textile/leather
processing (Ethomeen S/15; Hetoxamine S5; Mazeen S5); waxes and oils
(Hetoxamine S5; Varonic L205)
Industrial cleaners: metal processing surfactants (Hetoxamine S5)

POE (5) soya amine (cont'd.)

PROPERTIES:
Form:
   Liquid (Accomeen S5; Hetoxamine S5; Mazeen S5; Varonic L205)
   Clear liquid (Ethomeen S/15)
Color:
   Gardner 10 (Accomeen S5)
   Gardner 14 (Mazeen S5)
   Gardner 14 max. (Ethomeen S/15)
Odor:
   Amine-type (Accomeen S5)
Composition:
   99% active (Accomeen S5)
   100% active (Varonic L205)
   100% conc. (Mazeen S5)
Solubility:
   Sol. in acetone (cloudy > 50 C) (Ethomeen S/15); partly sol. (Mazeen S5)
   Sol. in benzene (Ethomeen S/15; Mazeen S5)
   Sol. in carbon tetrachloride (> 25 C) (Ethomeen S/15)
   Sol. in dioxane (Ethomeen S/15)
   Sol. in isopropanol (Ethomeen S/15; Hetoxamine S5; Mazeen S5)
   Partly sol. in min. oil (Hetoxamine S5; Mazeen S5)
   Sol. in organic solvents (Accomeen S5)
   Sol. in Stoddard solvent (> 75 C) (Ethomeen S/15)
   Disp. in water (Accomeen S5); forms gel or dispersion in water (Ethomeen S/15); gels
     in water (Hetoxamine S5; Mazeen S5)
Ionic Nature:
   Nonionic (Accomeen S5; Varonic L205)
   Cationic (Ethomeen S/15; Hetoxamine S5; Mazeen S5)
M.W.:
   480 (Hetoxamine S5)
Sp.gr.:
   0.95 (Accomeen S5; Ethomeen S/15)
   0.951 (Mazeen S5)
Density:
   7.9 lb/gal (Accomeen S5)
Surface Tension:
   33 dynes/cm (0.1% sol'n.) (Ethomeen S/15; Mazeen S5)
TOXICITY/HANDLING:
   Corrosive (Accomeen S5)
   Skin irritant, severe eye irritant (Ethomeen S/15)

SYNONYMS:
PEG-10 soyamine (CTFA)
PEG 500 soya amine

STRUCTURE:

$$R—N \Big\langle \begin{matrix} (CH_2CH_2O)_xH \\ (CH_2CH_2O)_yH \end{matrix}$$

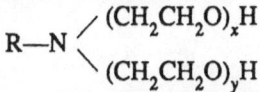

where R represents the soya radical and
avg. $(x + y) = 10$

CAS No.:
61791-24-0 (generic)
RD No.: 977063-23-2

TRADENAME EQUIVALENTS:
Accomeen S10 [Armstrong]
Ethomeen S/20 [Armak]
Mazeen S10 [Mazer]

CATEGORY:
Emulsifier, surfactant, antistat, dispersant, rewetting agent, lubricant

APPLICATIONS:
Cosmetic industry preparations: (Mazeen S10)
Farm products: herbicides (Mazeen S10); insecticides/pesticides (Mazeen S10)
Industrial applications: lubricating/cutting oils (Mazeen S10); printing inks (Mazeen S10); textile/leather processing (Ethomeen S/20; Mazeen S10)

PROPERTIES:
Form:
Liquid (Accomeen S10; Mazeen S10)
Clear liquid (Ethomeen S/20)
Color:
Gardner 10 (Accomeen S10)
Gardner 14 (Mazeen S10)
Gardner 14 max. (Ethomeen S/20)
Odor:
Amine (Accomeen S10)
Composition:
99% active (Accomeen S10)
100% conc. (Mazeen S10)
Solubility:
Sol. in acetone (Mazeen S10); sol. (cloudy) (Ethomeen S/20)
Sol. in benzene (Ethomeen S/20; Mazeen S10)
Sol. in carbon tetrachloride (> 25 C) (Ethomeen S/20)
Sol. in dioxane (Ethomeen S/20)
Sol. in isopropanol (Ethomeen S/20; Mazeen S10)
Partly sol. in min. oil (Mazeen S10)

POE (10) soya amine *(cont'd.)*

Sol. in organic solvents (Accomeen S10)
Sol. in Stoddard solvent (> 50C) (Ethomeen S/20)
Sol. in water (Accomeen S10; Mazeen S10); sol. in water (forms visc. sol'n.)
(Ethomeen S/20)
Ionic Nature:
Nonionic (Accomeen S10)
Cationic (Ethomeen S/20; Mazeen S10)
M.W.:
710 (Mazeen S10)
Sp.gr.:
1.02 (Accomeen S10; Ethomeen S/20; Mazeen S10)
Density:
8.5 lb/gal (Accomeen S10)
Surface Tension:
40 dynes/cm (0.1% sol'n.) (Ethomeen S/20; Mazeen S10)
43 dynes/cm (0.1% sol'n.) (Accomeen S10)
TOXICITY/HANDLING:
Corrosive (Accomeen S10)
Skin irritant, severe eye irritant (Ethomeen S/20)

POE (15) soya amine

SYNONYMS:
PEG-15 soyamine (CTFA)
PEG (15) soya amine
STRUCTURE:

$$R{-}N \Big\langle \begin{array}{l} (CH_2CH_2O)_xH \\ (CH_2CH_2O)_yH \end{array}$$

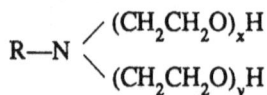

where   R represents the soya radical and
avg. $(x + y) = 15$
CAS No.:
61791-24-0 (generic)
RD No.: 977063-34-5
TRADENAME EQUIVALENTS:
Ethomeen S/25 [Armak]
Hetoxamine S15 [Heterene]
Mazeen S15 [Mazer]
CATEGORY:
Emulsifier, surfactant, antistat, dispersant, desizing agent, rewetting agent, lubricant

APPLICATIONS:
Cosmetic industry preparations: (Mazeen S15)
Farm products: agricultural oils/sprays (Hetoxamine S15); herbicides (Mazeen S15); insecticides/pesticides (Mazeen S15)
Industrial applications: lubricating/cutting oils (Mazeen S15); printing inks (Mazeen S15); textile/leather processing (Ethomeen S/25; Hetoxamine S15; Mazeen S15); waxes and oils (Hetoxamine S15)
Industrial cleaners: metal processing surfactants (Hetoxamine S15)

PROPERTIES:
Form:
Liquid (Hetoxamine S15; Mazeen S15)
Clear liquid (Ethomeen S/25)
Color:
Gardner 12 max. (Hetoxamine S15)
Gardner 18 (Mazeen S15)
Gardner 18 max. (Ethomeen S/25)
Composition:
95% tertiary amine min. (Hetoxamine S15)
100% conc. (Mazeen S15)
Solubility:
Sol. in acetone (Mazeen S15); sol. (cloudy) (Ethomeen S/25)
Sol. in benzene (Ethomeen S/25; Mazeen S15)
Sol. in carbon tetrachloride (> 25 C) (Ethomeen S/25)
Sol. in dioxane (Ethomeen S/25)
Sol. in isopropanol (Ethomeen S/25; Hetoxamine S15; Mazeen S15)
Insol. in min. oil (Hetoxamine S15)
Sol. in Stoddard solvent (> 75 C) (Ethomeen S/25)
Sol. in water (Ethomeen S/25; Hetoxamine S15; Mazeen S15)
Ionic Nature:
Cationic (Ethomeen S/25; Hetoxamine S15; Mazeen S15)
Equiv. Wt.:
895–960 (Hetoxamine S15)
Sp.gr.:
1.04 (Ethomeen S/25; Mazeen S15)
Surface Tension:
43 dynes/cm (0.1% sol'n.) (Ethomeen S/25; Mazeen S15)
TOXICITY/HANDLING:
Skin irritant, severe eye irritant (Ethomeen S/25)

POE (2) stearyl amine

SYNONYMS:
PEG-2 stearamine (CTFA)
PEG 100 stearyl amine

STRUCTURE:

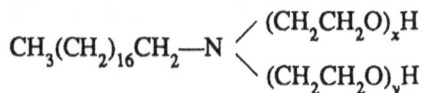

$$CH_3(CH_2)_{16}CH_2—N \begin{array}{l} (CH_2CH_2O)_xH \\ (CH_2CH_2O)_yH \end{array}$$

where avg. $(x + y) = 2$

CAS No.:
RD No.: 977063-45-8

TRADENAME EQUIVALENTS:
Chemeen 18/2 [Chemax]
Ethomeen 18/12 [Armak]
Hetoxamine ST2 [Heterene]

CATEGORY:
Emulsifier, antistat, lubricant, desizing agent, softener, dispersant

APPLICATIONS:
Farm products: (Hetoxamine ST2)
Industrial applications: glass processing (Chemeen 18/2); metal buffing (Chemeen 18/2); rubber (Chemeen 18/2); textile/leather processing (Chemeen 18/2; Ethomeen 18/12; Hetoxamine ST2); waxes and oils (Hetoxamine ST2)
Industrial cleaners: metal processing surfactants (Hetoxamine ST2)

PROPERTIES:

Form:
Solid (Chemeen 18/2; Ethomeen 18/12; Hetoxamine ST2)

Color:
Gardner 7 max. (Ethomeen 18/12)

Composition:
95% tertiary amine (Ethomeen 18/12)

Solubility:
Sol. in acetone (< 25 C) (Ethomeen 18/12)
Sol. in benzene (Ethomeen 18/12)
Sol. in carbon tetrachloride (Ethomeen 18/12)
Sol. in isopropanol (Ethomeen 18/12; Hetoxamine ST2)
Sol. in min. oil (Hetoxamine ST2)
Insol. in water (Ethomeen 18/12)

Ionic Nature:
Cationic (Ethomeen 18/12)

M.W.:
362 (Chemeen 18/2)
388 (Hetoxamine ST2)

Sp.gr.:
0.96 (Ethomeen 18/12)

TOXICITY/HANDLING:

Corrosive, skin irritant, severe eye irritant (Ethomeen 18/12)

POE (5) stearyl amine

SYNONYMS:

PEG-5 stearamine (CTFA)

PEG (5) stearyl amine

STRUCTURE:

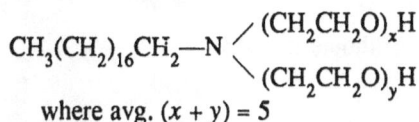

$$CH_3(CH_2)_{16}CH_2-N \diagup \begin{matrix} (CH_2CH_2O)_xH \\ \\ (CH_2CH_2O)_yH \end{matrix}$$

where avg. $(x + y) = 5$

CAS No.:

RD No. 977065-71-6

TRADENAME EQUIVALENTS:

Chemeen 18-5 [Chemax]

Ethomeen 18/15 [Armak]

Hetoxamine ST5 [Heterene]

CATEGORY:

Emulsifier, dispersant, antistat, lubricant, desizing agent, softener, water repellant

APPLICATIONS:

Farm products: agricultural oils/sprays (Hetoxamine ST5)

Industrial applications: glass processing (Chemeen 18-5); metalworking (Chemeen 18-5); rubber (Chemeen 18-5); textile/leather processing (Chemeen 18-5; Ethomeen 18/15; Hetoxamine ST5; Hetoxamine ST5); waxes (Hetoxamine ST5)

Industrial cleaners: metal processing surfactants (Hetoxamine ST5)

PROPERTIES:

Form:

Liquid to paste (Ethomeen 18/15)

Solid (Chemeen 18-5; Hetoxamine ST5)

Color:

Gardner 8 max. (Ethomeen 18/15)

Solubility:

Sol. in acetone (cloudy @ < 50 C) (Ethomeen 18/15)

Sol. in benzene (Ethomeen 18/15)

Sol. cloudy in carbon tetrachloride (Ethomeen 18/15)

Sol. in dioxane (< 75 C) (Ethomeen 18/15)

Sol. in isopropanol (Hetoxamine ST5); (< 25 C) (Ethomeen 18/15)

Sol. in min. oil (Hetoxamine ST5)

Sol. cloudy in Stoddard solvent (Ethomeen 18/15)

POE (5) stearyl amine *(cont'd.)*

Forms gel in water (Ethomeen 18/15)
Ionic Nature:
Cationic (Ethomeen 18/15)
Nonionic (Hetoxamine ST5)
M.W.:
520 (Hetoxamine ST5)
540 (Chemeen 18-5)
Sp.gr.:
0.98 (Ethomeen 18/15)
Surface Tension:
34 dynes/cm (0.1% sol'n.) (Ethomeen 18/15)
TOXICITY/HANDLING:
Corrosive, skin irritant, severe eye irritant (Ethomeen 18/15)

POE (50) stearyl amine

SYNONYMS:
PEG-50 stearamine (CTFA)
PEG (50) stearyl amine
STRUCTURE:

$$CH_3(CH_2)_{16}CH_2{-}N \left\langle \begin{array}{l} (CH_2CH_2O)_xH \\ (CH_2CH_2O)_yH \end{array} \right.$$

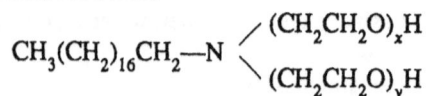

where avg. $(x + y) = 50$
CAS No.:
RD No.: 977065-74-9
TRADENAME EQUIVALENTS:
Chemeen 18-50 [Chemax]
Ethomeen 18/60 [Armak]
Hetoxamine ST-50 [Heterene]
Icomeen 18-50 [BASF Wyandotte]
Trymeen SAM-50 [Emery]
CATEGORY:
Emulsifier, surfactant, antistat, lubricant, leveling agent, dispersant, desizing agent, water repellent
APPLICATIONS:
Farm products: agricultural oils/sprays (Hetoxamine ST-50)
Industrial applications: dyes and pigments (Trymeen SAM-50); glass processing (Chemeen 18-50; Trymeen SAM-50); rubber (Chemeen 18-50; Trymeen SAM-50); textile/leather processing (Chemeen 18-50; Ethomeen 18/60; Hetoxamine ST-50); waxes and oils (Hetoxamine ST-50)

Industrial cleaners: metal buffing (Chemeen 18-50; Hetoxamine ST-50; Trymeen SAM-50)

PROPERTIES:

Form:
Solid (Ethomeen 18/60; Hetoxamine ST-50; Trymeen SAM-50)
Cast solid (Icomeen 18-50)

Color:
White (Ethomeen 18/60)
Gardner 4 (Trymeen SAM-50)
Gardner 8 max. (Hetoxamine ST-50)
Gardner 9 max. (Icomeen 18-50)

Composition:
100% active (Icomeen 18-50; Trymeen SAM-50)

Solubility:
Sol. in acetone (< 100 C) (Ethomeen 18/60)
Sol. in benzene (< 100 C) (Ethomeen 18/60)
Sol. in carbon tetrachloride (Ethomeen 18/60)
Sol. in dioxane (< 100 C) (Ethomeen 18/60)
Sol. in isopropanol (Hetoxamine ST-50); (< 100 C) (Ethomeen 18/60)
Insol. in min. oil (Hetoxamine ST-50)
Sol. in perchloroethylene (@ 5%) (Trymeen SAM-50)
Sol. in water (Ethomeen 18/60; Hetoxamine ST-50; Icomeen 18-50); (@ 5%) (Trymeen SAM-50)

Ionic Nature:
Cationic (Ethomeen 18/60; Hetoxamine ST-50; Trymeen SAM-50)
Mildly cationic (Icomeen 18-50)

M.W.:
2380 (Chemeen 18-50)

Equiv. Wt.:
2350–2500 (Hetoxamine ST-50)

Sp.gr.:
1.07 (Icomeen 18-50)
1.12 (Ethomeen 18/60)

M.P.:
50 C (Trymeen SAM-50)

Cloud Pt.:
> 100 C (Trymeen SAM-50)

HLB:
17.7 (Icomeen 18-50)
17.8 (Trymeen SAM-50)

pH:
9 (5% aq.) (Icomeen 18-50)

POE (50) stearyl amine (cont'd.)

Surface Tension:
49 dynes/cm (0.1% sol'n.) (Ethomeen 18/60)
TOXICITY/HANDLING:
Mild skin and eye irritant (Ethomeen 18/60)
STD. PKGS.:
55-gal (450 lb net) steel drums (Icomeen 18-50)

POE (2) tallow amine

SYNONYMS:
PEG-2 tallow amine (CTFA)
PEG 100 tallow amine
STRUCTURE:

$$R-N \begin{cases} (CH_2CH_2O)_xH \\ (CH_2CH_2O)_yH \end{cases}$$

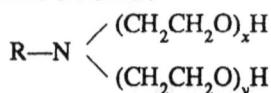

where   R represents the tallow radical and
avg. $(x + y) = 2$
CAS No.:
61791-26-2 (generic)
RD No.: 977063-46-9
TRADENAME EQUIVALENTS:
Alkaminox T2 [Alkaril]
Chemeen T-2 [Chemax]
Ethomeen T/12 [Armak]
Hetoxamine T2 [Heterene]
Mazeen T2 [Mazer]
CATEGORY:
Emulsifier, dispersant, antistat, lubricant, softener, antiprecipitant, leveling and migrating agent, desizing agent, rewetting agent
APPLICATIONS:
Cosmetic industry preparations: (Mazeen T2)
FARM products: (Hetoxamine T2); herbicides (Mazeen T2); insecticides/pesticides (Mazeen T2)
Household detergents: carpet & upholstery shampoos (Chemeen T-2)
Industrial applications: dyes and pigments (Chemeen T-2); lubricating/cutting oils (Mazeen T2); plastics (Alkaminox T2); printing inks (Mazeen T2); textile/leather processing (Alkaminox T2; Chemeen T-2; Ethomeen T/12; Hetoxamine T2; Mazeen T2); waxes and oils (Hetoxamine T2)
Industrial cleaners: metal processing surfactants (Hetoxamine T2)

POE (2) tallow amine (cont'd.)

PROPERTIES:
Form:
  Liquid (Hetoxamine T2; Mazeen T2)
  Semiliquid (Chemeen T-2)
  Paste (Alkaminox T2; Ethomeen T/12)
Color:
  Gardner 8 max. (Alkaminox T2; Ethomeen T/12)
  Gardner 11 (Mazeen T2)
Composition:
  95% tertiary amine (Ethomeen T/12)
  100% conc. (Mazeen T2)
Solubility:
  Sol. in acetone (Ethomeen T/12; Mazeen T2)
  Sol. in benzene (Ethomeen T/12; Mazeen T2)
  Sol. in carbon tetrachloride (Ethomeen T/12)
  Sol. in isopropanol (Ethomeen T/12; Hetoxamine T2; Mazeen T2)
  Sol. in min. acid (Alkaminox T2)
  Sol. in min. oil (Hetoxamine T2; Mazeen T2)
  Sol. in most organic solvents (Alkaminox T2)
  Sol. in Stoddard solvent (Ethomeen T/12)
  Sol. in water (> 50 C) (Ethomeen T/12); insol. (Alkaminox T2)
Ionic Nature:
  Cationic (Ethomeen T/12; Hetoxamine T2)
M.W.:
  350 (Chemeen T-2; Hetoxamine T2)
Sp.gr.:
  0.916 (Mazeen T2)
  0.92 (Alkaminox T2; Ethomeen T/12)
Surface Tension:
  29 dynes/cm (0.1% sol'n.) (Mazeen T2)
TOXICITY/HANDLING:
  Corrosive, skin irritant, severe eye irritant (Ethomeen T/12)

POE (5) tallow amine

SYNONYMS:
  PEG-5 tallow amine (CTFA)
  PEG (5) tallow amine

POE (5) tallow amine (cont'd.)

STRUCTURE:

$$R-N \begin{cases} (CH_2CH_2O)_xH \\ (CH_2CH_2O)_yH \end{cases}$$

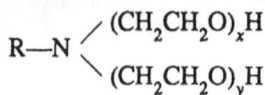

where   R represents the tallow radical and
avg. $(x + y) = 5$

CAS No.:
61791-26-2 (generic)

TRADENAME EQUIVALENTS:
Alkaminox T-5 [Alkaril]
Chemeen T-5 [Chemax]
Ethomeen T/15 [Armak]
Hetoxamine T5 [Heterene]
Katapol PN-430 [GAF]
Mazeen T5 [Mazer]

CATEGORY:
Emulsifier, dispersant, corrosion inhibitor, antistat, lubricant, softener, antiprecipitant, leveling and migrating agent, desizing agent, water repellent, rewetting agent

APPLICATIONS:
Cosmetic industry preparations: (Mazeen T5)
Farm products: agricultural oils/sprays (Hetoxamine T5); herbicides (Mazeen T5); insecticides/pesticides (Mazeen T5)
Household detergents: carpet & upholstery shampoos (Chemeen T-5)
Industrial applications: dyes and pigments (Alkaminox T-5; Chemeen T-5); industrial processing (Alkaminox T-5); inks (Mazeen T5); lubricating/cutting oils (Mazeen T5); metalworking (Katapol PN-430); paint mfg. (Alkaminox T-5); textile/leather processing (Alkaminox T-5; Chemeen T-5; Ethomeen T/15; Hetoxamine T5; Mazeen T5); waxes (Hetoxamine T5)
Industrial cleaners: metal processing surfactants (Hetoxamine T5)

PROPERTIES:
Form:
Liquid (Chemeen T-5)
Clear liquid (Alkaminox T-5; Ethomeen T/15)
Liquid to paste (Hetoxamine T5; Mazeen T5)
Slightly viscous liquid (Katapol PN-430)
Color:
Brown (Katapol PN-430)
Gardner 8 max. (Alkaminox T-5; Ethomeen T/15)
Gardner 12 (Mazeen T5)
Composition:
99% active (Katapol PN-430)
100% conc. (Alkaminox T-5; Chemeen T-5; Mazeen T5)

Solubility:
    Sol. in acetone (Mazeen T5); (cloudy) (Ethomeen T/15)
    Sol. in benzene (Mazeen T5); (cloudy) (Ethomeen T/15)
    Sol. cloudy in carbon tetrachloride (Ethomeen T/15)
    Sol. cloudy in dioxane (Ethomeen T/15)
    Sol. in isopropanol (Hetoxamine T5; Mazeen T5); (cloudy) (Ethomeen T/15)
    Sol. in kerosene (10% with slight haze) (Katapol PN-430)
    Sol. in min. acid (Alkaminox T-5)
    Sol. in min. oil (Hetoxamine T5; Mazeen T5)
    Sol. in most organic solvents (Alkaminox T-5)
    Sol. cloudy in Stoddard solvent (Ethomeen T/15)
    Sol. in water (Hetoxamine T5); gels in water (Mazeen T5); gels in water @ 30–80 C
      (Ethomeen T/15); insol. (Alkaminox T-5)
Ionic Nature:
    Cationic (Alkaminox T-5; Ethomeen T/15; Hetoxamine T5; Katapol PN-430; Mazeen
      T5)
M.W.:
    480 (Mazeen T5)
    490 (Chemeen T-5)
Sp.gr.:
    0.97 (Alkaminox T-5; Ethomeen T/15)
HLB:
    9.0 (Alkaminox T-5)
pH:
    8.0–10.0 (10% aq. disp.) (Katapol PN-430)
Surface Tension:
    34 dynes/cm (0.1% sol'n.) (Ethomeen T/15; Mazeen T5)
TOXICITY/HANDLING:
    Corrosive, skin irritant, severe eye irritant (Ethomeen T/15)

POE (15) tallow amine

SYNONYMS:
    PEG-15 tallow amine (CTFA)
    PEG (15) tallow amine
STRUCTURE:

$$R-N \Big\langle \begin{array}{l} (CH_2CH_2O)_xH \\ (CH_2CH_2O)_yH \end{array}$$

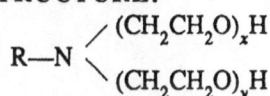

      where R represents the tallow radical and
             avg. $(x + y) = 15$

POE (15) tallow amine *(cont'd.)*

CAS No.:
61791-26-2 (generic)
RD No.: 977063-36-7

TRADENAME EQUIVALENTS:
Alkaminox T-15 [Alkaril]
Chemeen T-15 [Chemax]
Crodamet 1.T15 [Croda Chem. Ltd.]
Ethomeen T/25 [Armak]
Hetoxamine T15 [Heterene]
Icomeen T-15 [BASF Wyandotte]
Mazeen T-15 [Mazer]
Teric 17M15 [ICI Australia Ltd.]
Trymeen TAM-15 [Emery]

Distilled:
Katapol PN-730 [GAF]

CATEGORY:
Emulsifier, surfactant, dispersant, antistat, lubricant, dispersant, softener, antiprecipitant, leveling agent, dye assistant, migrating agent, desizing agent, water repellent, wetting agent, rewetting agent, dewatering agent, intermediate, corrosion inhibitor

APPLICATIONS:
Cosmetic industry preparations: (Mazeen T-15)
Farm products: agricultural oils/sprays (Hetoxamine T15); herbicides (Mazeen T-15); insecticides/pesticides (Mazeen T-15)
Household detergents: carpet & upholstery shampoos (Chemeen T-15)
Industrial applications: construction (Teric 17M15); dyes and pigments (Chemeen T-15; Katapol PN-730; Trymeen TAM-15); electronics (Teric 17M15); lubricating/cutting oils (Mazeen T-15; Teric 17M15); metalworking (Teric 17M15); printing inks (Mazeen T-15); textile/leather processing (Alkaminox T-15; Chemeen T-15; Ethomeen T/25; Hetoxamine T15; Katapol PN-730; Mazeen T-15; Trymeen TAM-15); waxes and oils (Alkaminox T-15; Hetoxamine T15)
Industrial cleaners: metal processing surfactants (Hetoxamine T15)

PROPERTIES:
Form:
Liquid (Chemeen T-15; Icomeen T-15; Mazeen T-15; Teric 17M15; Trymeen TAM-15)
Clear liquid (Alkaminox T-15; Ethomeen T/25; Katapol PN-730)
Liquid to paste (Hetoxamine T15)

Color:
Amber (Katapol PN-730)
Gardner 8 (Trymeen TAM-15)
Gardner 8 max. (Alkaminox T-15; Ethomeen T/25; Icomeen T-15)
Gardner 18 (Mazeen T-15)

POE (15) tallow amine *(cont'd.)*

Composition:
99% active min. (Katapol PN-730)
100% active (Icomeen T-15; Teric 17M15; Trymeen TAM-15)
100% conc. (Crodamet 1.T15; Mazeen T-15)

Solubility:
Sol. in acetone (Mazeen T-15); sol. cloudy (Ethomeen T/25)
Sol. in benzene (Mazeen T-15; Teric 17M15); sol. cloudy (< 100 C) (Ethomeen T/25)
Disp. in butyl stearate (@ 5%) (Trymeen TAM-15)
Sol. cloudy in carbon tetrachloride (Ethomeen T/25)
Sol. cloudy in dioxane (Ethomeen T/25)
Sol. in ethanol (Teric 17M15)
Sol. in ethyl acetate (Teric 17M15)
Sol. in glycerol trioleate (@ 5%) (Trymeen TAM-15)
Sol. in isopropanol (Hetoxamine T15; Mazeen T-15); sol. cloudy (Ethomeen T/25)
Partly sol. in kerosene (Teric 17M15)
Sol. in min. acid (Alkaminox T-15)
Partly sol. in min. oil (Teric 17M15)
Sol. in olein (Teric 17M15)
Sol. in most organic solvents (Alkaminox T-15)
Partly sol. in paraffin oil (Teric 17M15)
Sol. in perchloroethylene (Teric 17M15); sol. (@ 5%) (Trymeen TAM-15)
Disp. in Stoddard solvent (@ 5%) (Trymeen TAM-15)
Sol. in veg. oil (Teric 17M15)
Sol. in water (Alkaminox T-15; Hetoxamine T15; Icomeen T-15; Katapol PN-730; Mazeen T-15; Teric 17M15); sol. (@ 5%) (Trymeen TAM-15); sol. cloudy (Ethomeen T/25)

Ionic Nature:
Cationic (Crodamet 1.T15; Ethomeen T/25; Hetoxamine T15; Katapol PN-730; Mazeen T-15; Trymeen TAM-15)
Mildy cationic (Icomeen T-15)
Nonionic (Teric 17M15)

M.W.:
925 (Hetoxamine T15; Icomeen T-15)
930 (Chemeen T-15)

Sp.gr.:
1.028 (Mazeen T-15; Teric 17M15)
1.03 (Alkaminox T-15; Ethomeen T/25; Icomeen T-15)

Density:
8.5 lb/gal (Trymeen TAM-15)

Visc.:
175 cs (Trymeen TAM-15)
428 cps (20 C) (Teric 17M15)

POE (15) tallow amine (cont'd.)

M.P.:
  -19 ± 2 C (Teric 17M15)
Cloud Pt.:
  > 100 C (Trymeen TAM-15); (1% in hard water) (Teric 17M15)
HLB:
  13.3 (Teric 17M15)
  14.1 (Icomeen T-15)
  14.2 (Trymeen TAM-15)
Stability:
  Good in hard or saline waters and in reasonable concs. of acids or alkalis (Teric 17M15)
pH:
  8–10 (1% aq.) (Teric 17M15)
  9 (5% aq.) (Icomeen T-15)
Surface Tension:
  40 dynes/cm (0.01%, 20 C) (Teric 17M15)
  41 dynes/cm (0.1% sol'n.) (Ethomeen T/25; Icomeen T-15; Mazeen T-15)
TOXICITY/HANDLING:
  Skin irritant, severe eye irritant (Ethomeen T/25)
  May cause skin and eye irritation; spillages are slippery (Teric 17M15)
STD. PKGS.:
  55-gal (450 lb net) steel drums (Icomeen T-15)

POE (15) tallow aminopropylamine

SYNONYMS:
  PEG-15 tallow aminopropylamine (CTFA)
  PEG (15) tallow aminopropylamine
  N,N′, N′-POE (15)-N-tallow-1,3-diaminopropane
STRUCTURE:

$$R-N-(CH_2)_3-N \begin{cases} (CH_2CH_2O)_xH \\ (CH_2CH_2O)_yH \end{cases}$$
$$| \\ (CH_2CH_2O)_zH$$

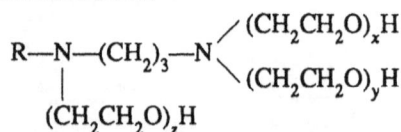

  where R represents the tallow radical and
      avg. $(x + y + z) = 15$
CAS No.:
  RD No.: 977066-71-9
TRADENAME EQUIVALENTS:
  Chemeen DT-15 [Chemax]
  Ethoduomeen T/25 [Akzo Chemie]

POE (15) tallow aminopropylamine *(cont'd.)*

CATEGORY:
Emulsifier, corrosion inhibitor, dyeing assistant
APPLICATIONS:
Farm products: agricultural oils/sprays (Chemeen DT-15)
Industrial applications: construction (Chemeen DT-15); petroleum industry (Ethoduomeen T/25); textile/leather processing (Chemeen DT-15); water treatment (Ethoduomeen T/25)
PROPERTIES:
Form:
Liquid (Chemeen DT-15; Ethoduomeen T/25)
Color:
Gardner 18 min. (Ethoduomeen T/25)
Composition:
95% active (Ethoduomeen T/25)
M.W.:
1020 (Chemeen DT-15)
Sp.gr.:
1.02 (Ethoduomeen T/25)
M.P.:
25 C max. (Ethoduomeen T/25)
Flash Pt.:
238 C (COC) (Ethoduomeen T/25)
TOXICITY/HANDLING:
Skin irritant, severe eye irritant (Ethoduomeen T/25)
STORAGE/HANDLING:
Avoid contact with strong oxidizing agents (Ethoduomeen T/25)
STD. PKGS.:
180 kg net bung-type drums (Ethoduomeen T/25)

POE (3) tridecyl ether

SYNONYMS:
PEG (3) tridecyl ether
PEG-3 tridecyl ether
POE (3) tridecyl alcohol
Trideceth-3 (CTFA)
Tridecyl alcohol, ethoxylated (3 EO)
Tridecyloxypoly (ethyleneoxy) ethanol (3 EO)
STRUCTURE:
$C_{13}H_{27}(OCH_2CH_2)_nOH$
where avg. $n = 3$

POE (3) tridecyl ether *(cont'd.)*

CAS No.:
 4403-12-7; 24938-91-8 (generic)
 RD No.: 977058-48-2
TRADENAME EQUIVALENTS:
 Chemal TDA-3 [Chemax Inc.]
 Emulphogene BC-420 [GAF]
 Hetoxol TD3 [Heterene]
 Iconol TDA-3 [BASF Wyandotte]
 Lipal 3TD [PVO Int'l.]
 Macol TD-3 [Mazer]
 Siponic TD-3 [Alcolac]
 Trycol TDA-3 [Emery]
 Volpo T3 [Croda Chem. Ltd.]
CATEGORY:
 Surfactant, emulsifier, detergent, wetting agent, dispersant, intermediate, foam stabilizer, solubilizer, leveling agent, emollient, lubricant, antifoam
APPLICATIONS:
 Bath products: bath oils (Hetoxol TD3)
 Cleansers: germicidal skin cleanser (Lipal 3TD)
 Cosmetic industry preparations: (Macol TD-3; Siponic TD-3); creams and lotions (Hetoxol TD3; Lipal 3TD; Siponic TD-3); hair preparations (Lipal 3TD); makeup (Lipal 3TD); shampoos (Hetoxol TD3); shaving preparations (Lipal 3TD)
 Household detergents: (Emulphogene BC-420; Hetoxol TD3; Siponic TD-3)
 Industrial applications: (Macol TD-3; Siponic TD-3); dyes and pigments (Hetoxol TD3); silicone products (Hetoxol TD3); textile/leather processing (Hetoxol TD3; Trycol TDA-3); waxes and oils (Emulphogene BC-420; Hetoxol TD3)
 Industrial cleaners: metal processing surfactants (Siponic TD-3); textile cleaning (Hetoxol TD3; Siponic TD-3)
 Pharmaceutical applications: (Macol TD-3); antiperspirant/deodorant (Lipal 3TD); ointments (Lipal 3TD)
PROPERTIES:
Form:
 Liquid (Chemal TDA-3; Hetoxol TD3; Iconol TDA-3; Lipal 3TD; Macol TD-3; Siponic TD-3; Trycol TDA-3; Volpo T3)
 Cloudy, slightly viscous but pourable liquid (Emulphogene BC-420)
Color:
 Water-white (Volpo T3)
 APHA 70 max. (Iconol TDA-3)
 Gardner 1 (Lipal 3TD)
 Gardner 1 max. (Trycol TDA-3)
 VCS 2 max. (50 C) (Emulphogene BC-420)
Odor:
 Mild, pleasant (Emulphogene BC-420)

POE (3) tridecyl ether (cont'd.)

Composition:
  100% active (Emulphogene BC-420; Iconol TDA-3; Lipal 3TD; Siponic TD-3; Trycol TDA-3)
  100% conc. (Macol TD-3)

Solubility:
  Sol. in arachis oil (Volpo T3)
  Sol. in butyl stearate (@ 5%) (Trycol TDA-3); partly sol. (Volpo T3)
  Sol. in ethanol (Volpo T3)
  Sol. in glycerol trioleate (@ 5%) (Trycol TDA-3)
  Sol. in isopropanol (Hetoxol TD3; Lipal 3TD)
  Sol. in kerosene (Volpo T3)
  Sol. in min. oil (Hetoxol TD3; Lipal 3TD; Volpo T3); disp. (Trycol TDA-3)
  Sol. in oil (Emulphogene BC-420)
  Sol. in oleic acid (Volpo T3)
  Sol. in oleyl alcohol (Volpo T3)
  Sol. in peanut oil (Lipal 3TD)
  Disp. in perchloroethylene (Trycol TDA-3)
  Sol. in propylene glycol (Lipal 3TD)
  Sol. in Stoddard solvent (@ 5%) (Trycol TDA-3)
  Sol. in toluene (Emulphogene BC-420)
  Sol. in trichlorethylene (Volpo T3)
  Disp. in water (Iconol TDA-3; Lipal 3TD; Trycol TDA-3); insol. (Emulphogene BC-420)

Ionic Nature:
  Nonionic (Emulphogene BC-420; Hetoxol TD3; Iconol TDA-3; Lipal 3TD; Macol TD-3; Siponic TD-3; Trycol TDA-3)

M.W.:
  325 (Iconol TDA-3)

Sp.gr.:
  0.92–0.94 (Emulphogene BC-420)
  0.95 (Iconol TDA-3)

Density:
  7.8 lb/gal (Trycol TDA-3)

Visc.:
  30 cps (Iconol TDA-3)
  30 cs (Trycol TDA-3)

Pour Pt.:
  –15 C (Iconol TDA-3)
  –7 to –1 C (Emulphogene BC-420)

Flash Pt.:
  > 200 F (PMCC) (Emulphogene BC-420)

Fire Pt.:
  > 165 C (Emulphogene BC-420)

POE (3) tridecyl ether (cont'd.)

Cloud Pt.:
  < 25 C (Trycol TDA-3); (1% aq.) (Iconol TDA-3)
  40–43 C (1% sol'n.) (Lipal 3TD)
HLB:
  7.9 (Hetoxol TD3; Macol TD-3; Siponic TD-3)
  8.0 (Iconol TDA-3; Trycol TDA-3; Volpo T3)
  8.0 ± 1 (Lipal 3TD)
Acid No.:
  1.0 max. (Volpo T3)
Iodine No.:
  1.0 max. (Volpo T3)
Hydroxyl No.:
  163–178 (Chemal TDA-3)
  165–175 (Hetoxol TD3; Lipal 3TD)
  170–180 (Volpo T3)
Stability:
  Stable to sulfuric acid and alkalis (Emulphogene BC-420)
  Stable over a wide pH range (Lipal 3TD)
  Stable to hydrolysis by acids and alkalis (Trycol TDA-3)
  Stable to alkalis and strong mineral acids (Volpo T3)
pH:
  6.0–7.5 (5% aq.) (Iconol TDA-3); (3%) (Volpo T3)
  6.5–8.5 (10% sol'n.) (Emulphogene BC-420)
Biodegradable: (Volpo T3)
STORAGE/HANDLING:
  Conc. mineral acids will react chemically with product (Trycol TDA-3)
STD. PKGS.:
  55-gal (420 lb net) steel drums (Iconol TDA-3)

POE (6) tridecyl ether

SYNONYMS:
  PEG-6 tridecyl ether
  PEG 300 tridecyl ether
  POE (6) tridecyl alcohol
  Trideceth-6 (CTFA)
STRUCTURE:
  $C_{13}H_{27}(OCH_2CH_2)_nOH$
    where avg. $n = 6$

POE (6) tridecyl ether (cont'd.)

CAS No.:
 24938-91-8 (generic)
 RD No.: 977058-49-3

TRADENAME EQUIVALENTS:
 Ahcowet DQ-114 [ICI United States]
 Chemal TDA-6 [Chemax]
 Emulphogene BC-610 [GAF]
 Ethosperse TDA-6 [Glyco]
 Hetoxol TD6 [Heterene]
 Lipal 6TD [PVO Int'l.]
 Lipocol TD-6 [Lipo]
 Macol TD-6 [Mazer]
 Renex 36 [ICI United States]
 Siponic TD-6 [Alcolac]
 Trycol TDA-6 [Emery]

CATEGORY:
 Wetting agent, detergent, emulsifier, dispersant, foam stabilizer, solubilizer, detergent, intermediate, leveling agent, scouring agent

APPLICATIONS:
 Bath products: bath oils (Hetoxol TD6)
 Cleansers: germicidal skin cleanser (Lipal 6TD)
 Cosmetic industry preparations: (Ethosperse TDA-6; Macol TD-6); creams and lotions (Hetoxol TD6; Lipal 6TD); hair preparations (Lipal 6TD); makeup (Lipal 6TD); perfumery (Lipocol TD-6); shampoos (Hetoxol TD6); shaving preparations (Lipal 6TD)
 Degreasers: (Siponic TD-6; Trycol TDA-6)
 Farm products: agricultural oils/sprays (Siponic TD-6)
 Household detergents: (Hetoxol TD6); alkaline cleaners (Emulphogene BC-610; Renex 36); dishwashing (Renex 36); mechanical dishwashing (Emulphogene BC-610)
 Industrial applications: (Ethosperse TDA-6; Macol TD-6); dyes and pigments (Hetoxol TD6; Renex 36); lubricating/cutting oils (Trycol TDA-6); paint mfg. (Renex 36); paper mfg. (Emulphogene BC-610); textile/leather processing (Ahcowet DQ-114; Hetoxol TD6); wood pulping (Emulphogene BC-610)
 Industrial cleaners: (Siponic TD-6); bottle cleaners (Renex 36); drycleaning compositions (Siponic TD-6); metal processing surfactants (Renex 36; Siponic TD-6); solvent cleaners (Siponic TD-6); textile cleaning (Ahcowet DQ-114; Emulphogene BC-610; Trycol TDA-6)
 Pharmaceutical applications: (Ethosperse TDA-6; Macol TD-6); antiperspirant/deodorant (Lipal 6TD); ointments (Lipal 6TD)

PROPERTIES:
Form:
 Liquid (Ahcowet DQ-114; Chemal TDA-6; Ethosperse TDA-6; Hetoxol TD6; Lipal

POE (6) tridecyl ether (cont'd.)

  6TD; Lipocol TD-6; Macol TD-6; Trycol TDA-6)
  Clear, oily liquid (Siponic TD-6)
  Cloudy to clear, thin oily liquid (Renex 36)
  Cloudy, slightly viscous liquid (Emulphogene BC-610)
Color:
  Colorless (Renex 36)
  Pale straw (Ethosperse TDA-6)
  VCS 2 max. (@ 50 C) (Emulphogene BC-610)
  < Gardner 1 (Trycol TDA-6)
  Gardner 1 (Lipal 6TD)
Composition:
  99% active (Siponic TD-6)
  100% active (Emulphogene BC-610; Ethosperse TDA-6; Lipal 6TD; Trycol TDA-6)
  100% conc. (Ahcowet DQ-114; Lipocol TD-6; Macol TD-6; Renex 36)

Solubility:
  Sol. in acetone (Emulphogene BC-610; Ethosperse TDA-6; Renex 36)
  Sol. in aniline (Renex 36)
  Sol. in butyl stearate (@ 5%) (Trycol TDA-6)
  Sol. in carbon tetrachloride (Renex 36)
  Sol. in Cellosolve (Renex 36)
  Sol. in ethanol (Emulphogene BC-610; Ethosperse TDA-6; Renex 36)
  Sol. in ethyl acetate (Ethosperse TDA-6; Renex 36)
  Sol. in glycerol trioleate (@ 5%) (Trycol TDA-6)
  Disp. cloudy in 5% aq. sol'n. of hydrochloric acid (Renex 36)
  Sol. in isopropanol (Hetoxol TD6; Lipal 6TD)
  Sol. in methanol (Emulphogene BC-610; Ethosperse TDA-6)
  Sol. in min. oil (Lipal 6TD); disp. (Emulphogene BC-610); disp. (@ 5%) (Trycol TDA-6)
  Sol. in oil (Siponic TD-6)
  Sol. in perchlorethylene (@ 5%) (Trycol TDA-6)
  Sol. in propylene glycol (Lipal 6TD)
  Disp. cloudy in 5% aq. sol'n. of sodium hydroxide (Renex 36)
  Sol. in Stoddard solvent (@ 5%) (Trycol TDA-6)
  Disp. cloudy in 5% aq. sol'n. of sulfuric acid (Renex 36)
  Sol. in toluene (Emulphogene BC-610; Renex 36)
  Sol. in toluol (Ethosperse TDA-6)
  Sol. in water (Hetoxol TD6; Lipal 6TD); disp. (Emulphogene BC-610; Ethosperse TDA-6); disp. cloudy (Renex 36); disp. (@ 5%) (Trycol TDA-6)

Ionic Nature:
  Nonionic (Ahcowet DQ-114; Emulphogene BC-610; Ethosperse TDA-6; Hetoxol TD6; Lipal 6TD; Lipocol TD-6; Macol TD-6; Renex 36; Siponic TD-6; Trycol TDA-6)

Sp.gr.:
0.97–0.99 (Emulphogene BC-610)
0.98 (Ethosperse TDA-6; Siponic TD-6)
1.0 (Renex 36)
Density:
8.2 lb/gal (Siponic TD-6; Trycol TDA-6)
Visc.:
50 cs (Trycol TDA-6)
80 cps (Ethosperse TDA-6; Renex 36; Siponic TD-6)
Pour Pt.:
4.5–7.5 C (Emulphogene BC-610)
Flash Pt.:
> 200 F (PMCC) (Emulphogene BC-610)
> 300 F (Renex 36)
355 F (Siponic TD-6)
Fire Pt.:
> 165 C (Emulphogene BC-610)
> 300 F (Renex 36)
Cloud Pt.:
< 25 C (Trycol TDA-6)
66–69 C (1% sol'n.) (Lipal 6TD)
< 32 F (1% aq.) (Renex 36)
HLB:
11.0 (Ethosperse TDA-6)
11.0 ± 1 (Lipal 6TD)
11.3 (Hetoxol TD6)
11.4 (Ahcowet DQ-114; Renex 36; Macol TD-6; Trycol TDA-6)
Acid No.:
1.0 max. (Ethosperse TDA-6; Renex 36)
> 2 (Siponic TD-6)
Hydroxyl No.:
110–125 (Chemal TDA-6; Lipal 6TD)
115–125 (Hetoxol TD6)
118–133 (Ethosperse TDA-6; Renex 36)
Stability:
Stable to sulfuric acid and alkalis (Emulphogene BC-610)
Stable over a wide pH range (Lipal 6TD)
Stable to hydrolysis by acids and alkalis (Trycol TDA-6)
pH:
6.0 (1% sol'n.) (Renex 36; Siponic TD-6)
6.0–8.0 (10% sol'n.) (Emulphogene BC-610)
Surface Tension:
27 dynes/cm (0.01% sol'n.) (Renex 36)

POE (6) tridecyl ether *(cont'd.)*

STORAGE/HANDLING:
Conc. min. acids will react chemically with product (Trycol TDA-6)
STD. PKGS.:
55-gal drums, tank trailers, truckloads (Siponic TD-6)

POE (12) tridecyl ether

SYNONYMS:
PEG-12 tridecyl ether
PEG 500 tridecyl ether
Trideceth-12 (CTFA)
STRUCTURE:
$C_{13}H_{27}(OCH_2CH_2)_nOH$
where avg. $n = 12$
CAS No.:
24938-91-8 (generic)
RD No.: 977058-51-7
TRADENAME EQUIVALENTS:
Ahcowet DQ-145 [ICI United States]
Chemal TDA-12 [Chemax]
Hetoxol TD-12 [Heterene]
Lipocol TD-12 [Lipo]
Renex 30 [ICI United States]
Siponic TD-12 [Alcolac]
CATEGORY:
Emulsifier, wetting agent, detergent, dispersant, foam stabilizer, solubilizer, leveling agent, intermediate, defoamer, conditioning agent
APPLICATIONS:
Bath products: bath oils (Hetoxol TD-12)
Cosmetic industry preparations: creams and lotions (Hetoxol TD-12; Lipocol TD-12; Siponic TD-12); shampoos (Hetoxol TD-12)
Household detergents: (Hetoxol TD-12); laundry detergent (Renex 30)
Industrial applications: dyes and pigments (Hetoxol TD-12; Lipocol TD-12); textile/leather processing (Ahcowet DQ-145; Hetoxol TD-12); waxes and oils (Hetoxol TD-12)
Industrial cleaners: bottle cleaners (Renex 30); metal processing surfactants (Renex 30); textile cleaning (Renex 30)
Pharmaceutical applications: antiperspirant/deodorant (Lipocol TD-12); depilatories (Lipocol TD-12)

PROPERTIES:
Form:
Liquid (Ahcowet DQ-145; Chemal TDA-12; Hetoxol TD-12; Siponic TD-12)
Clear to hazy liquid (precipitates on standing) (Renex 30)
Paste (Lipocol TD-12)

Color:
Colorless (Renex 30)
White (Lipocol TD-12)

Composition:
70% conc. (Siponic TD-12)
100% active (Lipocol TD-12)
100% conc. (Ahcowet DQ-145; Renex 30)

Solubility:
Sol. in lower alcohols (Renex 30)
Sol. in butyl Cellosolve (Renex 30)
Sol. in carbon tetrachloride (Renex 30)
Sol. in ethylene glycol (Renex 30)
Sol. in isopropanol (Hetoxol TD-12)
Sol. in propylene glycol (Renex 30)
Slightly sol. in veg. oils (Renex 30)
Sol. in water (Hetoxol TD-12; Renex 30)
Sol. in xylene (Renex 30)

Ionic Nature:
Nonionic (Ahcowet DQ-145; Hetoxol TD-12; Lipocol TD-12; Renex 30; Siponic TD-12)

Sp.gr.:
1.0 (Renex 30)

Visc.:
60 cps (Renex 30)

Pour Pt.:
55 F (Renex 30)

Flash Pt.:
> 300 F (Renex 30)

Fire Pt.:
> 300 F (Renex 30)

Cloud Pt.:
183 F (1% aq.) (Renex 30)

HLB:
14.5 (Ahcowet DQ-145; Hetoxol TD-12; Renex 30; Siponic TD-12)
14.6 ± 1 (Lipocol TD-12)

Acid No.:
1.0 max. (Lipocol TD-12; Renex 30)

POE (12) tridecyl ether *(cont'd.)*

Hydroxyl No.:
  70–85 (Lipocol TD-12)
  72–87 (Chemal TDA-12; Hetoxol TD-12)
  75–85 (Renex 30)
Stability:
  Acid and alkaline stable (Lipocol TD-12)
pH:
  6.0 (1% in dist. water) (Renex 30)
Surface Tension:
  28 dynes/cm (0.01% sol'n.) (Renex 30)
TOXICITY/HANDLING:
  Causes eye irritation; avoid contact with eyes (Renex 30)

Potassium lauryl sulfate *(CTFA)*

SYNONYMS:
  Sulfuric acid, monododecyl ester, potassium salt
EMPIRICAL FORMULA:
  $C_{12}H_{26}O_4S \cdot K$
STRUCTURE:
  $CH_3(CH_2)_{10}CH_2OSO_3K$
CAS No.:
  4706-78-9
TRADENAME EQUIVALENTS:
  Conco Sulfate P [Continental]
CATEGORY:
  Detergent, wetting agent
APPLICATIONS:
  Bath products: bubble bath (Conco Sulfate P)
  Cleansers: cleansing lotions (Conco Sulfate P)
  Cosmetic industry preparations: shampoos (Conco Sulfate P)
  Household detergents: (Conco Sulfate P); carpet & upholstery shampoos (Conco Sulfate P)
PROPERTIES:
Form:
  Paste (Conco Sulfate P)
Color:
  White (Conco Sulfate P)
Odor:
  Pleasant (Conco Sulfate P)

Potassium lauryl sulfate *(cont'd.)*

Composition:
   29.0–32.5% active (Conco Sulfate P)
Visc.:
   225 cps max. (Conco Sulfate P)
Cloud Pt.:
   2.5 C (Conco Sulfate P)
pH:
   7.0–8.0 (10% sol'n.) (Conco Sulfate P)

PPG-3-Isosteareth-9 (CTFA)

SYNONYMS:
   Polyoxyethylene (9) polyoxypropylene (3) isostearyl ether
   Polyoxypropylene (3) polyoxyethylene (9) isostearyl ether
STRUCTURE:

$$R(OCHCH_2)_x(OCH_2CH_2)_yOH$$
$$|$$
$$CH_3$$

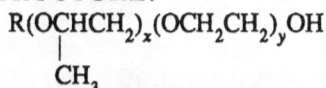

   where R represents the isostearyl radical,
         avg. $x = 3$ and
         avg. $y = 9$
CAS No.:
   RD No.: 977068-36-2
TRADENAME EQUIVALENTS:
   Arosurf 66-PE12 [Sherex]
CATEGORY:
   Detergent, emulsifier, dispersant, spreading agent
APPLICATIONS:
   Bath products: bath oils (Arosurf 66-PE12)
   Cosmetic industry preparations: (Arosurf 66-PE12); conditioners (Arosurf 66-PE12);
      hair dressings (Arosurf 66-PE12); shampoos (Arosurf 66-PE12); shaving prepara-
      tions (Arosurf 66-PE12); skin creams and lotions (Arosurf 66-PE12); toiletries
      (Arosurf 66-PE12)
PROPERTIES:
Form:
   Liquid (Arosurf 66-PE12)
Color:
   Gardner 1 (Arosurf 66-PE12)
Composition:
   100% conc. (Arosurf 66-PE12)

PPG-3-Isosteareth-9 *(cont'd.)*

Ionic Nature:
   Nonionic (Arosurf 66-PE12)
M.P.:
   −10 C (Arosurf 66-PE12)
HLB:
   12.2 (Arosurf 66-PE12)
Stability:
   Good (Arosurf 66-PE12)
pH:
   7 (1% sol'n. in DW) (Arosurf 66-PE12)
TOXICITY/HANDLING:
   Toxic orally and irritating to eyes at 100%; nontoxic orally but irritating to eyes @ 5% sol'n. (Arosurf 66-PE12)

Propylene glycol monolaurate

SYNONYMS:
   Dodecanoic acid, 2-hydroxypropyl ester
   Dodecanoic acid, monoester with 1,2-propanediol
   2-Hydroxypropyl dodecanoate
   Propylene glycol laurate (CTFA)
EMPIRICAL FORMULA:
   $C_{15}H_{30}O_3$
STRUCTURE:

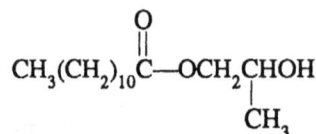

CAS No.:
   142-55-2; 27194-74-7
TRADENAME EQUIVALENTS:
   Cithrol PGML N/E [Croda Chem. Ltd.]
   Drewmulse PGML [PVO Int'l.]
   Hodag PGML [Hodag]
   Kessco Propylene Glycol Monolaurate E [Armak]
   Schercemol PGML [Scher]
Self-emulsifying grades:
   Cithrol PGML S/E [Croda Chem. Ltd.]

Propylene glycol monolaurate *(cont'd.)*

CATEGORY:

Emulsifier, coemulsifier, stabilizer, wetting agent, lubricant, antistat, emollient, chemical specialty, corrosion inhibitor, base, solubilizer, coupling agent, solvent, plasticizer, defoaming agent

APPLICATIONS:

Cosmetic industry preparations: (Cithrol PGML N/E, PGML S/E; Hodag PGML; Schercemol PGML); perfumery (Schercemol PGML); shampoos (Schercemol PGML)

Farm products: agricultural oils/sprays (Schercemol PGML); insecticides/pesticides (Schercemol PGML)

Food applications: flavors (Schercemol PGML)

Household detergents: liquid detergents (Schercemol PGML)

Industrial applications: (Cithrol PGML N/E, PGML S/E); dyes and pigments (Schercemol PGML); plastics (Schercemol PGML)

Pharmaceutical applications: (Cithrol PGML N/E, PGML S/E; Hodag PGML); sunscreens (Schercemol PGML)

PROPERTIES:

Form:

Liquid (Cithrol PGML N/E, PGML S/E; Drewmulse PGML; Hodag PGML; Kessco Propylene Glycol Monolaurate E)

Clear liquid (Schercemol PGML)

Color:

Gardner 5 max. (Schercemol PGML)

Odor:

Mild (Schercemol PGML)

Composition:

100% conc. (Cithrol PGML N/E, PGML S/E; Hodag PGML)

Solubility:

Sol. in alcohols (Schercemol PGML)

Sol. in aliphatic hydrocarbons (Schercemol PGML)

Sol. in aromatic hydrocarbons (Schercemol PGML)

Sol. in chlorinated hydrocarbons (Schercemol PGML)

Sol. in esters (Schercemol PGML)

Sol. in glycol ethers (Schercemol PGML)

Disp. in glycols (Schercemol PGML)

Sol. in ketones (Schercemol PGML)

Sol. in min. oil (Schercemol PGML)

Sol. in most organic solvents (Schercemol PGML)

Disp. in polyols (Schercemol PGML)

Disp. in triols (Schercemol PGML)

Sol. in veg. oil (Schercemol PGML)

Insol. in water (Schercemol PGML)

Propylene glycol monolaurate (*cont'd.*)

Ionic Nature:
Nonionic (Cithrol PGML N/E; Kessco Propylene Glycol Monolaurate E)
Anionic (Cithrol PGML S/E)
M.W.:
258 theoret. (Schercemol PGML)
Sp.gr.:
0.905 ± 0.01 (Schercemol PGML)
Density:
7.45 lb/gal (Schercemol PGML)
F.P.:
5 C max.(Schercemol PGML)
Flash Pt.:
160 C min. (OC) (Schercemol PGML)
HLB:
3.2 (Kessco Propylene Glycol Monolaurate E)
4.0 (Hodag PGML)
Acid No.:
5.0 max. (Schercemol PGML)
Iodine No.:
Nil (Schercemol PGML)
Saponification No.:
225–240 (Schercemol PGML)
Ref. Index:
1.4425 (Schercemol PGML)
pH:
7.0 (10% disp.) (Schercemol PGML)
STD. PKGS.:
55-gal (425 lb net) epoxy-lined, bung-head steel drums (Schercemol PGML)

Propylene glycol monooleate

SYNONYMS:
9-Octadecenoic acid, monoester with 1,2-propanediol
Propylene glycol oleate (CTFA)
EMPIRICAL FORMULA:
$C_{21}H_{40}O_3$
STRUCTURE:

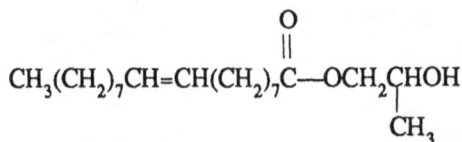

Propylene glycol monooleate *(cont'd.)*

CAS No.:
1330-80-9
TRADENAME EQUIVALENTS:
Atlas G-2185 [ICI United States]
Cithrol PGMO N/E [Croda Chem. Ltd.]
Radiasurf 7206 [Oleofina S.A.]
Self-emulsifying grades:
Cithrol PGMO S/E [Croda Chem. Ltd.]
CATEGORY:
Emulsifier, coemulsifier, stabilizer, wetting agent, lubricant, antistat, opacifier, dispersant, w/o emulgent, scouring and detergent aid, defoamer, plasticizer, rust inhibitor
APPLICATIONS:
Cosmetic industry preparations: (Cithrol PGMO N/E, PGMO S/E; Radiasurf 7206)
Farm products: insecticides/pesticides (Radiasurf 7206)
Food applications: food emulsifying (Atlas G-2185)
Industrial applications: (Cithrol PGMO N/E, PGMO S/E); dyes and pigments (Radiasurf 7206); lubricating/cutting oils (Radiasurf 7206); paint mfg. (Radiasurf 7206); plastics (Radiasurf 7206); polishes and waxes (Radiasurf 7206); printing inks (Radiasurf 7206); textile/leather processing (Radiasurf 7206)
Pharmaceutical applications: (Cithrol PGMO N/E, PGMO S/E; Radiasurf 7206)
PROPERTIES:
Form:
Liquid (Atlas G-2185; Cithrol PGMO N/E, PGMO S/E; Radiasurf 7206)
Color:
Amber (Radiasurf 7206)
Yellow (Atlas G-2185)
Composition:
100% conc. (Cithrol PGMO N/E, PGMO S/E)
Solubility:
Sol. in benzene (@ 10%) (Radiasurf 7206)
Sol. cloudy in hexane (@ 10%) (Radiasurf 7206)
Sol. in isopropanol (@ 10%) (Radiasurf 7206)
Sol. cloudy in min. oil (@ 10%) (Radiasurf 7206)
Sol. in common organic solvents (Atlas G-2185)
Sol. in trichlorethylene (@ 10%) (Radiasurf 7206)
Sol. in veg. oil (Atlas G-2185); (@ 10%) (Radiasurf 7206)
Ionic Nature:
Nonionic (Cithrol PGMO N/E; Radiasurf 7206)
Anionic (Cithrol PGMO S/E)
M.W.:
460 avg. (Radiasurf 7206)
Sp.gr.:
0.900 (37.8 C) (Radiasurf 7206)

305

Propylene glycol monooleate *(cont'd.)*

Visc.:
   18.80 cps (37.8 C) (Radiasurf 7206)
Congealing Pt.:
   40 F or less (Atlas G-2185)
Flash Pt.:
   157 C (COC) (Radiasurf 7206)
Cloud Pt.:
   –9 C (Radiasurf 7206)
HLB:
   2.3 avg. (Radiasurf 7206)
   3.2 (Atlas G-2185)
Acid No.:
   2.0 max. (Radiasurf 7206)
Iodine No.:
   74–80 (Radiasurf 7206)
Saponification No.:
   172–180 (Radiasurf 7206)
STD. PKGS.:
   190-kg net bung drums or bulk (Radiasurf 7206)

Propylene glycol monostearate

SYNONYMS:
   Octadecanoic acid, monoester with 1,2-propanediol
   Propylene glycol stearate (CTFA)
EMPIRICAL FORMULA:
   $C_{21}H_{42}O_3$
STRUCTURE:

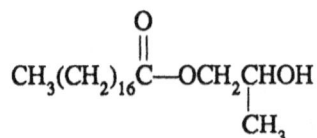

CAS No.:
   1323-39-3
TRADENAME EQUIVALENTS:
   Aldo PGHMS [Lonza]
   Aldo PMS [Glyco]
   Cithrol PGMS N/E [Croda Chem. Ltd.]
   Cyclochem PGMS [Cyclo Chem.]

Propylene glycol monostearate (cont'd.)

TRADENAME EQUIVALENTS *(cont'd.):*
Drewlene 10 [PVO Int'l.]
Drewmulse PGMS [Drew Produtos Quimicos]
Durpro 107-55 [Durkee/SCM Corp.]
Emerest 2380 [Emery]
Grindtek PGMS-90 [Grindsted]
Hodag PGMS [Hodag]
Homotex PS-90 [Kao Corp.]
Kessco Propylene Glycol Monostearate Pure [Armak]
Lipo PGMS [Lipo]
Mapeg PGMS [Mazer]
Mazol PGMS [Mazer]
Monosteol [Gattefosse]
Nikkol PMS-1C [Nikko]
PMS-33 [Hefti Ltd.]
Radiasurf 7201 [Oleofina S.A.]
Schercemol PGMS [Scher]
Tegin P412 [Th. Goldschmidt A.G.]
Self-emulsifying grades:
Cerasynt PA [Van Dyk]
Cithrol PGMS S/E [Croda Chem. Ltd.]
Emerest 2381 [Emery]
Lexemul P [Inolex]
Nikkol PMS-1CSE [Nikko]
Tegin P, P-411SE [Goldschmidt Chem.]
Witconol RHP [Witco/Organics]

CATEGORY:
Emulsifier, coemulsifier, stabilizer, emollient, lubricant, thickener, conditioner, foam modifier, solubilizer, plasticizer, occlusive agent, opacifier, pearlescent, spreading agent, dispersant, wetting aid, antistat, w/o emulgent, scouring and detergent aid, defoamer, corrosion inhibitor

APPLICATIONS:
Bath products: bath oils (Lipo PGMS)
Cosmetic industry preparations: (Aldo PGHMS, PMS; Cithrol PGMS N/E, PGMS S/E; Emerest 2380, 2381; Hodag PGMS; Mapeg PGMS; Monosteol; Nikkol PMS-1C; PMS-33; Radiasurf 7201; Schercemol PGMS; Tegin P); conditioners (Aldo PMS; Schercemol PGMS); creams and lotions (Cyclochem PGMS; Durpro 107-55; Emerest 2381; Kessco Propylene Glycol Monostearate Pure; Lexemul P; Lipo PGMS; Schercemol PGMS; Tegin P-411SE); hair preparations (Aldo PMS; Schercemol PGMS); hand care products (Kessco Propylene Glycol Monostearate Pure); makeup (Cyclochem PGMS; Durpro 107-55; Kessco Propylene Glycol Monostearate Pure; PMS-33); personal care products (Schercemol PGMS); perfumery (Kessco Propylene Glycol Monostearate Pure); shampoos (Aldo PMS; Drewmulse

Propylene glycol monostearate (cont'd.)

PGMS; PMS-33; Schercemol PGMS); toiletries (Aldo PMS)

Farm products: insecticides/pesticides (Radiasurf 7201)

Food applications: (Homotex PS-90; Mazol PGMS; PMS-30); food emulsifying (Drewlene 10; Mazol PGMS)

Household detergents: (Aldo PMS); dishwashing (Emerest 2381)

Industrial applications: (Aldo PGHMS; Cithrol PGMS N/E, PGMS S/E; Witconol RHP); dyes and pigments (Drewmulse PGMS; Radiasurf 7201; Tegin P-411SE); industrial processing (Mapeg PGMS); lubricating/cutting oils (Mapeg PGMS; Radiasurf 7201); metalworking (Mapeg PGMS); paint mfg. (Radiasurf 7201); plastics (Radiasurf 7201); polishes and waxes (Radiasurf 7201); printing inks (Radiasurf 7201); textile/leather processing (Mapeg PGMS; Radiasurf 7201)

Pharmaceutical applications: (Aldo PGHMS, PMS; Cithrol PGMS N/E, S/E; Hodag PGMS; Mapeg PGMS; Nikkol PMS-1C; PMS-33; Radiasurf 7201); antiperspirant/deodorant (Aldo PMS)

PROPERTIES:

Form:

Paste (Witconol RHP)

Soft solid (Cyclochem PGMS)

Solid (Cithrol PGMS N/E, PGMS S/E; Drewlene 10; Emerest 2380; Hodag PGMS; Mapeg PGMS; Mazol PGMS; Monosteol; Nikkol PMS-1C, PMS-1CSE; PMS-33; Schercemol PGMS; Tegin P)

Block (Grindtek PGMS-90)

Beads (Aldo PGHMS; Drewlene 10; Emerest 2381; Homotex PS-90)

Flake (Aldo PMS; Cerasynt PA; Drewlene 10; Drewmulse PGMS; Kessco Propylene Glycol Monostearate Pure; Lexemul P)

Solid wax (Lipo PGMS)

Waxy (Tegin P-411SE, P412)

Color:

White (Aldo PMS; Lipo PGMS)

Whitish (Grindtek PGMS-90)

Off-white (Cyclochem PGMS)

White-cream (Mazol PGMS; Schercemol PGMS)

Light yellow (Mapeg PGMS)

Yellow (Drewlene 10)

Gardner 2 (Emerest 2381)

Odor:

Slight, typical (Schercemol PGMS)

Composition:

48% active (Drewmulse PGMS)

55% PGME min. (Durpro 107-55)

100% active (Emerest 2381)

100% conc. (Aldo PGHMS; Cerasynt PA; Cithrol PGMS N/E, PGMS S/E; Emerest 2380; Hodag PGMS; Homotex PS-90; Lexemul P; Mazol PGMS; Nikkol PMS-1C,

Propylene glycol monostearate *(cont'd.)*

PMS-1CSE; PMS-33; Tegin P, P-411SE; Witconol RHP)

Solubility:
Sol. in alcohols (Schercemol PGMS)
Sol. in aromatic hydrocarbons (Schercemol PGMS)
Sol. in chlorinated hydrocarbons (Schercemol PGMS)
Sol. in esters (Schercemol PGMS)
Sol. in ethanol (Grindtek PGMS-90); sol. hot (Aldo PMS); disp. (Mazol PGMS)
Disp. in glycerol (Schercemol PGMS)
Disp. in glycols (Schercemol PGMS)
Sol. in isopropanol (Emerest 2381; Mapeg PGMS)
Sol. in min. oil (Mapeg PGMS; Schercemol PGMS); sol. hot (Aldo PMS); disp. (Emerest 2381)
Sol. in most organic solvents (Schercemol PGMS)
Sol. warm in paraffin oil (Grindtek PGMS-90)
Sol. warm in peanut oil (Grindtek PGMS-90)
Disp. in polyols (Schercemol PGMS)
Sol. in propylene glycol (Mapeg PGMS; Mazol PGMS); sol. warm (Grindtek PGMS-90)
Sol. in soybean oil (Mapeg PGMS); partly sol. (Mazol PGMS)
Sol. in toluene (Grindtek PGMS-90)
Sol. in toluol (Mapeg PGMS); disp. hot (Emerest 2381)
Sol. hot in veg. oil (Aldo PMS)
Disp. in water (Emerest 2381); insol. (Aldo PMS; Mapeg PGMS; Schercemol PGMS)
Sol. warm in white spirit (Grindtek PGMS-90)

Ionic Nature:
Nonionic (Aldo PGHMS, PMS; Cerasynt PA; Cithrol PGMS N/E; Cyclochem PGMS; Drewmulse PGMS; Durpro 107-55; Emerest 2380; Hodag PGMS; Homotex PS-90; Mazol PGMS; Nikkol PMS-1C, PMS-1CSE; Radiasurf 7201; Schercemol PGMS; Witconol RHP)
Anionic (Cithrol PGMS S/E; Emerest 2381; Tegin P)

M.W.:
326 theoret. (Schercemol PGMS)

F.P.:
40 C (Emerest 2381)

M.P.:
33.5–38.5 C (Kessco Propylene Glycol Monostearate Pure)
35 C (Schercemol PGMS)
36 C (Cyclochem PGMS; Mapeg PGMS)
46–52 C (Durpro 107-55)
110 F (Drewlene 10)

Flash Pt.:
> 170 C (OC) (Schercemol PGMS)
390 F (COC) (Kessco Propylene Glycol Monostearate Pure)

Propylene glycol monostearate *(cont'd.)*

HLB:
   1.8 (Emerest 2380)
   2.2 (Durpro 107-55)
   2.3 (Mapeg PGMS)
   2.8 (Tegin P-411SE)
   3.0 (Aldo PGHMS)
   3.0 ± 1 (Lipo PGMS)
   3.4 (Drewlene 10; Hodag PGMS; Homotex PS-90; Mazol PGMS)
   3.5 (Grindtek PGMS-90; Nikkol PMS-1C; PMS-33)
   4.0 (Emerest 2381; Lexemul P; Monosteol; Nikkol PMS-1CSE)
   4.4 (Tegin P)

Acid No.:
   3.0 max. (Aldo PMS; Kessco Propylene Glycol Monostearate Pure)
   4.0 max. (Schercemol PGMS)
   5.0 max. (Cyclochem PGMS; Mapeg PGMS; Mazol PGMS)
   6.0 max. (Lipo PGMS)
   20 (Emerest 2381)

Iodine No.:
   0.5 max. (Kessco Propylene Glycol Monostearate Pure)
   1.0 max. (Aldo PMS; Schercemol PGMS)
   2.0 (Mazol PGMS)
   6.0 max. (Mapeg PGMS)

Saponification No.:
   170 (Emerest 2381)
   170–180 (Mazol PGMS)
   170–185 (Aldo PMS)
   170–190 (Drewlene 10)
   172–182 (Mapeg PGMS)
   180 ± 5 (Schercemol PGMS)
   180–192 (Lipo PGMS)
   186 (Cyclochem PGMS)

Sodium capryl lactylate (CTFA)

EMPIRICAL FORMULA:
$C_{16}H_{28}O_6 \cdot Na$

STRUCTURE:

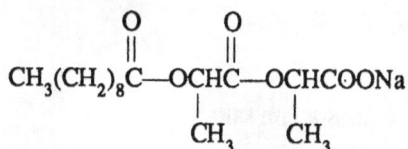

CAS No.:
RD No.: 977067-37-0

TRADENAME EQUIVALENTS:
Pationic 122A [Patco]

CATEGORY:
Surfactant, microbicide

PROPERTIES:

Form:
Clear viscous liquid (Pationic 122A)

Color:
Amber (Pationic 122A)

Composition:
100% conc. (Pationic 122A)

Solubility:
Sol. in isopropanol (@ 1%) (Pationic 122A)
Sol. in isopropyl myristate (@ 1%) (Pationic 122A)
Sol. in propylene glycol (@ 1%) (Pationic 122A)
Sol. hazy in distilled water (@ 1%) (Pationic 122A)

Ionic Nature: Anionic

Visc.:
24 cps (2% in dist. water) (Pationic 122A)

HLB:
11.3 (Pationic 122A)

Acid No.:
65–85 (Pationic 122A)

Saponification No.:
235–265 (Pationic 122A)

pH:
5.25 (2% aq.) (Pationic 122A)

311

Sodium capryl lactylate *(cont'd.)*

Surface Tension:
   24.72 dynes/cm (0.1% conc.) (Pationic 122A)
TOXICITY/HANDLING:
   Moderate skin irritant in conc. form (Pationic 122A)
STORAGE/HANDLING:
   Store under cool, dry conditions (Pationic 122A)

Sodium cocoyl glutamate (CTFA)

SYNONYMS:
   L-Glutamic acid, N-coco acyl derivatives, monosodium salts
   Monosodium N-cocoyl-L-glutamate
   Sodium salt of coconut acid amide of glutamic acid
STRUCTURE:
   HOOC—CH$_2$CH$_2$CH—COONa
   $\qquad\qquad\qquad$ |
   $\qquad\qquad$ NH—CR
   $\qquad\qquad\qquad\quad$ ||
   $\qquad\qquad\qquad\quad$ O

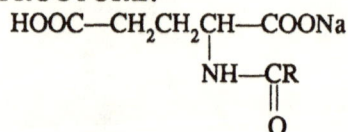

   where RCO⁻ represents the coconut acid radical
CAS No.:
   68187-32-6
TRADENAME EQUIVALENTS:
   Acylglutamate CS-11 [Ajinomoto]
   Amisoft GS-11 [Ajinomoto]
CATEGORY:
   Detergent, emulsifier, foaming agent, emollient, bactericide
APPLICATIONS:
   Cosmetic industry preparations: (Amisoft GS-11); personal care products (Acyl-glutamate CS-11)

PROPERTIES:
Form:
   Powder (Acylglutamate CS-11; Amisoft GS-11)
Composition:
   100% active (Acylglutamate CS-11; Amisoft GS-11)
Ionic Nature:
   Anionic (Acylglutamate CS-11)
Stability:
   Good in hard water (Acylglutamate CS-11)
Biodegradable: (Acylglutamate CS-11)

312

Sodium cocoyl glutamate *(cont'd.)*

TOXICITY/HANDLING:
Nonskin irritation even for infants and eczema cases (Acylglutamate CS-11)

Sodium dodecylbenzenesulfonate *(CTFA)*

SYNONYMS:
Benzenesulfonic acid, dodecyl-, sodium salt
Dodecylbenzene sodium sulfonate
Dodecylbenzene sulfonate, sodium salt
Dodecylbenzenesulfonic acid, sodium salt
Sodium lauryl benzene sulfonate

EMPIRICAL FORMULA:
$C_{18}H_{30}O_3S \cdot Na$

STRUCTURE:

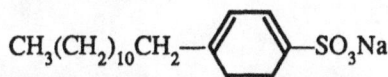

$$CH_3(CH_2)_{10}CH_2 - \langle \bigcirc \rangle - SO_3Na$$

CAS No.:
25155-30-0

TRADENAME EQUIVALENTS:
Arylan S Flake, SBC 25, SC30, SX Flake [Lankro]
Bio Soft D-35X, D-40, D-60, D-62 [Stepan]
Calsoft F-90, L-40, L-60 [Pilot]
Conco AAS-35, AAS-40, AAS-45S, AAS-65, AAS-90 [Continental]
Cycloryl DDB40 [Cyclo]
Elfan WA, WA Pulver [Akzo Chemie]
Hetsulf 40, 40X, 60S [Heterene]
Kadif 50 Flakes [Witco Chem. Ltd.]
Manro BA25, BA30 [Manro]
Manro SDBS30 [Manro]
Marlon A350, A360, A365, A375, A390, A396 [Chem. Werke Huls]
Marlon AFR [Chem. Werke Huls] (modified)
Mercol 15, 25, 30, Special [Capital City]
Merpisap AE50, AE60, AE70 [Kempen]
Merpisap AP80W, AP85W, AP90, AP90P [Kempen]
Merpisap KH30 [Kempen]
Nacconol 35SL, 40F, 90F [Stepan]
Nacconol 40G, 90G [Stepan]
Nansa 1106/P, 1169/P [Albright & Wilson/Detergents]
Nansa HS80/S [Albright & Wilson/Detergents]

Sodium dodecylbenzenesulfonate *(cont'd.)*

TRADENAME EQUIVALENTS *(cont'd.):*
  Nansa HS80 Soft, HS85/S [Albright & Wilson/Marchon]
  Nansa SB25, SB62 [Albright & Wilson/Detergents]
  Nansa SL30 [Albright & Wilson/Marchon]
  Nansa SS30, SS60 [Albright & Wilson/Marchon]
  Reworyl NKS50, NKS100 [Rewo Chem. Werke]
  Richonate 40B, 45B, 45BX, 50BD, 60B [Richardson]
  Richonate 1850, 1850U, C-50H [Richardson]
  Siponate DS-4, DS-10 [Alcolac]
  Sterling AB-40, AB-80 [Canada Packers]
  Sterling LA Paste 55%, 60% [Canada Packers]
  Sul-fon-ate AA-9, AA-10 [Tennessee/Cities Service]
  Sul-fon-ate LA-10 [Tennessee]
  Surco 55S, 60S [Onyx]
  Tairygent CB-1, CB-2 [Formosa Chem & Fibre]
  Witconate 60B, 90 Flakes [Witco/Organics]
  Witconate 1250 Slurry [Witco/Organics]
  Witconate C-50H [Witco/Organics]
CATEGORY:
  Wetting agent, emulsifier, detergent, foaming agent, dispersant, intermediate, penetrant, surfactant, stabilizer, solubilizer
APPLICATIONS:
  Automobile cleaners: (Conco AAS-45S); car shampoo (Marlon AFR; Sterling AB-40, AB-80)
  Bath products: bubble bath (Calsoft L-40; Nacconol 35SL)
  Cleansers: (Marlon AFR)
  Cosmetic industry preparations: (Conco AAS-45S; Mercol 15, 25, 30, Special; Nacconol 35SL; Sul-fon-ate AA-9, AA-10; Witconate 60B, 1250 Slurry); shampoos (Calsoft L-60; Nacconol 35SL, 90F); toiletries (Witconate 60B, 1250 Slurry)
  Degreasers: (Calsoft F-90)
  Farm products: agricultural oils/sprays (Calsoft L-40, L-60; Nacconol 40G, 90G; Nansa HS80/S); wettable powders (Nansa HS80/S); fertilizers (Nacconol 35SL, 90F; Sul-fon-ate AA-9, AA-10); herbicides (Nacconol 40F, 40G, 90G); insecticides/pesticides (Mercol 15, 25, 30, Special; Nacconol 40F, 40G, 90F, 90G; Nansa HS80 Soft, HS85/S; Sul-fon-ate AA-9, AA-10)
  Food applications: food cleaning (Mercol 15, 25, 30, Special); dairy cleaners (Conco AAS-45S; Nacconol 35SL; Sul-fon-ate AA-9, AA-10); fruit/vegetable washing (Calsoft L-40, L-60; Kadif 50 Flakes; Nacconol 35SL; Siponate DS-4, DS-10; Sul-fon-ate AA-9, AA-10); food packaging (Siponate DS-4, DS-10)
  Household detergents: (Calsoft L-40, L-60; Conco AAS-45S; Kadif 50 Flakes; Manro BA25, BA30, SDBS30; Marlon A350, A360, A365, A375; Mercol 15, 25, 30, Special; Merpisap AE50, AE60, AE70, AP80W, AP85W, AP90, AP90P, KH30; Nacconol 35SL, 40G, 90G; Nansa SL30, SS30, SS60; Reworyl NKS50, NKS100;

Sodium dodecylbenzenesulfonate *(cont'd.)*

Richonate 40B, 45B, 45BX, 50BD, 60B, 1850, 1850U, C-50H; Sterling AB-40, AB-80); all-purpose cleaner (Calsoft F-90; Marlon A390, A396; Nacconol 40F; Sterling LA Paste 55%, 60%; Sul-fon-ate AA-9, AA-10); built detergents (Calsoft L-60; Kadif 50 Flakes); carpet & upholstery shampoos (Calsoft L-40, L-60; Conco AAS-45S; Kadif 50 Flakes; Nacconol 35SL); detergent base (Arylan SBC25, SC30, SX Flake; Cycloryl DDB40; Hetsulf 40, 60S; Sterling LA Paste 55%, 60%; Surco 55S, 60S; Witconate 60B, 90 Flakes, 1250 Slurry); dishwashing (Conco AAS-45S; Elfan WA; Kadif 50 Flakes; Marlon AFR; Nacconol 40F; Richonate 40B, 45B, 45BX, 50BD, 60B, 1850, 1850U, C-50H); hard surface cleaner (Calsoft F-90; Nansa HS80/S); heavy-duty cleaner (Calsoft L-60; Elfan WA, WA Pulver; Hetsulf 60S; Nacconol 40F; Nansa HS80 Soft, HS80/S, HS85/S; Siponate DS-10); laundry detergent (Calsoft F-90; Kadif 50 Flakes; Mercol 15, 25, 30, Special; Nacconol 35SL, 40F, 40G, 90F, 90G; Nansa HS80 Soft, HS85/S; Sul-fon-ate AA-9, AA-10); light-duty cleaners (Hetsulf 60S; Nacconol 40F); liquid detergents (Bio Soft D-35X, D-40, D-60, D-62; Conco AAS-35; Hetsulf 40X; Manro BA25, BA30, SDBS30; Marlon A350, A360, A365, A375; Merpisap AE50, AE60, AE70, KH30; Reworyl NKS50; Sterling LA Paste 55%, 60%); paste detergents (Marlon A350, A360, A365, A375; Merpisap AE50, AE60, AE70; Reworyl NKS50); powdered detergents (Arylan SX Flake; Merpisap AE50, AE60, AE70, AP80W, AP85W, AP90, AP90P; Nansa HS80/S; Reworyl NKS100; Sterling AB-40, AB-80)

Industrial applications: cement/concrete industry (Calsoft L-40, L-60; Nacconol 40F, 40G, 90G; Sul-fon-ate AA-9, AA-10); construction (Calsoft L-40, L-60; Nacconol 40F, 40G, 90G; Nansa HS80/S); dyes and pigments (Calsoft L-40, L-60; Kadif 50 Flakes; Mercol 15, 25, 30, Special; Nacconol 35SL, 40F, 40G, 90F, 90G; Sul-fon-ate AA-9, AA-10); electroplating (Nacconol 40F, 40G, 90G; Sul-fon-ate AA-9, AA-10); fire fighting (Mercol 15, 25, 30, Special; Nacconol 40F, 40G, 90G; Sul-fon-ate AA-9, AA-10); metalworking (Mercol 15, 25, 30, Special; Nacconol 40F, 40G, 90F, 90G; Nansa HS80 Soft, HS80/S, HS85/S); mining (Nacconol 40F, 40G, 90G; Sul-fon-ate AA-9, AA-10); ore flotation (Calsoft L-40, L-60; Nacconol 40F, 40G, 90G; Sul-fon-ate AA-9, AA-10); paint mfg. (Mercol 15, 25, 30, Special); paper mfg. (Mercol 15, 25, 30, Special; Nacconol 35SL, 40G, 90F, 90G; Nansa HS80/S; Sul-fon-ate AA-9, AA-10); petroleum industry (Siponate DS-10; Sul-fon-ate AA-9, AA-10); plastics (Nacconol 40F; Siponate DS-4, DS-10); polymers/ polymerization (Arylan S Flake, SBC25, SC30; Calsoft F-90, L-40, L-60; Elfan WA; Manro BA25, BA30, SDBS30; Nacconol 40F, 40G, 90G; Nansa 1106/P, 1169/P); rubber (Nacconol 40F; Sul-fon-ate AA-9, AA-10); textile/leather processing (Conco AAS-35, AAS-45S; Merpisap AP80W, AP90, AP90P, KH30; Nacconol 35SL, 40F, 90F; Reworyl NKS50, NKS100; Sul-fon-ate AA-9, AA-10; Witconate 90 Flakes)

Industrial cleaners: (Calsoft L-40; Conco AAS-35; Kadif 50 Flakes; Mercol 15, 25, 30, Special; Merpisap AE50, AE60, AE70, AP80W, AP85W, AP90, AP90P, KH30; Richonate 40B, 45B, 45BX, 50BD, 60B, 1850, 1850U, C-50H; Sterling LA Paste 55%, 60%; Witconate 60B, 90 Flakes, 1250 Slurry, C50H); aircraft cleaners (Conco

315

Sodium dodecylbenzenesulfonate (cont'd.)

AAS-45S); all-purpose cleaners (Marlon A390, A396); drycleaning compositions (Richonate 40B, 45B, 45BX, 50BD, 60B, 1850, 1850U, C-50H); institutional cleaners (Kadif 50 Flakes; Siponate DS-10); metal processing surfactants (Calsoft L-60; Kadif 50 Flakes; Nacconol 40F, 40G, 90F, 90G; Sul-fon-ate AA-9, AA-10; Witconate 90 Flakes); sanitation (Nacconol 40G, 90G); street cleaners (Sul-fon-ate AA-9, AA-10); strippers (Cycloryl DDB40); textile cleaning (Calsoft L-40, L-60; Elfan WA, WA Pulver; Mercol 15, 25, 30, Special; Nacconol 40G, 90G; Nansa HS80/S)

PROPERTIES:

Form:

Liquid (Arylan SBC25, SC30, SX Flake; Bio Soft D-35X; Conco AAS-35; Cycloryl DDB40; Hetsulf 40X; Nansa SL30, SS30; Siponate DS-4; Witconate 60B, 1250 Slurry, C50H)

Clear liquid (Bio Soft D-40; Manro BA25, BA30; Nacconol 35SL); @ 20 C (Nansa 1106/P)

Cloudy liquid (@ 20 C) (Nansa 1169/P)

Hazy, pasty liquid, clear above R.T. (Calsoft L-60)

Clear viscous liquid (Calsoft L-40; Manro SDBS30)

Slurry (Bio Soft D-60, D-62; Hetsulf 60S; Surco 55S, 60S)

Liquid to paste (Conco AAS-45S; Marlon A350, A360, A365, A375)

Paste (Elfan WA; Hetsulf 40; Merpisap AE50, AE60, AE70, KH30; Nansa SS60; Sterling LA Paste 55%, 60%; Reworyl NKS50; Tairygent CB-1, CB-2)

Beads (Elfan WA Pulver; Mercol 15, 25, 30, Special)

Flake (Arylan S Flake; Calsoft F-90; Kadif 50 Flakes; Nacconol 40F, 90F; Nansa HS85/S; Siponate DS-10; Sterling AB-80; Sul-fon-ate AA-9, AA-10, LA-10; Witconate 90 Flakes)

Granular powder (Nacconol 40G, 90G)

Powder (Elfan WA Pulver; Marlon A390, A396; Merpisap AP80W, AP85W, AP90, AP90P; Nansa HS80 Soft, HS80/S; Reworyl NKS100; Sterling AB-40)

Color:

Nearly water-white (Calsoft L-60)

White (Mercol 15, 25, 30; Merpisap AE50, AE60, AE70, AP85W, AP90, AP90P; Nacconol 40F; Sul-fon-ate AA-9, AA-10, LA-10; Tairygent CB-1, CB-2)

White to tan (Conco AAS-45S)

Light (Bio Soft D-60; Kadif 50 Flakes)

Light cream (Nacconol 40F, 90G)

Cream (Arylan S Flake; Nacconol 40G; Nansa HS80 Soft, HS80/S, HS85/S, SS60)

Light straw (Bio Soft D-35X; Nacconol 35SL)

Golden (Nansa SS30)

Golden yellow (@ 20 C) (Nansa 1106/P, 1169/P)

Straw-yellow (Bio Soft D-62)

Pale amber (Manro BA25, BA30)

Amber (Manro SDBS30)

Sodium dodecylbenzenesulfonate *(cont'd.)*

Light/pale yellow (Calsoft L-40; Elfan WA Pulver; Nansa SL30)
Yellow (Cycloryl DDB40; Merpisap KH30)
Yellowish (Elfan WA)
Dark brown (Siponate DS-4)
Gardner 3 (30% solids) (Richonate 40B, 45B, 45BX, 50BD, 60B, 1850, 1850U, C-50H)

Odor:

Bland (Bio Soft D-40, D-62)
Odorless (Calsoft L-60; Sul-fon-ate AA-9)
Practically odorless (Kadif 50 Flakes)
Faint (Nacconol 35SL, 40F, 90F)
Mild (Manro BA25, BA30, SDBS30)
Neutral (Mercol 15, 25, 30, Special)
Typical (Conco AAS-45S)

Composition:

20% active (Mercol Special)
23% active (Siponate DS-4)
25% active min. (Manro BA25)
25% conc. (Arylan SBC 25; Nansa SB25)
29% active min. (Manro BA30, SDBS30)
30% active (Arylan SC30; Merpisap KH30; Nansa SL30)
30% active min. (Nansa SS30)
30 ± 1% active (Nansa 1106/P, 1169/P)
31.5% conc. (Conco AAS-35)
34% active min. (Bio Soft D-35X)
35% active (Hetsulf 40X)
35–38% active (Nacconol 35SL)
36% active (Richonate 40B)
37% active (Richonate 50BD, 1850, 1850U)
38% active (Hetsulf 40)
38–40% active (Cycloryl DDB40)
38–42% active (Nacconol 40G)
40% active (Bio Soft D-40; Mercol 15, 25, 30; Nacconol 40F; Richonate 45B; Sterling AB-40)
40% conc. (Conco AAS-40)
41–43% active (Conco AAS-45S)
42% active (Richonate 45BX)
42% solids (Calsoft L-40)
44% active (Richonate C-50H)
49–52% active (Surco 55S)
50% active (Elfan WA; Kadif 50 Flakes; Marlon A350; Merpisap AE50)
50% active, 21% water (Bio Soft D-60)
50% conc. (Reworyl NKS50)

Sodium dodecylbenzenesulfonate *(cont'd.)*

55% active (Sterling LA Paste 55%)
57% active (Hetsulf 60S; Richonate 60B; Sterling LA Paste 60%)
58–61% active (Surco 60S)
60% active (Bio Soft D-62; Marlon A360; Merpisap AE60)
60% active min. (Nansa SS60)
60% conc. (Conco AAS-65)
60 ± 2% active (Tairygent CB-1, CB-2)
60% solids (Calsoft L-60)
62% conc. (Nansa SB62)
65% active (Marlon A365)
70% active (Merpisap AE70)
75% active (Marlon A375)
75% conc. (Arylan SX Flake)
78 ± 3% active (Nansa HS80 Soft)
80% active (Elfan WA Pulver; Sterling AB-40)
80% conc. (Merpisap AP80W)
80 ± 3% active (Nansa HS80/S)
82% active (Arylan S Flake)
85% active (Merpisap AP85W)
85 ± 3% active (Nansa HS85/S)
88% active (Merpisap AP90, AP90P)
90% active (Calsoft F-90; Marlon A390; Nacconol 40F; Sul-fon-ate AA-9)
90% active min. (Nacconol 90G)
96% active (Marlon A396; Sul-fon-ate AA-10)
97.4% sodium dodecylbenzene sulfonate (Sul-fon-ate LA-10)
98% active (Siponate DS-10)
98% conc. (Reworyl NKS100)

Solubility:
Very sol. in water (Sul-fon-ate LA-10); sol. in water (Bio Soft D-35X, D-40, D-60, D-62; Kadif 50 Flakes; Mercol 15, 25, 30, Special; Nacconol 35SL, 40F, 90F; Witconate 60B, 90 Flakes, 1250 Slurry, C50H); 250 g/l @ 20 C (Elfan WA); 100 g/l @ 20 C (Elfan WA Pulver); 15 g/100 ml (Nacconol 90G); 2 g/100 ml (Nacconol 40G)

Ionic Nature:
Anionic

Sp.gr.:
1.02 (25/20 C) (Surco 55S, 60S)
1.04 (20 C) (Manro BA25)
1.05 (20 C) (Manro BA30)
1.05 (30 C) (Manro SDBS30)
1.08 (Elfan WA)
1.11 (20 C) (Nacconol 35SL)

Sodium dodecylbenzenesulfonate *(cont'd.)*

Density:
   0.15 g/cc (Mercol 15)
   0.25 g/cc (Mercol 25)
   0.30 g/cc (Mercol 30, Special)
   0.5 g/cc (Nansa HS85/S)
   0.55 g/cc (Nansa HS80 Soft)
   0.65 g/cc (Nansa HS80/S)
   1.0 g/cc (20 C) (Nansa 1106/P, 1169/P)
   300 g/l (Elfan WA Pulver)
   18.7–22.0 lb/ft^3 (untamped) (Nacconol 40F)
   22.0 lb/ft^3 (untamped) (Nacconol 90F)
   30 lb/ft^3 (untamped) (Nacconol 90G)
   35 lb/ft^3 (untamped) (Nacconol 40G)
   8.5 lb/gal (Richonate C-50H)
   8.7 lb/gal (Bio Soft D-40; Calsoft L-60)
   8.9 lb/gal (Bio Soft D-35X; Richonate 1850, 1850U)
   9.0 lb/gal (Richonate 40B)
   9.1 lb/gal (Bio Soft D-62; Richonate 45BX, 50BD)
   9.12 lb/gal (20 C) (Nacconol 35SL)
   9.2 lb/gal (Richonate 45B, 60B)
Visc.:
   103 cps (Nacconol 35SL)
   250 cps (20 C) (Manro BA25)
   ≈ 250 cs (20 C) (Nansa SS30)
   1000 cps (20 C) (Manro BA30)
   1000 cps (30 C) (Manro SDBS30)
   2500 cs (20 C) (Nansa 1106/P)
   2500–3500 cs (20 C) (Nansa SL30)
   5500 cs (20 C) (Nansa 1169/P)
F.P.:
   < 0 C (Elfan WA)
M.P.:
   5 C (Elfan WA)
Flash Pt.:
   > 200 F (Surco 55S, 60S)
Cloud Pt.:
   0 C (Nansa SS30)
   0 C max. (Nacconol 35SL)
   10 C (Bio Soft D-35X)
   11 C (quick cool) (Nansa 1106/P)
   15 C (Nansa SL30)
   17 C (Conco AAS-45S)
   25 C (quick cool) (Nansa 1169/P)

Sodium dodecylbenzenesulfonate *(cont'd.)*

29 C (Bio Soft D-40)

Clear Pt.:

18.0–18.5 C (Conco AAS-45S)

Stability:

Good (Bio Soft D-35X)

Stable to acids and alkalis (Conco AAS-45S; Nacconol 35SL; Nansa SL30, SS30, SS60)

Stable to mild acids or alkalis, and sol'ns. of electrolytes (Sul-fon-ate AA-9, AA-10)

Electrolyte tolerant (Witconate C50H)

Stable to water hardness (Conco AAS-45S; Nacconol 35SL; Sul-fon-ate AA-9, AA-10)

Stable to oxidizing and reducing agents (Nacconol 35SL, 40F, 40G)

Decomposed by strong oxidants (Sul-fon-ate LA-10)

Stable to reducing agents; oxidizing agents may cause some darkening (Nacconol 90F, 90G)

Stable to mineral and organic acids, dilute alkalis (Nacconol 40F, 40G, 90F)

Excessive amounts of conc. alkalis and salt may cause salting out of organic constituents (Nacconol 40G, 90G)

pH:

6.0 (Sul-fon-ate LA-10)

6.0–7.5 (Nacconol 35SL); (1% aq.) (Nacconol 80G); (10% aq.) (Nacconol 90F)

6.0–6.8 (2% sol'n.) (Nansa SS30)

6.4–7.6 (1% aq.) (Nacconol 40F)

6.5–7.5 (Richonate 40B, 45B)

7 ± 1 (1% sol'n.) (Tairygent CB-1, CB-2)

7.0–7.5 (2% sol'n.) (Nansa SL30)

7.0–8.0 (Richonate 45BX; Surco 55S, 60S); (1% sol'n.) (Sul-fon-ate AA-9, AA-10); (5% aq.) (Hetsulf 40X, 60S); (10% aq.) (Manro BA25, BA30)

7.0–8.5 (Mercol 15, 25, 30, Special; Richonate C-50H)

7.0–9.0 (2% sol'n.) (Nansa SS60); (10% aq.) (Manro SDBS30)

7.4 (Calsoft L-60)

7.5 (Calsoft L-40); (10% sol'n.) (Siponate DS-10)

7.5–8.5 (Richonate 60B, 1850); (5% aq.) (Hetsulf 40)

8.0 (1% aq.) (Kadif 50 Flakes)

8.0 ± 1 (5% aq.) (Nansa 1106/P, 1169/P)

8.0–9.0 (Elfan WA Pulver; Richonate 50BD, 1850U)

8.5 ± 1.0 (1% sol'n.) (Nansa HS80 Soft)

9.0 ± 1 (Elfan WA)

9.0–11.0 (1% aq.) (Nansa HS80/S, HS85/S)

Surface Tension:

30.6 dynes/cm (86 F) (Sul-fon-ate AA-9)

31 dynes/cm (0.1% conc.) (Nacconol 40F)

35 dynes/cm (0.1% conc.) (Nacconol 90F)

Sodium dodecylbenzenesulfonate *(cont'd.)*

37.4 dynes/cm (0.1% conc.) (Bio Soft D-60)
41.8 dynes/cm (0.1% conc.) (Nacconol 35SL)
Biodegradable: (Bio Soft D-35X, D-40, D-60, D-62; Calsoft F-90; Hetsulf 40, 40X, 60S; Manro BA25, BA30, SDBS30; Mercol 15, 25, 30, Special; Merpisap AE60, AP90, AP90P, KH30; Nacconol 40G, 90G; Nansa HS80/S; Richonate , 40B, 45B, 45BX, 50BD, 60B, 1850, 1850U; Sul-fon-ate LA-10); 90% min. (Nacconol 35SL, 40F, 90F); > 80% (Nansa HS80 Soft)

TOXICITY/HANDLING:

Eye irritant; avoid prolonged exposure to dust; do not ingest (Mercol 15, 25, 30, Special)

Avoid prolonged contact with skin and inhalation of dust (Nacconol 40G, 90G; Nansa HS80 Soft)

Toxic orally; irritating to eyes and skin; nontoxic dermally; combustion and/or decomposition will release sulfur dioxide (Sul-fon-ate LA-10)

STORAGE/HANDLING:

Store in tanks between 80–90 C (Elfan WA)

Hygroscopic—store dry (Elfan WA Pulver)

Moderately hygroscopic—store in a cool, dry place (Nacconol 40G, 90G)

Store in a cool, dry place (Nacconol 90F)

Store in a cool, dry place; avoid moisture pickup (Kadif 50 Flakes)

STD. PKGS.:

Bags and drums (Mercol 15, 25, 30, Special)

15 kg net paper bags (Kadif 50 Flakes)

25 kg paper bag (Elfan WA Pulver)

25 kg net multiply paper bag with polythene liner (Nansa HS80 Soft)

40 kg net bags (Merpisap AP85W)

40 kg net poly bags (Merpisap AP90, AP90P)

120 kg net poly drums (Merpisap AE60)

125 kg drums, 20 ton tanks (Elfan WA)

125 kg net drums (Merpisap AE50)

200 kg net drums (Merpisap AE70)

200 kg net iron drums or road tankers (Merpisap KH30)

45 gal drums or road tankers (Manro BA25, BA30, SDBS30)

51 gal Leverpak fiber drums (Sul-fon-ate AA-9, AA-10)

55 gal Liquipak drums (Nacconol 35SL)

55 gal Leverpak drums with aluminum barrier (Nacconol 40F)

55 gal (450 lb net) Liquipak drums, bulk tank car, tank truck (Calsoft L-40, L-60)

Sodium hydrogenated tallow glutamate (CTFA)

SYNONYMS:
Monosodium n-hydrogenated-tallowyl-l-glutamate
Sodium salt of hydrogenated tallow acid amide of glutamic acid

STRUCTURE:

HOOCCH$_2$CH$_2$CHCOONa
         |
         HN—CR
             ||
             O

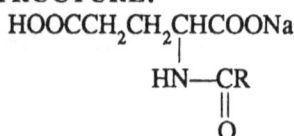

where RCO$^-$ represents the hydrogenated tallow radicals

CAS No.:
RD No. 977067-45-0

TRADENAME EQUIVALENTS:
Acylglutamate HS-11 [Ajinomoto]
Amisoft HS-11 [Ajinomoto]

CATEGORY:
Detergent, emulsifier, foaming agent, emollient, bactericide

APPLICATIONS:
Cosmetic industry preparations: (Amisoft HS-11); personal care products (Acylglutamate HS-11)

PROPERTIES:
Form:
Powder (Acylglutamate HS-11; Amisoft HS-11)
Composition:
100% active (Acylglutamate HS-11; Amisoft HS-11)
Ionic Nature:
Anionic (Acylglutamate HS-11)
Stability:
Good in hard water (Acylglutamate HS-11)
Biodegradable: (Acylglutamate HS-11)
TOXICITY/HANDLING:
Nonskin irritation even for infants and eczema cases (Acylglutamate HS-11)

Sodium lauroyl glutamate (CTFA)

SYNONYMS:
Glutamic acid, N-(1-oxododecyl)- monosodium salt
Monosodium N-lauroyl-L-glutamate
N-(1-Oxododecyl) glutamic acid, monosodium salt
Sodium salt of the lauric acid amide of glutamic acid

EMPIRICAL FORMULA:
C$_{17}$H$_{31}$NO$_5$ • Na

Sodium lauroyl glutamate *(cont'd.)*

STRUCTURE:

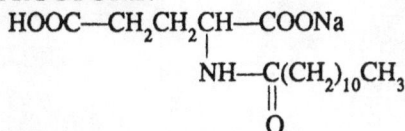

HOOC—CH₂CH₂CH—COONa

NH—C(CH₂)₁₀CH₃

O

CAS No.:
29923-31-7 (L-form); 29923-34-0 (DL form); 42926-22-7 (L-form)

TRADENAME EQUIVALENTS:
Acylglutamate LS-11 [Ajinomoto]
Amisoft LS-11 [Ajinomoto]

CATEGORY:
Detergent, emulsifier, foaming agent, emollient, bactericide

APPLICATIONS:
Cosmetic industry preparations: (Amisoft LS-11); personal care products (Acylglutamate LS-11)

PROPERTIES:

Form:
Powder (Acylglutamate LS-11; Amisoft LS-11)

Composition:
100% active (Acylglutamate LS-11; Amisoft LS-11)

Ionic Nature:
Anionic (Acylglutamate LS-11)

Stability:
Good in hard water (Acylglutamate LS-11)

Biodegradable: (Acylglutamate LS-11)

TOXICITY/HANDLING:
Nonskin irritation even for infants and eczema cases (Acylglutamate LS-11)

Sodium lauroyl lactylate

SYNONYMS:
Dodecanoic acid, 2-(1-carboxyethoxy)-1-methyl-2-oxoethyl ester, sodium salt

EMPIRICAL FORMULA:
$C_{18}H_{32}O_6 \cdot Na$

STRUCTURE:

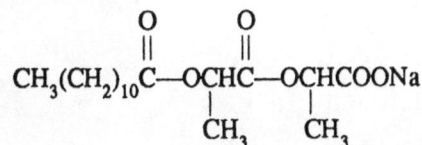

CH₃(CH₂)₁₀C—OCHC—OCHCOONa

O   O

CH₃   CH₃

Sodium lauroyl lactylate *(cont'd.)*

CAS No.:
13557-75-0
RD No.: 977067-51-8
TRADENAME EQUIVALENTS:
Pationic 138C [C.J. Patterson]
CATEGORY:
Emulsifier, foam booster
APPLICATIONS:
Cleansers: facial cleansers (Pationic 138C)
Cosmetic industry preparations: personal care products (Pationic 138C)
PROPERTIES:
Form:
Waxy (Pationic 138C)
Composition:
100% conc. (Pationic 138C)
Ionic Nature: Anionic
HLB:
14.4 (Pationic 138C)

Sodium lauryl sulfate *(CTFA)*

SYNONYMS:
1-Dodecanol, hydrogen sulfate, sodium salt
Sulfuric acid, monododecyl ester, sodium salt
EMPIRICAL FORMULA:
$NaC_{12}H_{25}SO_4$
STRUCTURE:
$CH_3(CH_2)_{10}CH_2OSO_3Na$
CAS No.
151-21-3
TRADENAME EQUIVALENTS:
Akyposal NLS [Chemax]
Alkasurf WAQ [Alkaril]
Alscoap LN-40, LN-90 [Toho]
Arsul WAQ [Arjay]
Avirol SL-2010 [Henkel]
Berol 474 [Berol]
Calfoam SLS-30 [Pilot]
Carsonol SLS, SLS Paste B, SLS Special [Lonza]
Cedepon LS-30, LS-30P [Domtar/CDC]

TRADENAME EQUIVALENTS *(cont'd.):*

Conco Sulfate WA, WA Dry, WA Special, WAG, WAS, WB-45, WBS-45, WR, WX [Continental]
Condanol NLS28, NLS30, NLS90 [Dutton & Reinisch]
Cycloryl 21 [Cyclo]
Cycloryl 21LS, 21SP [Cyclo]
Drewpon 100 [Drew Produtos]
Drewpon AS, CN [Drew Produtos]
Duponol C, ME Dry, QC, WA Dry, WA Paste, WAQE [DuPont]
Elfan 200 Pulver, 260 S [Akzo Chemie]
Elfan 240 [Akzo Chemie]
Emal O, 10 [Kao Corp.]
Emersal 6400, 6402, 6403, 6410 [Emery]
Empicol 0045 [Albright & Wilson/Detergents]
Empicol 0185, 0266, 0303 [Albright & Wilson/Detergents]
Empicol 0919 [Albright & Wilson]
Empicol LM [Albright & Wilson/Marchon]
Empicol LM/T [Albright & Wilson]
Empicol LM45 [Albright & Wilson/Marchon]
Empicol LMV [Albright & Wilson/Marchon]
Empicol LMV/T [Albright & Wilson]
Empicol LS30, LS30B, LS30P [Albright & Wilson]
Empicol LX, LX28 [Albright & Wilson/Marchon]
Empicol LXS95 [Albright & Wilson]
Empicol LXV [Albright & Wilson/Marchon]
Empicol LY28/S [Albright & Wilson]
Empicol LZ, LZ34 [Albright & Wilson/Marchon]
Empicol LZ/D [Albright & Wilson/Marchon]
Empicol LZ/E, LZG30, LZGV [Albright & Wilson/Marchon]
Empicol LZGV/C [Albright & Wilson]
Empicol LZP, LZV [Albright & Wilson/Marchon]
Empicol LZV/E [Albright & Wilson/Marchon]
Empicol WA [Albright & Wilson/Marchon]
Empicol WAK [Albright & Wilson]
Empimin LR28 [Albright & Wilson]
Equex S, SP, SW [Procter & Gamble]
Gardinol WA Paste [Ronsheim & Moore]
Hartenol LAS-30, LES-60 [Hart]
Incronol SLS [Croda Surfactants]
Jordanol SL300 [Jordan]
Lakeway 101-10, 101-10P,N, 101-11, 101-19 [Bofors Lakeway]
Lonzol LS-300 [Lonza]
Manro SLS28 [Manro]

Sodium lauryl sulfate (cont'd.)

TRADENAME EQUIVALENTS *(cont'd.):*
Maprofix 563, 563-SD, LCP, LK USP, MM, WAC, WAC-LA, WAM, WA Paste, WAQ [Onyx/Millmaster]
Mars SLS-A95, SLS-30 [Mars]
Merpinal LM40, LM50, LS40, LS50 [Elektrochem Fabrik Kempen]
Michelene LS-90 [M. Michel]
Neopon LS [Nippon Nyukazai]
Nikkol SLS [Nikko]
Norfox SLS [Norman, Fox & Co.]
Nutrapon W, WAC, WAQ [Clough]
Polyfac SLS-30, Polyfac SLS-30 Special [Westvaco]
Polystep B-3, B-5, B-24 [Stepan]
Rewopol NLS 15/L, NLS 30/L [Rewo]
Rewopol NLS 28 [Dutton & Reinisch]
Rewopol NLS 90 [Rewo]
Rewopol NLS 90-Powder [Dutton & Reinisch]
Richonol 2310, 6522, A, A Powder, C [Richardson]
Sactipon 2S3 [Lever Industriel]
Sactol 2S3 [Lever Industriel]
Sandoz Sulfate WA Dry, WAG, WAS, WA Special [Sandoz Colors & Chem.]
Sermul EA150 [Servo BV]
Sipex LCP, SB, SD, UB [Alcolac]
Sipon LS, LSB [Alcolac]
Standapol WA-AC, WAQ-115, WAQ-LC, WAQ-LCX [Henkel]
Standapol WAQ Special [Henkel/Canada]
Standapol WAS-100 [Henkel]
Steinapol NLS28, NLS90-Powder [Dutton & Reinisch]
Stepanol ME Dry, WA100, WAC, WA Extra, WA Paste, WAQ, WA Special [Stepan]
Sterling WA Paste, WA Powder, WAQ-CH, WAQ-Cosmetic, WAQ-LO, WAQ P [Canada Packers]
Sulfatol 33 Pasta, 33 Pasta 13 [Aarhus Olie]
Sulfatol 33 Paste [Aarhus Olie]
Sulfopon 101, 101 Special, 102, 103 [Henkel KGaA]
Sulfopon K-35 [Henkel KGaA]
Sulfopon WA-3 [Henkel/Canada]
Sulfopon WA30 [Henkel Argentina]
Sulfopon WAQLCX, WAQ Special [Henkel/Canada]
Sulfotex LCX, WA [Henkel]
Sulphonated Lorol Paste [Ronsheim & Moore]
Swascol 1PC, 3L, 1-P [Swastik]
Texapon K-12 [Henkel]
Texapon K-1294 [Henkel KGaA]
Texapon K-1296 [Henkel/Canada]

Sodium lauryl sulfate (cont'd.)

TRADENAME EQUIVALENTS *(cont'd.)*:
Texapon L-100 [Henkel]
Texapon LS Highly Conc. Needles [Henkel KGaA]
Texapon OT Highly Conc. Needles [Henkel KGaA]
Texapon V Highly Conc. Needles [Henkel]
Texapon Z Granules [Henkel KGaA]
Texapon Z Highly Conc. Needles [Henkel]
Texapon Z Highly Conc. Powder [Henkel]
Ultra Sulfate SL-1 [Witco/Organics]
Witcolate 6522, A, A Powder [Witco/Organics]
Zoharpon LAS, LAS Special [Zohar Detergent]

CATEGORY:
Wetting agent, detergent, emulsifier, thickener; also antipitting agent, base, cleaning agent, dispersant, process aid, foaming agent, lubricant, penetrant, raw material, solubilizer, surfactant, suspending agent

APPLICATIONS:
Automobile cleaners: car shampoo (Carsonol SLS, SLS Paste B, SLS Special; Mars SLS-30, SLS-A95; Sipex SB; Stepanol ME Dry, WAC, WA Extra, WA Paste, WA Special, WAQ)

Bath products: bubble bath (Alkasurf WAQ; Calfoam SLS-30; Carsonol SLS, SLS Paste B, SLS Special; Conco Sulfate WA Dry, WAG, WAS, WA Special, WB-45, WBS-45; Condanol NLS28, NLS90; Drewpon 100; Elfan 240; Empicol 0045, LX, LX28, LXV, LZ, LZ34, LZV; Jordanol SL300; Lakeway 101-10, 101-19; Lonzol LS-300; Maprofix WAC, WAM, WAQ; Mars SLS30, SLS-A95; Rewopol NLS28, NLS90 Powder; Sipon LSB; Standapol WA-AC, WAQ-115, WAQ-LC, WAQ Special, WAS-100; Steinapol NLS28, NLS90-Powder; Stepanol ME Dry, WAC, WA Extra, WA Paste, WA Special, WAQ; Sterling WA Paste, WAQ-CH, WAQ-Cosmetic, WAQ-LO, WAQ-P; Sulfatol 33MO, 33 Pasta, 33 Pasta 13; Sulfopon 101, 101 Special, WA-3, WA-30, WAQ Special; Sulfotex WA; Texapon K-12, LS Highly Conc. Needles, OT Highly Conc. Needles, Z Granules; Ultra Sulfate SL-1; Witcolate A Powder)

Cleansers: (Conco Sulfate WA; Standapol WA-AC, WAQ-115, WAQ-LC, WAQ Special, WAS-100); body cleansers (Ultra Sulfate SL-1); cleansing creams (Mars SLS30, SLS-A95; Stepanol ME Dry); hand cleanser (Cycloryl 21LS; Polyfac SLS30)

Cosmetic industry preparations: (Alkasurf WAQ; Calfoam SLS-30; Carsonol SLS, SLS Paste B, SLS Special; Conco Sulfate WB-45, WBS-45; Condanol NLS28, NLS90; Duponol C; Elfan 200 Pulver, 260 S; Emersal 6402, 6403, 6410; Hartenol LAS-30, LES-60; Manro SLS28; Mars SLS-30, SLS-A95; Norfox SLS; Rewopol NLS28, NLS90-Powder; Sipon LS, LSB; Steinapol NLS28, NLS90-Powder; Stepanol ME Dry, WAC, WA Extra, WA Paste, WA Special, WAQ; Sterling WA Paste, WAQ-P; Sulfatol 33 Paste; Sulfopon WA-3, WAQ Special; Swascol 3L; Texapon K-12, V Highly Conc. Needles, Z Highly Conc. Needles, Z Highly Conc.

Powder; Witcolate A); conditioners (Calfoam SLS-30); cosmetic base (Calfoam SLS-30; Empicol LM, LMV; Neopon LS); creams and lotions (Alkasurf WAQ; Calfoam SLS-30; Carsonol SLS, SLS Paste B, SLS Special; Duponol C; Neopon LS; Stepanol ME Dry, WAC, WA Extra, WA Paste, WA Special, WAQ; Texapon K12-96); personal care products (Empicol 0045; Lakeway 101-10, 101-10P, 101-19, 101-N; Lonzol LS-300; Maprofix LK (USP), WAC-LA; Nutrapon W, WAC, WAQ; Sandoz Sulfate WA Dry, WAG, WAS, WA Special); shampoo base (Alscoap LN-40, LN-90; Emersal 6402, 6403, 6410; Empicol LS30B, LS30P, WAK; Sactol 2S3); shampoos (Alkasurf WAQ; Calfoam SLS-30; Carsonol SLS Paste B, SLS Special; Conco Sulfate WA, WA Dry, WAG, WAS, WA Special, WB-345, WBS-45; Condanol NLS28, NLS30, NLS90; Cycloryl 21, 21LS; Drewpon 100, AN, AS; Duponol QC, WA Paste; Elfan 240, 260S; Emal O, 10; Emersal 6400; Empicol 0045, 0919, LM45, LM/T, LMV/T, LS30P, LX, LX28, LXV, LZ, LZ34, LZV, WAK; Equex S, SW; Gardinol WA Paste; Incronol SLS; Jordanol SL300; Lakeway 101-10, 101-19; Lonzol LS-300; Manro SLS28; Maprofix MM, WA Paste, WAC, WAM, WAQ; Mars SLS-30, SLS-A95; Neopon LS; Polyfac SLS30, SLS-30 Special; Rewopol NLS28, NLS90 Powder; Sandoz Sulfate WAS, WA Special; Sipon LS, LSB; Standapol WA-AC, WAQ-115, WAQ-LC, WAQ Special, WAS-100; Steinapol NLS28, NLS90 Powder; Stepanol ME Dry, WAC, WA Extra, WA Paste, WA Special, WAQ; Sterling WA Paste, WA Powder, WAQ-CH, WAQ-Cosmetic, WAQ-LO, WAQ-P; Sulfatol 33MO, 33 Paste, 33 Pasta, 33 Pasta 13; Sulfopon 101, 101 Special, 103, WA-3, WA30, WAQ Special; Sulphonated Lorol Paste; Swascol 1PC, 3L; Texapon LS Highly Conc. Needles, OT Highly Conc. Needles, Z Granules; Ultra Sulfate SL-1; Zoharpon LAS, LAS Special); shaving preparations (Carsonol SLS, SLS Paste B, SLS Special; Duponol C; Emal O, 10; Sipon LS, LSB; Stepanol ME Dry, WAC, WA Extra, WA Paste, WA Special, WAQ; Witcolate A Powder); toiletries (Empicol 0919, LX, LX28, LXV, LZ, LZ34, LZP, LZV; Manro SLS28; Sulfatol 33MO, 33 Pasta, 33 Pasta 13; Witcolate A)

Farm products: (Empicol 0185); insecticides/pesticides (Drewpon 100; Stepanol ME Dry, WAC, WA Extra, WA Paste, WA Special, WAQ)

Food applications: food additives (Maprofix 563, 563SD); dairy cleaners (Stepanol ME Dry, WAC, WA Extra, WA paste, WA Special, WAQ); fruit/vegetable washing (Carsonol SLS, SLS Paste B, SLS Special; Maprofix 563, 563SD; Sipex SB; Stepanol ME Dry, WAC, WA Extra, WA Paste, WA Special, WAQ); food packaging (Sipex SB)

Household detergents: (Akyposal NLS; Alscoap LN-40, LN-90; Carsonol SLS, SLS Paste B, SLS Special; Emersal 6402, 6403, 6410; Empicol LM45, LS30; Hartenol LES-60; Jordanol SL300; Lonzol LS-300; Maprofix LCP, LK (USP); Merpinal LM50, LS40, LS50; Nikkol SLS; Norfox SLS; Sactol 2S3; Sandoz Sulfate WA Dry, WAG, WAS, WA Special; Stepanol ME Dry, WAC, WA Extra, WA Paste, WA Special, WAQ; Texapon OT Highly Conc. Needles; Sulfopon 102; Ultra Sulfate SL-1); all-purpose cleaner (Lonzol LS-300; Michelene LS-90); built detergents (Duponol WA Dry); carpet & upholstery shampoos (Alkasurf WAQ; Calfoam SLS-

30; Carsonol SLS, SLS Paste B, SLS Special; Conco Sulfate WA, WA Dry, WAG, WAS, WA Special, WB-45, WBS-45; Emersal 6400, 6402, 6410; Empicol 0919, LS30P, LX, LX28, LXV; Equex SW; Incronol SLS; Lakeway 101-10, 101-19; Lonzol LS-300; Maprofix MM, WA Paste; Mars SLS-30, SLS-A95; Sandoz Sulfate WA Dry, WAG, WAS, WA Special; Sipex SD; Standapol WAQ-LCX; Stepanol ME Dry, WAC, WA Extra, WA Paste, WA Special, WAQ; Sulfatol 33MO, 33 Paste, 33 Pasta, 33 Pasta 13; Sulfopon WAQ LCX, WAQ Special; Sulfotex LCX); detergent base (Empicol LM, LMV; Rewopol NLS28, NLS90, NLS90-Powder; Steinapol NLS28, NLS90-Powder; Sulfopon K35); dishwashing (Mars SLS-30, SLS-A95; Richonol 2310, A, A Powder; Sulfatol 33MO, 33 Pasta, 33 Pasta 13); hard surface cleaner (Alkasurf WAQ; Sulfotex LCX); laundry detergent (Alkasurf WAQ; Hartenol LAS-30); fine fabric detergent (Calfoam SLS-30; Gardinol WA Paste); light-duty cleaners (Elfan 240; Emersal 6400; Empicol LS30; Richonol 2310, 6522, A, A Powder, C; Sulfatol 33MO, 33 Pasta, 33 Pasta 13; Sulfopon 101, 101 Special, 103); liquid detergents (Polyfac SLS-30 Special; Sulfotex WA; Swascol 3L)

Industrial applications: carpet backing (Empicol LY28/S, LZ, LZ-34, WA); dyes and pigments (Conco Sulfate WA Dry, WA Special, WAG, WAS, WB-45, WBS-45; Hartenol LAS-30, LES-60; Maprofix LK (USP)); electroplating (Empicol 0185; Sipex SD); industrial processing (Carsonol SLS, SLS Paste B, SLS Special; Conco Sulfate WA Dry, WA Special, WAG, WAS, WB-45, WBS-45; Cycloryl 21SP; Duponol ME Dry, WA Dry; Empicol LS30, LS30P, LZG30, LZGV, LZGV/C, WAK; Incronol SLS; Lonzol LS-300; Maprofix LCP, MM, WA Paste, WAC-LA; Mars SLS-30, SLS-A95; Nutrapon W, WAC, WAQ; Stepanol ME Dry, WAC, WA Extra, WA Paste, WA Special, WAQ; Sulfatol 33MO, 33 Paste, 33 Pasta, 33 Pasta 13; Witcolate A); ore flotation (Conco Sulfate WB-45, WBS-45); photography (Empicol LY28/S); plastics (Empicol LX, LX28, LXV, LZ, LZ34, LZ/E, LZG30, LZGV, LZGV/C, LZP, LZV, LZV/E); polymers/polymerization (Arsul WAQ; Avirol SL-2010; Berol 474; Calfoam SLS-30; Carsonol SLS, SLS Paste B, SLS Special; Conco Sulfate WB-45, WBS-45; Condanol NLS28, NLS30, NLS90; Duponol WAQE; Elfan 260S; Emal 10; Emersal 6400; Empicol LX, LX28, LXS95, LXV, LY28/S, LZ, LZ34, LZ/E, LZG30, LZGV, LZGV/C, LZP, LZV, LZV/E; Equex SP; Incronol SLS; Lakeway 101-10P,N, 101-11, 101-19; Lonzol LS-300; Manro SLS28; Maprofix LCP, WAC-LA; Polyfac SLS-30 Special; Polystep B3, B24; Rewopol NLS15/L, NLS28, NLS30/L, NLS-90Powder; Sermul EA150; Sipex LCP, SB, UB; Steinapol NLS28, NLS-90Powder; Stepanol ME Dry, WA-100, WAC, WA Extra, WA Paste, WA Special, WAQ; Sterling WAQ-CH, WAQ-LO; Sulfopon 102; Swascol 1P; Ultra Sulfate SL-1; Witcolate A Powder); printing inks (Empicol 0185, LZ/E, LZV/E; Lakeway 101-10P,N, 101-19); rubbers (Empicol LX, LX28, LXV, LZ, LZ34, LZ/E, LZG30, LZGV, LZGV/C, LZP, LZV, LZV/E); foamed rubber (Calfoam SLS-30; Empicol LX, LX28, WA); textile/leather processing (Hartenol LAS-30; Jordanol SL300; Sipex SD; Stepanol ME Dry, WAC, WA Extra, WA Paste, WA Special, WAQ; Ultra Sulfate SL-1)

Sodium lauryl sulfate *(cont'd.)*

Industrial cleaners: (Carsonol SLS, SLS Paste B, SLS Special; Conco Sulfate WB-45, WBS-45; Duponol ME Dry, WA Dry; Empicol LS30, LS30P, LZG30, LZGV, LZGV/C, WA, WAK; Incronol SLS; Lonzol LS-300; Maprofix LCP, MM, WA Paste, WAC-LA; Mars SLS-30, SLS-A95; Nutrapon W, WAC, WAQ; Stepanol ME Dry, WAC, WA Extra, WA Paste, WA Special, WAQ; Sulfatol 33MO, 33 Paste, 33 Pasta, 33 Pasta 13; Ultra Sulfate SL-1; Witcolate A); drycleaning compositions (Alkasurf WAQ); institutional cleaners (Alkasurf WAQ; Hartenol LAS-30, LES-60); metal processing surfactants (Sipex SB, SD; Stepanol ME Dry, WAC, WA Extra, WA Paste, WA Special, WAQ); railway cleaners (Stepanol ME Dry, WAC, WA Extra, WA paste, WA Special, WAQ); sanitizers (Alkasurf WAQ; Mars SLS-30, SLS-A95); specialty detergents (Cycloryl 21; Empicol 0919, LS30P, WAK; Equex S; Maprofix WAC, WAM, WAQ); textile cleaning (Elfan 260S; Hartenol LAS-30, LES-60; Merpinal LM50, LS40, LS50; Sipex SB)

Pet shampoos: (Carsonol SLS, SLS Paste B, SLS Special)

Pharmaceutical applications: (Empicol LXS95; Stepanol ME Dry, WA-100, WAC, WA Extra, WA Paste, WA Special, WAQ; Sterling WA Powder; Zoharpon LAS); dental preparations (Alscoap LN-90; Conco Sulfate WA Dry, WA Special, WAG, WAS, WB-45, WBS-45, WR, WX; Drewpon 100; Duponol C; Elfan 200 Pulver; Emal O, 10; Empicol 0045, 0185, 0266, 0303, 0919, LM45, LM/T, LMV/T, LX, LX28, LXS95, LXV, LZ, LZ34, LZP, LZV; Maprofix 563, 563SD; Sandoz Sulfate WA Dry, WAG; Sipon LS; Stepanol WA-100; Swascol 1P, 1PC; Texapon K12, K12-94, L-100, Z Granules; Zoharpon LAS, LAS Special); medical preparations (Duponol C; Sipon LSB); mouthwash (Maprofix 563, 563SD; Texapon K12-94, L-100); ointments (Sipon LS, LSB); tablet mfg. (Empicol 0185; Texapon K12-96)

PROPERTIES:
Form:

Liquid (Akyposal NLS; Alkasurf WAQ; Alscoap LN-40; Arsul WAQ; Avirol SL-2010; Calfoam SLS-30, SLS Special; Cedepon LS-30, LS-30P; Conco Sulfate WAS, WB-45, WBS-45; Condanol NLS28; Cycloryl 21, 21LS, 21SP; Drewpon CN; Duponol WAQE; Elfan 240; Emersal 6400, 6402, 6403, 6410; Empicol 0919, LX28, LY28/S (@ 30 C); Equex S; Hartenol LAS-30, LES-60; Jordanol SL300; Lakeway 101-10, 101-11, 101-19; Maprofix LCP, WAC, WAC-LA, WAM, WAQ; Neopon LS; Nutrapon W; Polyfac SLS30, SLS-30 Special; Polystep B-24; Re-wopol NLS15/L, NLS30/L, NLS28; Richonol 6522, C, A; Sandoz WAS; Sipex UB; Standapol WAQ-LC, WAQ-LCX; Steinapol NLS28; Sterling WAQ-CH, WAQ-LO; Sulfatol 33MO, 33 Pasta 13; Sulfotex LCX, WA; Sulfopon 102, WAQ Special; Swascol 3L; Ultra Sulfate SL-1; Witcolate 6522, A)

Clear liquid (Carsonol SLS; Empicol 0919, LS30, LS30B, LS30P, WA, WAK; Equex SP; Lonzol LS-300; Manro SLS28; Mars SLS-30, SLS-A95; Polystep B-5; Sipex LSB, SB; Stepanol WAC, WA Extra, WA Special, WAQ)

Viscous liquid (Duponol QC; Empicol LS30, LS30B; Equex SW; Incronol SLS; Nutrapon WAC, WAQ; Sipon LS; Standapol WAQ-115, WAQ Special, WAS-100; Stepanol WAQ; Sterling WA Paste, WAQ-Cosmetic, WAQ-P)

330

Liquid/paste (Elfan 260S; Sulfopon 101 Special)

Extrusion (Michelene LS-90)

Paste (Calfoam SLS-30, SLS Paste B; Conco Sulfate WA; Condanol NLS30; Cycloryl
  21; Drewpon AS, CN; Empicol LM45, LX28, LY28/S (@ 20 C), LZ34, LZG30;
  Incronol SLS; Maprofix MM, WA Paste; Merpinal LM50; Norfox SLS; Richonol
  6522, C; Sactipon 2S3; Sactol 2S3; Sandoz Sulfate WA Special; Standapol WA-
  AC; Sulfatol 33 Paste, 33 Pasta, 33 Pasta 13; Sulfopon 101, 103, K35, WA-3, WA-
  30; Sulphonated Lorol Paste; Swascol 3L)

Smooth paste (Conco Sulfate WA Special)

Soft paste (Berol 474; Merpinal LS40, LS50)

Clear paste (Stepanol WA Paste)

Beads (Duponol C; Elfan 200 Pulver)

Granular (Maprofix 563; Texapon Z Granules)

Powder (Alscoap LN-90; Conco Sulfate WA Dry, WR, WX; Condanol NLS90;
  Drewpon 100; Duponol ME Dry, WA Dry; Elfan 200 Pulver; Emal O, 10; Empicol
  0045, 0185, 0266, 0303, LM, LX, LXS95, LZ/D, LZ/E, LZP; Lakeway 101-10P;
  Maprofix 563, 563SD, LK (USP); Nikkol SLS; Polystep B-3; Rewopol NLS90,
  NLS90-Powder; Richonol A Powder; Sandoz Sulfate WA Dry; Sipex SD; Steina-
  pol NLS90-Powder; Stepanol ME Dry, WA-100; Sterling WA Powder; Swascol
  1P; Texapon K-12, K12-94, K12-96, L-100, Z Highly Conc. Powder; Witcolate A
  Powder)

Spray-dried powder (Empicol LM/T, LZ)

Crystalline or amorphous powder (Swascol 1PC)

Needles (Conco Sulfate WAG; Emal 10; Empicol LMV, LMV/T, LXV, LZGV,
  LZGV/C, LZV, LZV/E; Lakeway 101-10N; Sandoz Sulfate WAG; Texapon LS
  Highly Conc. Needles, OT Highly Conc. Needles, V Highly Conc. Needles, V
  Highly Conc. Needles)

Color:

Colorless (Conco Sulfate WAS; Incronol SLS)

Water-white (Calfoam SLS-30; Duponol QC; Maprofix WAC-LA; Standapol WAQ-
  115, WAQ-LC, WAQ Special, WAS-100; Stepanol WA Paste, WA Special, WAQ;
  Sulfopon WAQ Special; Sulfotex LCX)

White (Calfoam SLS30; Conco Sulfate WA Special, WA Dry, WAG, WB-45, WBS-
  45; Condanol NLS90; Duponol C, ME Dry, WA Dry; Elfan 200 Pulver; Empicol
  0045, 0185, LM45, LMV/T, LM/T, LX, LXS95, LXV, LZ, LZ34, LZ/E, LZP, LZV,
  LZV/E; Incronol SLS; Lakeway 101-10, 101-10P,N, 101-11, 101-19; Merpinal
  LM50, LS40, LS50; Polystep B-3; Rewopol NLS90-Powder; Richonol A Powder,
  C; Sandoz Sulfate WA Dry, WAG, WA Special; Sipex SD; Steinapol NLS90-
  Powder; Stepanol ME Dry, WA-100; Sulfatol 33 Paste; Texapon K-12, K-1296, L-
  100, V Highly Conc. Needles, Z Highly Conc. Needles, Z Highly Conc. Powder)

White pearlescent (Alkasurf WAQ; Standapol WA-AC; Sulfopon WA-3)

White to cream (Gardinol WA Paste)

Cream (Duponol ME Dry; Empicol LZG-30, LZGV, LZBV/C)

Sodium lauryl sulfate (cont'd.)

Off-white (Condanol NLS30)
Light (Hartenol LAS-30, LES-60; Maprofix WAC)
Low (Carsonol SLS, SLS Paste B, SLS Special)
Pale (Polystep B-5, B-24; Sipon LSB)
Pale/light yellow (Condanol NLS28, NLS90; Cycloryl 21, 21LS; Duponol WAQE;
   Empicol 0919, LS30, LS30B, LS30P, LX28, LY28/S, WA, WAK; Equex SP, SW;
   Lakeway 101-10, 101-11, 101-19; Manro SLS28; Mars SLS-30; Polyfac SLS30,
   SLS-30 Special; Rewopol NLS28, NLS90-Powder; Sandoz Sulfate WAS; Steina-
   pol NLS28, NLS90-Powder; Stepanol WAC)
Yellow (Cycloryl 21SP; Elfan 260S; Mars SLS-A95)
Gardner 1 (Emersal 6400)
Gardner 2 (Richonol 2310, 6522, A)
Hazen < 50 (Berol 474)
Odor:
Bland, clean, characteristic (Duponol QC)
Clean, pleasant (Equex SP, SW)
Mild (Calfoam SLS-30; Manro SLS28)
Low (Carsonol SLS Paste B, SLS Special)
Mildy fatty (Duponol C)
Pleasant (Conco Sulfate WA Special, WAS, WB-45, WBS-45; Equex SP, SW)
Typical (Alkasurf WAQ; Conco Sulfate WA Dry, WAG; Condanol NLS30; Duponol
   QC)
Composition:
15% conc. (Rewopol NLS 15/L)
25% active (Sulfatol 33 Pasta 13)
27% active (Empicol WA)
27.5–30.5% active (Sandoz Sulfate WAS, WA Special)
28% active (Calfoam SLS-30; Condanol NLS28; Empicol LX28; Empimin LR28;
   Manro SLS28; Polystep B-24; Rewopol NLS28; Sipon LS; Steinapol NLS28)
28–30% active (Cycloryl 21, 21LS, 21SP; Drewpon AS, CN; Incronol SLS; Lonzol
   LS-300; Maprofix LCP, WAC, WAC-LA, WAM, WAQ; Stepanol WAC, WA
   Extra, WA Paste, WAQ, WA Special; Sulfopon WA-3, WAQ Special)
28.5% active (Equex SW; Richonol A, C)
28.5–33.0% active (Conco Sulfate WAS, WA Special)
29% active (Emersal 6403; Equex S; Lakeway 101-10, 101-11; Mars SLS-30; Polyfac
   SLS-30; Polystep B-5; Richonol 6522; Sipex SB; Sipon LSB)
$29.0 \pm 1.0\%$ active (Empicol LY28/S)
29.5% active (Polyfac SLS-30 Special)
29–30% active (Duponol QC)
29–30.5% active (Duponol WAQE)
29–31% active (Sulfopon WA30)
29.8% active (Equex SP)
30% active (Alkasurf WAQ; Arsul WAQ; Carsonol SLS, SLS Paste B, SLS Special;

Sodium lauryl sulfate (cont'd.)

Cedepon LS-30; Condanol NLS30; Elfan 260S; Emersal 6400, 6402, 6410; Empicol LZG30; Hartenol LAS-30; Rewopol NLS 30/L; Sipex LCP, UB; Swascol 3L; Ultra Sulfate SL-1)

30% solids (Mars SLS-A95)
30.0 ± 1% active (Empicol 0919, LS30, LS30B, LS30P)
34% active (Empicol LZ34)
35.0 ± 1.0% active (Empicol WAK)
38% active (Sulfatol 33 Pasta)
39–41% active (Berol 474; Sulfatol 33 Paste)
40% active (Merpinal LM40, LS40)
40% conc. (Gardinol WA Paste)
41.5% active (Maprofix MM, WA Paste)
44% active (Duponol WA Dry)
45% active (Empicol LM45; Lakeway 101-19)
45.0–47.5% active (Conco Sulfate WB-45, WBS-45)
50% active (Merpinal LM50, LS50)
60% active (Hartenol LES-60)
85% active (Condanol NLS90; Empicol LZGV, LXV, LZV; Rewopol NLS 90-Powder; Steinapol NLS90-Powder)
85–89% active (Texapon Z Highly Conc. Needles)
89% active (Empicol LZ, LZP)
90% active (Drewpon 100; Elfan 200 Pulver; Empicol LMV/T, LX, LZ/E, LZV/E; Lakeway 101-10P,N; Maprofix LK (USP); Michelene LS-90; Texapon K-12, V Highly Conc. Needles, Z Highly Conc. Powder)
90–96% active (Duponol C, ME Dry; Sandoz Sulfate WA Dry, WAG)
92–98% active (Conco Sulfate WA Dry, WAG)
93% active (Richonol A Powder; Stepanol ME Dry)
94% active (Empicol 0045, 0185, LM/T, LXS95)
95% active (Maprofix 563-SD; Sipex SD)
96% active (Swascol 1PC)
> 96% active (Texapon K-1296)
97% active (Maprofix 563)
97.5% active (Polystep B-3)
98.5% active (Stepanol WA-100)
99% active (Texapon L-100)
Solubility:
Sol. in ethanol (Berol 474; Duponol ME Dry, QC, WA Dry, WAQE)
Disp. in glycerol trioleate (Emersal 6400)
Sol. in methanol (Duponol ME Dry, QC, WA Dry, WAQE)
Sol. in propylene glycol (Berol 474)
Sol. in water (Alkasurf WAQ; Berol 474; Carsonol SLS, SLS Paste B, SLS Special; Duponol ME Dry, QC, WA Dry, WAQE; Emersal 6400; Stepanol ME Dry, WA-100, WAC, WA Extra, WA Paste, WA Special, WAQ; Witcolate A); @ 30 C (Elfan

333

Sodium lauryl sulfate *(cont'd.)*

260S); 10–15 g in 100 ml water @ 25 C (Duponol C)

Ionic Nature: Anionic

M.W.:
   292 (Empicol WA)
   300 (Condanol NLS28; Rewopol NLS-28, NLS90-Powder; Steinapol NLS28, NLS90-Powder)
   302 (Duponol ME Dry, WAQE)

Sp.gr.:
   0.35 (25/20 C) (Maprofix LK (USP))
   0.40 (25/20 C) (Maprofix 563SD)
   0.66 (25/20 C) (Maprofix 563)
   1.0 (Condanol NLS30)
   1.03 (Mars SLS-A95)
   1.04 (Elfan 260S; Empicol WA; Equex S, SP; Manro SLS28; Mars SLS-30)
   1.04 (25/20 C) (Maprofix WAC, WAC-LA)
   1.05 (Empicol LS30P, WA; Equex SW)
   1.05 (25/20 C) (Maprofix LCP, WAM, WAQ)
   1.05 ± 0.005 (Empicol 0919)
   1.09 (25/20 C) (Maprofix MM, WA Paste)

Density:
   1.6 lb/gal (Richonol A Powder)
   2.08 lb/gal (Texapon K-1296)
   2.6 lb/gal (Duponol WA Dry)
   3.3 lb/gal (Duponol C, ME Dry)
   8.5 lb/gal (Lakeway 101-10, 101-11)
   8.6 lb/gal (Duponol WAQE; Lakeway 101-19; Mars SLS-30, SLSA-95)
   8.7 lb/gal (Carsonol SLS, SLS Special; Emersal 6400; Equex S, SP; Polyfac SLS-30 Special; Richonol 6522, A)
   8.73 lb/gal (Sulfotex LCX)
   8.8 lb/gal (Carsonol SLS Paste B; Equex SW)
   9.2 lb/gal (Richonol C)
   0.35 g/cm^3 (20 C) (Empicol 0045, 0185, LM/T, LX, LXS95, LZ, LZ/E, LZP)
   0.5 g/cm^3 (20 C) (Empicol LMV/T, LXV, LZGV, LZGV/C, LZV, LZV/E)
   0.95 g/cm^3 (20 C) (Empicol LM45)
   1.030 g/cm^3 (20 C) (Berol 474)
   1.05 g/cm^3 (20 C) (Empicol LX28, LY28/S, LZ34, LZG30)
   220 g/l (Elfan 200 Pulver)

Visc.:
   40 cps (Mars SLS-A95)
   50 cps (Brookfield) (Sipex LCP, UB)
   50–200 cps (Brookfield) (Equex S)
   90 cps (Brookfield @ 80 F) (Equex SP)
   100 cps max. (Brookfield) (Lakeway 101-19; Standapol WAQ-LC)

120 cps max. (Brookfield, 35 C, #2 spindle, 60 rpm) (Empicol 0919, LS30P)
150 cps (Sipex SB; Sipon LSB)
200 cps (20 C) (Manro SLS28)
200 cps (30 C) (Elfan 260S)
250 cps (Brookfield) (Lakeway 101-11)
350 cps (Brookfield) (Lakeway 101-10)
< 500 cps (80 F) (Duponol QC)
500 cps max. (27 C) (Duponol WAQE)
550 cps (Mars SLS-30)
750 ± 300 cps (Brookfield, 35 C, #2 spindle, 6 rpm) (Empicol LS30)
800 ± 200 cps (Brookfield, 35 C, #2 spindle, 6 rpm) (Empicol LS30B)
1000 cps max. (Standapol WAQ-115)
1000–3000 cps (Sandoz Sulfate WAS)
1000–5000 cps (Conco Sulfate WAS; Sandoz Sulfate WA Special)
1500 cps max. (Standapol WAQ Special, WAS-100)
19,000–31,000 cps (Conco Sulfate WA Special)
20,000 cps min. (Standapol WA-AC)
25,000 cps (Sipon LS)
40,000 cps (Brookfield, 80 F) (Equex SW)
28 cs (20 C) (Empicol LY28/S)
424 cSt (100 F) (Emersal 6400)

F.P.:
15 C (Elfan 260S)
Pour Pt.:
15 C (Emersal 6400)
Solidification Pt.:
34 F (Equex SP)
35 F (Equex S)
55 F (Equex SW)
Flash Pt.:
> 100 C (Berol 474)
> 200 F (Maprofix 563, 563SD, LCP, LK (USP), MM, WA Paste, WAC, WAC-LA, WAM, WAQ)
Cloud Pt.:
−1 C (Empicol LY28/S)
1 C (Sipex LCP)
2 C (Sulfotex LCX)
8 C max. (Standapol WAQ-LC)
12 C max. (Standapol WAQ-115)
14 C max. (Mars SLS-30)
15 C (Lakeway 101-11)
15 C max. (Standapol WAQ Special, WAS-100)
17–21 C (Conco Sulfate WA Special; Sandoz Sulfate WA Special)

Sodium lauryl sulfate *(cont'd.)*

    17–23 C (Conco Sulfate WAS; Sandoz Sulfate WAS)
    18 C (Lakeway 101-10; Sipex UB)
    18 C max. (Mars SLS-A95)
    19 C (Lakeway 101-19)
    20 C (Sipex SB; Sipon LS, LSB)
    25 C max. (Standapol WA-AC)
    > 100 C (5% saline) (Emersal 6400)
    37 F (1% sol'n.) (Duponol WA Dry)
    47 F (Conco Sulfate WB-45, WBS-45)
    50 F (Polyfac SLS-30 Special; Richonol 6522)
    54 F (Equex S)
    55 F (Equex SP)
    60 F (Richonol A)
    65 F (Duponol QC, WAQE; Polyfac SLS30)
    70 F (1% aq. sol'n.) (Duponol ME Dry)
    77 F (Equex SW)
Clear Pt.:
    19–26 C (Conco Sulfate WA Special)
    23–27 C (Conco Sulfate WAS)
    25 C (Berol 474)
HLB:
    40 (Sipex SD, UB; Sipon LS, LSB)
    > 40 (Emersal 6402, 6403, 6410)
Stability:
    Stable to weak bases and very weak acids (Berol 474)
    Stable to alkalis (Duponol C, ME Dry, WA Dry; Mars SLS-30, SLS-A95; Sipex UB)
    Good (Sipex SB)
    Good to anionic and nonionic surfactants, acids, lower alcohols, and electrolytes (Mars SLS-30, SLS-A95)
    Hydrolyzes in strong acid solutions (Duponol C, ME Dry, WA Dry)
    Stable to freezing, uv light, aging; stable to heat (150 F) for several hours (Duponol QC, WAQE)
    Unaffected by water hardness (Mars SLS-30, SLS-A95; Texapon V Highly Conc. Needles, Z Highly Conc. Needles, Z Highly Conc. Powder)
Storage Stability:
    Stable in normal storage (Duponol C, ME Dry, WA Dry)
pH:
    6.0–9.0 (1% sol'n.) (Texapon K12-96)
    6.5–7.5 (10% sol'n.) (Conco Sulfate WA Dry, WAG; Sandoz Sulfate WA Dry, WAG)
    6.5–8.0 (5% sol'n.) (Sulfopon WA30)
    6.8–7.5 (1% aq.) (Texapon Z Highly Conc. Needles)
    7.0 (1% aq.) (Texapon L-100)
    7.0–8.0 (Richonol C); (1% aq.) (Berol 474; Rewopol NLS28; Steinapol NLS28;

Texapon V Highly Conc. Needles); (10% aq.) (Manro SLS28)

7.0–8.5 (10% sol'n.) (Mars SLS-A95)

7.0–9.0 (1% solids) (Rewopol NLS90-Powder; Steinapol NLS90-Powder); (10% aq.) (Standapol WAQ-LC; Sulfotex LCX)

7.3–8.5 (10% sol'n.) (Lonzol LS-300)

7.4–8.6 (10% sol'n.) (Conco Sulfate WA Special, WAS; Sandoz Sulfate WAS, WA Special)

7.5 (3% sol'n.) (Sipex SD)

7.5 ± 0.7 (10% aq. sol'n.) (Empicol LS30P)

7.5–8.5 (Richonol 6522, A); (0.25% aq.) (Texapon Z Highly Conc. Powder); (10% sol'n.) (Carsonol SLS, SLS Special; Standapol WA-AC, WAQ-115, WAQ Special, WAS-100; Stepanol WAC, WA Extra, WA Paste, WA Special, WAQ)

7.5–9.0 (1% aq.) (Texapon K-12); (10% sol'n.) (Mars SLS-30)

7.5–9.5 (2% sol'n.) (Sulfatol 33 Paste)

7.5–10.0 (10% sol'n.) (Stepanol ME Dry, WA-100)

7.8 (10% sol'n.) (Equex S, SW)

8.0 (Calfoam SLS-30); (10% sol'n.) (Equex SP; Lakeway 101-10, 101-11, 101-19; Polyfac SLS-30, SLS-30 Special; Sipex LCP, SB, UB; Sipon LS, LSB)

8.0–8.5 (10% sol'n.) (Conco Sulfate WB-45, WBS-45; Incronol SLS)

8.0–9.5 (2% aq. sol'n.) (Empicol LZG30); (5% aq. sol'n.) (Empicol LX28)

8.0–10.0 (2% aq. sol'n.) (Empicol LM45)

8.4–8.8 (10% sol'n.) (Carsonol SLS Paste B)

8.5 ± 0.7 (10% sol'n.) (Empicol 0919, WAK)

8.5 ± 1.0 (10% sol'n.) (Empicol LS30, LS30B)

8.5–9.5 (2% aq. sol'n.) (Empicol LZ34); (5% aq. sol'n.) (Empicol LY28/S, WA)

8.5–10.0 (Richonol A Powder)

9.0–10.5 (1% aq. sol'n.) (Empicol LM/T, LMV/T)

9.5–10.5 (1% aq. sol'n.) (Empicol 0045, 0185, LX, LXV, LZ, LZGV, LZGV/C, LZP, LZV)

9.6–10.5 (1% aq. sol'n.) (Empicol LXS95)

10.0 (1%) (Lakeway 101-10P,N)

Surface Tension:

25 dynes/cm (0.1% sol'n.) (Berol 474)

33 dynes/cm (0.1% sol'n.) (Duponol WA Dry)

35 dynes/cm (0.1% sol'n.) (Duponol ME Dry)

Biodegradable: (Berol 474; Calfoam SLS-30; Carsonol SLS, SLS Paste B, SLS Special; Manro SLS28; Mars SLS-30, SLS-A95; Merpinal LM50, LS50; Sermul EA150; Sulfatol 33 Paste; Sulphonated Lorol Paste)

TOXICITY/HANDLING:

Degreasing effect on skin in concentrated form (Berol 474)

Avoid prolonged contact with skin (Empicol 0919, LS30, LS30B, LS30P, WAK; Incronol SLS)

Irritating to skin and eyes in concentrated form (Sulfotex LCX; Texapon K-12-96)

Sodium lauryl sulfate *(cont'd.)*

Irritating to nasal and oral mucosa in conc. form; dust causes sneezing (Texapon K-1296)

Avoid ingestion (Sulfotex LCX)

STORAGE/HANDLING:

Should not be heated above 140 F nor stored above 120 F; recommended handling temp. range is 70–110 F (Calfoam SLS-30)

Store at temps. of 50–120 F (Incronol SLS)

Avoid prolonged storage above 50 C (Empicol WAK)

Store in closed containers above 7 C (Sulfotex LCX)

Store in well-closed containers; corrosive to most metals (Duponol C)

Spillages are slippery (Empicol 0919, LS30, LS30B, LS30P, WAK; Incronol SLS)

Protect from moisture (Elfan 200 Pulver; Texapon K-1296)

Avoid contact with acids because of hydrolysis (Elfan 200 Pulver, 260S)

Product will set up to a firm off-white paste on storage under cool conditions; can be readily liquefied by warming with constant stirring (Empicol 0919)

STD. PKGS.:

45-gal drums or road tankers (Manro SLS28)

51-gal (150 lb net) Leverpak (Duponol C)

55-gal drums, tank trucks, rail cars (Polyfac SLS-30, SLS-30 Special)

55-gal (450 lb net) lined open-head steel drums; bulk tank car, tank truck (Calfoam SLS-30)

55-gal (450 lb net) Liquipak containers (Incronol SLS)

55 lb net fiber drums, 44 lb net bags (Texapon K-1296)

480 lb net fiber drums; bulk, tank wagons, rail cars (Sulfotex LCX)

25 kg net plastic pails (Empicol WAK)

125-kg drums, bulk (Elfan 260S)

200-kg net iron drums (Merpinal LM50, LS50)

200-kg net lined open-head mild steel drums (Empicol 0919, LS30, LS30B, LS30P, WAK; Merpinal LS40)

200 liter internally lacquered open-head steel drums (Sulfatol 33 Paste)

Drums (Hartenol LAS-30, LES-60)

Lacquered drums (Condanol NLS30)

Perma-Plex drums; tank trucks, tank cars (Duponol WAQE)

Sodium myristyl ether sulfate

SYNONYMS:

Sodium myreth sulfate (CTFA)

STRUCTURE:

$CH_3(CH_2)_{12}CH_2(OCH_2CH_2)_nOSO_3Na$

where $n = 1-4$

Sodium myristyl ether sulfate *(cont'd.)*

CAS No.:
25446-80-4
RD No.: 977059-34-9

TRADENAME EQUIVALENTS:
Cycloryl ME60 [Cyclo]
Standapol ES-40, ES-40 Conc. [Henkel]
Texapon K14S Special [Henkel]

CATEGORY:
Base, detergent, foaming agent

APPLICATIONS:
Bath products: (Standapol ES-40 Conc.; Texapon K14S Special); bubble bath (Cycloryl ME60); bath oils (Cycloryl ME60)
Cleansers: (Standapol ES-40 Conc.)
Cosmetic industry preparations: (Cycloryl ME60); shampoo base (Cycloryl ME60); shampoos (Standapol ES-40 Conc.; Texapon K14S Special)

PROPERTIES:
Form:
Liquid (Cycloryl ME60; Texapon K14S Special)
Clear liquid (Standapol ES-40 Conc.)
Color:
Light/pale yellow (Cycloryl ME60; Standapol ES-40)
Composition:
28% conc. (Texapon K14S Special)
58–60% active (Cycloryl ME60; Standapol ES-40 Conc.)
Ionic Nature:
Anionic (Cycloryl ME60; Standapol ES-40 Conc.; Texapon K14S Special)
Cloud Pt.:
10 C max. (Standapol ES-40 Conc.)
Stability:
Stable over a wide pH range (Standapol ES-40 Conc.)
pH:
7.5–9.0 (10% aq.) (Standapol ES-40 Conc.)

STORAGE/HANDLING:
Contains ethanol—use proper precautionary methods for storage and handling (Standapol ES-40 Conc.)

Sodium POE (7) lauryl ether sulfate

SYNONYMS:
Sodium laureth-7 sulfate (CTFA)
Sodium PEG (7) lauryl ether sulfate

Sodium POE (7) lauryl ether sulfate *(cont'd.)*

EMPIRICAL FORMULA:
$C_{26}H_{54}O_{11}S \cdot Na$

STRUCTURE:
$CH_3(CH_2)_{10}CH_2(OCH_2CH_2)_nOSO_3Na$
where avg. $n = 7$

CAS No.:
9004-82-4 (generic); 66197-75-9
RD No.: 977067-48-3

TRADENAME EQUIVALENTS:
Sipon ES-7 [Alcolac]
Steol CS-760 [Stepan]

CATEGORY:
Detergent, wetting agent, foaming agent, mild surfactant, solubilizer

APPLICATIONS:
Bath products: bubble bath (Sipon ES-7; Steol CS-760)
Cosmetic industry preparations: perfumery (Sipon ES-7); shampoos (Sipon ES-7; Steol CS-760); toiletries (Sipon ES-7)

PROPERTIES:

Form:
Liquid (Sipon ES-7; Steol CS-760)

Color:
Pale yellow (Steol CS-760)

Composition:
28% active (Sipon ES-7)
59% active; 12–15% ethnaol (Steol CS-760)

Solubility:
Sol. in water (Steol CS-760)

Ionic Nature:
Anionic (Steol CS-760)

Sp.gr.:
1.059 (Steol CS-760)

Visc.:
100 cps (Steol CS-760)

Cloud Pt.:
1 C (Steol CS-760)

pH:
8.0 (10%) (Steol CS-760)

Surface Tension:
46.2 dynes/cm (0.1%) (Steol CS-760)

Biodegradable: (Steol CS-760)

STORAGE/HANDLING:
Contains ethanol—use proper precautionary methods for storage and handling (Steol CS-760)

Sodium POE (12) lauryl ether sulfate

SYNONYMS:

Sodium laureth-12 sulfate (CTFA)

Sodium PEG 600 lauryl ether sulfate

EMPIRICAL FORMULA:

$C_{36}H_{74}O_{16}S \cdot Na$

STRUCTURE:

$CH_3(CH_2)_{10}CH_2(OCH_2CH_2)_nOSO_3Na$

where avg. $n = 12$

CAS No.:

9004-82-4 (generic); 66161-57-7

RD No.: 977069-66-1

TRADENAME EQUIVALENTS:

Polystep B-21, B-23 [Stepan]

Sipon ES-12 [Alcolac]

Standapol 125-E Conc., 130E [Henkel]

CATEGORY:

Detergent, foaming agent, surfactant, solubilizer

APPLICATIONS:

Bath products: (Standapol 125-E Conc., 130E); bubble bath (Sipon ES-12)

Cleansers: (Standapol 125-E Conc., 130E)

Cosmetic industry preparations: perfumery (Sipon ES-12); shampoos (Sipon ES-12; Standapol 125-E Conc., 130E); skin preparations (Standapol 125-E Conc., 130E); toiletries (Sipon ES-12)

Industrial applications: latex (Polystep B-21, B-23)

PROPERTIES:

Form:

Liquid (Sipon ES-12)

Clear liquid (Polystep B-21; Standapol 125-E Conc., 130E)

Hazy liquid (Polystep B-23)

Color:

Pale (Polystep B-21)

Amber (Polystep B-23)

Pale yellow (Standapol 130E)

Straw yellow (Standapol 125-E Conc.)

Composition:

28% active (Polystep B-21)

28–30% active (Standapol 130E)

58–60% active (Standapol 125-E Conc.)

60% active (Polystep B-23; Sipon ES-12)

Solubility:

Sol. in hydrophilic solvents (Standapol 125-E Conc., 130E)

Sodium POE (12) lauryl ether sulfate *(cont'd.)*

Sol. in water (Standapol 125-E Conc., 130E)

Cloud Pt.:

10 C max. (Standapol 125-E Conc.)

Gel Pt.:

10 C max. (Standapol 130E)

Stability:

Stable at pH 5–12 under normal temp. conditions (Standapol 125-E Conc., 130E)

pH:

7.5–9.0 (10% aq.) (Standapol 125-E Conc., 130E)

STORAGE/HANDLING:

Contains ethanol—use proper precautionary methods for storage and handling (Standapol 125-E Conc.)

Sodium stearoyl lactylate (CTFA)

SYNONYMS:

Octadecanoic acid, 2-(1-carboxyethoxy)-1-methyl-2-oxoethyl ester, sodium salt

Sodium salt of stearoyl-2-lactylic acid

Sodium stearyl-2-lactylate

EMPIRICAL FORMULA:

$C_{24}H_{44}O_6 \cdot Na$

STRUCTURE:

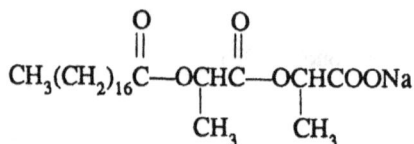

CAS No.

25383-99-7

TRADENAME EQUIVALENTS:

Admul SSL 2003, SSL 2004 [PPF Int'l.]

Artodan [Grinsted]

Emplex [Patco]

Grindtek FAL1 [Grinsted]

Lamegin NSL [Grunau]

Pationic 145A [Patco]

Pationic SSL [Patco]

Radiamuls SSL 2990 [Oleofina S.A.]

CATEGORY:
Emulsifier, stabilizer, viscosity builder

APPLICATIONS:
Cosmetic industry preparations: (Lamegin NSL; Pationic SSL)
Food applications: (Admul SSL 2003, SSL 2004; Emplex; Radiamuls SSL 2990); food emulsifying (Emplex)

PROPERTIES:

Form:
Flake (Admul SSL 2003; Pationic 145A; Radiamuls SSL 2990)
Powder (Admul SSL 2004; Emplex; Grindtek FAL1; Lamegin NSL; Pationic SSL; Radiamuls SSL 2990)

Color:
White (Pationic 145A)
Cream (Grindtek FAL1)
Light tan (Emplex; Pationic SSL)

Odor:
Mild caramel (Emplex)

Composition:
100% active (Artodan)
100% conc. (Pationic 145A, SSL; Radiamuls SSL 2990)

Solubility:
Sol. warm in ethanol (Grindtek FAL1)
Disp. in min. oil @ 1% conc. and 25 C (Pationic SSL)
Readily disp. in hot oil (Emplex)
Sol. warm in paraffin oil (Grindtek FAL1)
Sol. warm in propylene glycol (Grindtek FAL1)
Sol. warm in toluene (Grindtek FAL1)
Readily disp. in hot water (Emplex); disp. in dist. water @ 1% conc. and 25 C (Pationic 145A, SSL)
Sol. warm in white spirit (Grindtek FAL1)

Ionic Nature:
Anionic

Visc.:
175 cps (2% in dist. water) (Pationic SSL)
185 cps (2% in dist. water) (Pationic 145A)

M.P.:
41–46 C (Pationic SSL)
45–55 C (Radiamuls SSL 2990)
57–60 C (Pationic 145A)

HLB:
6.5 (Pationic SSL)
8.3 (Pationic 145A)
10 (estimate) (Grindtek FAL1)

Sodium stearoyl lactylate *(cont'd.)*

Acid No.:
  60–70 (Pationic SSL)
  60–80 (Pationic 145A; Emplex)
Ester No.:
  150–190 (Emplex)
Saponification No.:
  169–190 (Pationic 145A)
  210–235 (Pationic SSL)
pH:
  5.70 (2% aq.) (Pationic 145A)
  5.95 (2% aq.) (Pationic SSL)
Surface Tension:
  28.87 dynes/cm (0.1% conc.) (Pationic 145A)
  32.34 dynes/cm (0.1% conc. (Pationic SSL)
TOXICITY/HANDLING:
  Minor skin irritant in conc. form (Pationic 145A)
STORAGE/HANDLING:
  Store under cool, dry conditions in closed containers; avoid exposure to temps. above
    90 F (Emplex; Pationic 145A, SSL)
STD. PKGS.:
  25-kg net bags (standard pallets: 1 ton) (Radiamuls SSL 2990)
  50-lb net poly-lined cartons, 24 cartons/stretch-wrapped pallet (1200 lb) (Emplex)

Stearyl dimethyl amine

SYNONYMS:
  N,N-Dimethyl-1-octadecanamine
  Dimethyl stearamine (CTFA)
  Dimethyl stearylamine
  1-Octadecanamine, N,N-dimethyl-
  Octadecyl dimethyl amine
EMPIRICAL FORMULA:
  $C_{20}H_{43}N$
STRUCTURE:

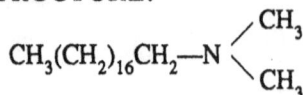

CAS No.:
  124-28-7

TRADENAME EQUIVALENTS:
Adma 8 [Ethyl Corp.]
Armeen DM18D [Akzo Chemie]
Nissan Tertiary Amine AB [Nippon Oils & Fats]
Distilled grades:
Adogen 342D [Sherex]
Crodamine 3.A18D [Croda Ltd.]
Kemamine T-9902D [Humko Div./Witco]
Lilamin 342D [Lilachim S.A.]

CATEGORY:
Intermediate, corrosion inhibitor, germicide, catalyst, emulsifier, extraction reagent

APPLICATIONS:
Household detergents: raw material (Armeen DM18D; Nissan Tertiary Amine AB)
Industrial applications: (Adma 8); dyes and pigments (Lilamin 342D); lubricating/ cutting oils (Nissan Tertiary Amine AB); metalworking (Crodamine 3.A18D); petroleum industry (Adogen 342D; Kemamine T-9902D; Lilamin 342D); plastics (Crodamine 3.A18D; Lilamin 342D; Nissan Tertiary Amine AB); rubber (Crodamine 3.A18D; Lilamin 342D); textile/leather processing (Crodamine 3.A18D)

PROPERTIES:
Form:
Liquid (Adma 8; Armeen DM18D; Kemamine T-9902D; Lilamin 342D)
Liquid or semisolid (Nissan Tertiary Amine AB)
Paste (Crodamine 3.A18D)
Color:
Light yellow (Nissan Tertiary Amine AB)
Yellow (Armeen DM18D)
Gardner 2 max. (Lilamin 342D)
Odor:
Amine (Armeen DM18D)
Composition:
95% active (Lilamin 342D)
95% conc. (Kemamine T-9902D)
95% min. tertiary amine (Nissan Tertiary Amine AB)
98% active (Armeen DM18D)
100% conc. (Adma 8; Adogen 342D; Crodamine 3.A18D)
Solubility:
Sol. in oil (Adogen 342D)
Insol. in water (Armeen DM18D)
Ionic Nature:
Cationic (Adma 8; Armeen DM18D; Crodamine 3.A18D)
M.W.:
305 (Lilamin 342D)

Stearyl dimethyl amine *(cont'd.)*

Sp.gr.:
0.79 (Armeen DM18D)
F.P.:
20 C (Armeen DM18D)
B.P.:
145–160 C (3 mm Hg) (Armeen DM18D)
Flash Pt.:
164 (OC) (Lilamin 342D)
178 C (COC) (Armeen DM18D)
Iodine No.:
2 max. (Lilamin 342D; Nissan Tertiary Amine AB)
3 max. (Armeen DM18D)
TOXICITY/HANDLING:
Skin irritant, severe eye irritant (Armeen DM18D)
STORAGE/HANDLING:
Avoid contact with strong oxidizing agents and strong acids (Armeen DM18D)
STD. PKGS.:
13-kg can, 160-kg drum (Nissan Tertiary Amine AB)
200-liter bung-type steel drums (Armeen DM18D)

TRADENAME PRODUCTS AND
GENERIC EQUIVALENTS

Accomeen C2 [Armstrong]—POE (2) coconut amine
Accomeen C5 [Armstrong]—POE (5) coconut amine
Accomeen C10 [Armstrong]—POE (10) coconut amine
Accomeen C15 [Armstrong]—POE (15) coconut amine
Accomeen S5 [Armstrong]—POE (5) soya amine
Accomeen S10 [Armstrong]—POE (10) soya amine
Acconon 400 MO [Armstrong]—POE (8) monooleate
Accoquat 2C-75, 2C-75-H [Armstrong]—Dicoco dimethyl ammonium chloride
Accosperse 20 [Armstrong]—POE (20) sorbitan monolaurate
Accosperse 60 [Armstrong]—POE (20) sorbitan monostearate
Accosperse 80 [Armstrong]—POE (20) sorbitan monooleate
Acylglutamate CS-11 [Ajinomoto]—Sodium cocoyl glutamate
Acylglutamate CT-12 [Ajinomoto]—Monotriethanolamine N-cocoyl-L-glutamate
Acylglutamate HS-11 [Ajinomoto]—Sodium hydrogenated tallow glutamate
Acylglutamate LS-11 [Ajinomoto]—Sodium lauroyl glutamate
Adma 2 [Ethyl Corp.]—Lauryl dimethyl amine
Adma 6 [Ethyl Corp.]—Palmityl dimethyl amine
Adma 8 [Ethyl Corp.]—Stearyl dimethyl amine
Admul CSL 2007, CSL 2008 [PPF Int'l.]—Calcium stearoyl lactylate
Admul SSL 2003, SSL 2004 [PPF Int'l.]—Sodium stearoyl lactylate
Adogen 342D [Sherex]—Stearyl dimethyl amine, distilled
Adogen 462 [Sherex]—Dicoco dimethyl ammonium chloride
Ahco 3998 [ICI Americas]—POE (20) oleyl ether
Ahco 7166T [ICI United States]—POE (20) sorbitan tristearate
Ahco 7596D [ICI United States]—POE (4) sorbitan monolaurate
Ahco 7596T [ICI United States]—POE (20) sorbitan monolaurate
Ahco DFO-110 [ICI United States]—POE (20) sorbitan trioleate
Ahco DFO-150 [ICI United States]—POE (20) sorbitan monooleate
Ahco DFP-156 [ICI United States]—POE (20) sorbitan monopalmitate
Ahco DFS-96 [ICI United States]—POE (4) sorbitan monostearate
Ahco DFS-149 [ICI United States]—POE (20) sorbitan monostearate
Ahcovel Base N-15 [ICI United States]—Glyceryl monostearate, self-emulsifying
Ahcowet DQ-114 [ICI United States]—POE (6) tridecyl ether
Ahcowet DQ-145 [ICI United States]—POE (12) tridecyl ether
Akyposal ALS33 [Chem-Y]—Ammonium lauryl sulfate
Akyposal MLES35 [Chem-Y GmbH]—Monoethanolamine lauryl ether sulfate
Akyposal NLS [Chemax]—Sodium lauryl sulfate
Alconate CPA [Alcolac]—Disodium coco amide MIPA sulfosuccinate
Alconate L-3 [Alcolac]—Disodium lauryl ether sulfosuccinate

WHAT EVERY CHEMICAL TECHNOLOGIST WANTS TO KNOW
ABOUT...EMULSIFIERS AND WETTING AGENTS

Alconate LEA [Alcolac]—Disodium lauramido MEA-sulfosuccinate

Aldo HMO, MO, MO Tech. [Glyco]—Glyceryl monooleate

Aldo HMS, MS, MSLG [Glyco]—Glyceryl monostearate

Aldo MLD [Glyco]—Glyceryl monolaurate

Aldo MOD [Glyco]—Glyceryl monooleate (SE grade)

Aldo MSD [Glyco]—Glyceryl monostearate, self-emulsifying

Aldo PGHMS [Glyco]—Propylene glycol monostearate

Aldo PMS [Glyco]—Propylene glycol monostearate

Aldosperse MS-20, MS-20 FG [Glyco]—POE (20) glyceryl monostearate

Alkaminox C-2 [Alkaril]—POE (2) coconut amine

Alkaminox T-2 [Alkaril]—POE (2) tallow amine

Alkaminox T-5 [Alkaril]—POE (5) tallow amine

Alkaminox T-15 [Alkaril]—POE (15) tallow amine

Alkamuls 200-DO [Alkaril]—POE (4) dioleate

Alkamuls 200-DS [Alkaril]—POE (4) distearate

Alkamuls 200-ML [Alkaril]—POE (4) monolaurate

Alkamuls 200-MO [Alkaril]—POE (4) monooleate

Alkamuls 200-MS [Alkaril]—POE (4) monostearate

Alkamuls 400-DO [Alkaril]—POE (8) dioleate

Alkamuls 400-DS [Alkaril]—POE (8) distearate

Alkamuls 400-MC [Alkaril]—POE (8) monococoate

Alkamuls 400-ML [Alkaril]—POE (8) monolaurate

Alkamuls 400-MO [Alkaril]—POE (8) monooleate

Alkamuls 400-MS [Alkaril]—POE (8) monostearate

Alkamuls 600-DO [Alkaril]—POE (12) dioleate

Alkamuls 600-DS [Alkaril]—POE (12) distearate

Alkamuls 600-ML [Alkaril]—POE (12) monolaurate

Alkamuls 600-MO [Alkaril]—POE (12) monooleate

Alkamuls 600-MS [Alkaril]—POE (12) monostearate

Alkamuls AG-CA [Alkaril]—Calcium sulfate

Alkamuls GML, GML-45 [Alkaril]—Glyceryl monolaurate

Alkamuls GMO, GMO-45, GMO-45LG, GMO-55LG [Alkaril]—Glyceryl monooleate

Alkamuls GMS [Alkaril]—Glyceryl monostearate

Alkamuls PSML-4 [Alkaril]—POE (4) sorbitan monolaurate

Alkamuls PSML-20 [Alkaril]—POE (20) sorbitan monolaurate

Alkamuls PSMO-5 [Alkaril]—POE (5) sorbitan monooleate

Alkamuls PSMO-20 [Alkaril]—POE (20) sorbitan monooleate

Alkamuls PSMS-4 [Alkaril]—POE (4) sorbitan monostearate

Alkamuls PSMS-20 [Alkaril]—POE (20) sorbitan monostearate

Alkamuls PSTO-20 [Alkaril]—POE (20) sorbitan trioleate

Alkamuls PSTS-20 [Alkaril]—POE (20) sorbitan tristearate

Alkasurf ALS [Alkaril]—Ammonium lauryl sulfate

Alkasurf DLS [Alkaril]—Diethanolamine lauryl sulfate
Alkasurf EA-60 [Alkaril]—Ammonium lauryl ether sulfate
Alkasurf IPAM [Alkaril]—Isopropylamine dodecylbenzenesulfonate
Alkasurf LA Acid [Alkaril]—Dodecylbenzene sulfonic acid (linear)
Alkasurf LA-3 [Alkaril]—Pareth-25-3
Alkasurf LA-7 [Alkaril]—Pareth-25-7
Alkasurf LA-12 [Alkaril]—Pareth 25-12
Alkasurf LAN-1 [Alkaril]—PEG-1 lauryl ether
Alkasurf LAN-15 [Alkaril]—POE (15) lauryl ether
Alkasurf LAN-23 [Alkaril]—POE (23) lauryl ether
Alkasurf MLS [Alkaril]—Monoethanolamine lauryl sulfate
Alkasurf OA-2 [Alkaril]—POE (2) oleyl ether
Alkasurf OA-10 [Alkaril]—POE (10) oleyl ether
Alkasurf SS-L9ME [Alkaril]—Disodium lauramido MEA-sulfosuccinate
Alkasurf SS-LA-3 [Alkaril]—Disodium lauryl ether sulfosuccinate
Alkasurf SS-OA [Alkaril]—Disodium N-oleyl sulfosuccinamate
Alkasurf WADX [Alkaril]—Diethanolamine lauryl sulfate (modified)
Alkasurf WAQ [Alkaril]—Sodium lauryl sulfate
Alkazine O [Alkaril]—Oleyl imidazoline
Alrosol C [Ciba-Geigy]—Capric acid diethanolamide
Alrosol Conc. [Ciba-Geigy]—Capric acid diethanolamide
Alscoap LN-40, LN-90 [Toho]—Sodium lauryl sulfate
Amerchol Polysorbate 80 [Amerchol]—POE (20) sorbitan monooleate
Ameroxol OE-2 [Amerchol]—POE (2) oleyl ether
Ameroxol OE-10 [Amerchol]—POE (10) oleyl ether
Ameroxol OE-20 [Amerchol]—POE (20) oleyl ether
Amine O [Ciba-Geigy]—Oleyl imidazoline
Amisoft C-T-12 [Ajinomoto]—Monotriethanolamine N-cocoyl-L-glutamate
Amisoft GS-11 [Ajinomoto]—Sodium cocoyl glutamate
Amisoft HS-11 [Ajinomoto]—Sodium hydrogenated tallow glutamate
Amisoft LS-11 [Ajinomoto]—Sodium lauroyl glutamate
Amphoterge K [Lonza]—Cocoamphopropionate
Ardet LAS [Ardmore]—Dodecylbenzene sulfonic acid (linear)
Arlacel 989 [ICI United States]—POE (7) hydrogenated castor oil
Arlatone G [ICI United States]—POE (25) hydrogenated castor oil
Armac OD [Akzo Chemie]—Oleyl amine acetate
Armeen DM12D [Akzo Chemie]—Lauryl dimethyl amine
Armeen DM16D [Akzo Chemie]—Palmityl dimethyl amine
Armeen DM18D [Akzo Chemie]—Stearyl dimethyl amine
Armeen DMCD [Armak]—Coco dimethyl amine
Armeen DMMCD [Armak]—Coco dimethyl amine
Armeen O [Armak]—Oleyl amine

Armeen OD [Akzo Chemie]—Oleyl amine

Armotan PML20 [Akzo Chemie]—POE (20) sorbitan monolaurate

Armotan PMO20 [Akzo]—POE (20) sorbitan monooleate

Arosurf 66-E2 [Sherex]—POE (2) isostearyl ether

Arosurf 66-E10 [Sherex]—POE (10) isostearyl ether

Arosurf 66-E20 [Sherex]—POE (20) isostearyl ether

Arosurf 66-PE12 [Sherex]—PPG-3-isosteareth-9

Arquad 2C-75 [Armak]—Dicoco dimethyl ammonium chloride

Arquad 12-33, 12-50 [Armak]—Lauryl trimethyl ammonium chloride

Arquad DM14B-90 [Akzo Chemie]—Myristyl dimethyl benzyl ammonium chloride (di-hydrate)

Arsul DDB [Magna/Arjay]—Dodecylbenzene sulfonic acid (branched)

Arsul LAS [Magna/Arjay]—Dodecylbenzene sulfonic acid (linear)

Arsul WAQ [Arjay]—Sodium lauryl sulfate

Artodan [Grinsted]—Sodium stearoyl lactylate

Arylan CA [Lankro]—Calcium dodecylbenzene sulfonate

Arylan S Acid [Diamond Shamrock/Process]—Dodecylbenzene sulfonic acid (branched)

Arylan S Flake, SBC 25, SC30, SX Flake [Lankro]—Sodium dodecylbenzenesulfonate

Arylan SBC Acid [Lankro]—Dodecylbenzene sulfonic acid (linear)

Arylan SC Acid [Lankro]—Dodecylbenzene sulfonic acid (linear)

Atlas G-1702 [ICI United States]—POE (6) sorbitol beeswax

Atlas G-1726 [ICI United States]—POE (20) sorbitol beeswax

Atlas G-2162 [ICI United States]—POE (25) propylene glycol monostearate

Atlas G-2185 [ICI United States]—Propylene glycol monooleate

Atlas G-2198 [ICI United States]—POE (40) monostearate

Atmos 150 [ICI]—Glyceryl monostearate

Atmos 300 [ICI]—Glyceryl monooleate

Atmul 84, 84K, 124 [ICI]—Glyceryl monostearate

Avirol 200 [Henkel]—Ammonium lauryl sulfate

Avirol SL-2010 [Henkel]—Sodium lauryl sulfate

Barquat MS-100 [Lonza]—Myristyl dimethyl benzyl ammonium chloride (dihydrate)

Barquat MX-50, MX-80 [Lonza]—Myristyl dimethyl benzyl ammonium chloride

Berol 474 [Berol]—Sodium lauryl sulfate

Bio-Dac 50-22 [Bio-Lab]—Didecyl dimethyl ammonium chloride

Bio Soft D-35X, D-40, D-60, D-62 [Stepan]—Sodium dodecylbenzenesulfonate

Bio Soft LAS-97 [Stepan]—Dodecylbenzene sulfonic acid (linear)

Bio Soft S-100 [Stepan]—Dodecylbenzene sulfonic acid (linear)

Brij 30, 30 SP [ICI United States]—POE (4) lauryl ether

Brij 35, 35SP [ICI United States]—POE (23) lauryl ether

Brij 52 [ICI United States]—POE (2) cetyl ether

Brij 56 [ICI United States]—POE (10) cetyl ether

Brij 58 [ICI United States]—POE (20) cetyl ether
Brij 92, 93 [ICI United States]—POE (2) oleyl ether
Brij 96 [ICI United States]—POE (10) oleyl ether
Brij 97 [ICI United States]—POE (10) oleyl ether
Brij 98, 99 [ICI United States]—POE (20) oleyl ether
BTC 824 [Onyx]—Myristyl dimethyl benzyl ammonium chloride
BTC 1010 [Onyx]—Didecyl dimethyl ammonium chloride
BTCO 1010 [Onyx]—Didecyl dimethyl ammonium chloride
Calfoam SLS-30 [Pilot]—Sodium lauryl sulfate
Calimulse PRS [Pilot]—Isopropylamine dodecylbenzenesulfonate
Calsoft F-90, L-40, L-60 [Pilot]—Sodium dodecylbenzenesulfonate
Calsoft LAS-99 [Pilot]—Dodecylbenzene sulfonic acid (linear)
Capmul EMG [Capital City Products]—POE (20) glyceryl monostearate
Capmul GMO [Stokely-Van Camp]—Glyceryl monooleate
Capmul POE-S [Capital City]—POE (20) sorbitan monostearate
Carsonol ALES-4 [Carson]—Ammonium lauryl ether sulfate
Carsonol ALES-5 [Carson Chem.]—Ammonium POE (5) lauryl ether sulfate
Carsonol ALS, ALS Special [Carson]—Ammonium lauryl sulfate
Carsonol DLS [Carson]—Diethanolamine lauryl sulfate
Carsonol MLS [Carson]—Magnesium lauryl sulfate
Carsonol SES-A [Carson]—Ammonium lauryl ether sulfate
Carsonol SLS, SLS Paste B, SLS Special [Carson]—Sodium lauryl sulfate
Carsosulf UL-100 Acid [Carson]—Dodecylbenzene sulfonic acid (linear)
CE-55-2 [Hefti]—POE (2) cetyl ether
CE-55-20 [Hefti]—POE (20) cetyl ether
Cedepal SA-406 [Domtar]—Ammonium lauryl ether sulfate
Cedepon LA-30 [Domtar/CDC]—Ammonium lauryl sulfate
Cedepon LS-30, LS-30P [Domtar/CDC]—Sodium lauryl sulfate
Cerasynt 840 [Van Dyk]—POE (20) monostearate
Cerasynt 945, SD, WM [Van Dyk]—Glyceryl monostearate
Cerasynt PA [Van Dyk]—Propylene glycol monostearate, self-emulsifying
Cerasynt Q [Van Dyk]—Glyceryl monostearate, self-emulsifying
Cetats [Hexcel]—Cetyl trimethyl ammonium p-toluene sulfonate
Charlab LPC [Catawba-Charlab]—Lauryl pyridinium chloride
Chemal LA-4 [Chemax]—POE (4) lauryl ether
Chemal LA-12 [Chemax]—POE (12) lauryl ether
Chemal LA-23 [Chemax]—POE (23) lauryl ether
Chemal OA-4 [Chemax]—POE (4) oleyl ether
Chemal OA-10 [Chemax]—POE (10) oleyl ether
Chemal OA-20, OA-20/70 [Chemax]—POE (20) oleyl ether
Chemal TDA-3 [Chemax]—POE (3) tridecyl ether
Chemal TDA-6 [Chemax]—POE (6) tridecyl ether

Chemal TDA-12 [Chemax]—POE (12) tridecyl ether
Chemax E-200-ML [Chemax]—POE (6) monolaurate
Chemax E-400-ML [Chemax]—POE (8) monolaurate
Chemax E-1000-MO [Chemax]—POE (20) monooleate
Chemax E-1000-MS [Chemax]—POE (20) monostearate
Chemax HCO-5 [Chemax]—POE (5) hydrogenated castor oil
Chemax HCO-25 [Chemax]—POE (25) hydrogenated castor oil
Chemax PEG 200 DO [Chemax]—POE (4) dioleate
Chemax PEG 400 DO [Chemax]—POE (8) dioleate
Chemax PEG 600 DO [Chemax]—POE (12) dioleate
Chemax SBO [Chemax]—Butyl oleate, sulfated
Chemax TO-10 [Chemax]—POE (10) monotallate
Chemeen 18/2 [Chemax]—POE (2) stearyl amine
Chemeen 18/5 [Chemax]—POE (5) stearyl amine
Chemeen 18/50 [Chemax]—POE (50) stearyl amine
Chemeen C-2 [Chemax]—POE (2) coconut amine
Chemeen C-5 [Chemax]—POE (5) coconut amine
Chemeen C-10 [Chemax]—POE (10) coconut amine
Chemeen C-15 [Chemax]—POE (15) coconut amine
Chemeen DT-15 [Chemax]—POE (15) tallow aminopropylamine
Chemeen T-2 [Chemax]—POE (2) tallow amine
Chemeen T-5 [Chemax]—POE (5) tallow amine
Chemeen T-15 [Chemax]—POE (15) tallow amine
Chemfac PB-184 [Chemax]—POE (4) oleyl ether phosphate
Chemquat 12-33, 12-50 [Chemax]—Lauryl trimethyl ammonium chloride
Cindet GE [Cindet]—Isopropylamine dodecylbenzenesulfonate
Cithrol 2DO [Croda Chem. Ltd.]—POE (4) dioleate
Cithrol 2DS [Croda Chem. Ltd.]—POE (4) distearate
Cithrol 2ML (Croda)—POE (4) monolaurate
Cithrol 2MO [Croda]—POE (4) monooleate
Cithrol 2MS [Croda]—POE (4) monostearate
Cithrol 4DS [Croda]—POE (8) distearate
Cithrol 4ML [Croda]—POE (8) monolaurate
Cithrol 4MO [Croda]—POE (8) monooleate
Cithrol 4MS [Croda]—POE (8) monostearate
Cithrol 6DO [Croda Chem. Ltd.]—POE (12) dioleate
Cithrol 6DS [Croda Chem. Ltd.]—POE (12) distearate
Cithrol 6ML [Croda]—POE (12) monolaurate
Cithrol 6MO [Croda]—POE (12) monooleate
Cithrol 6MS [Croda]—POE (12) monostearate
Cithrol 10MS [Croda]—POE (20) monostearate
Cithrol 15MS [Croda]—POE (1500) monostearate

Cithrol 40MS [Croda]—POE (75) monostearate
Cithrol A [Croda]—POE (8) monooleate
Cithrol DGML N/E [Croda]—Diethylene glycol monolaurate
Cithrol DGML S/E [Croda]—Diethylene glycol monolaurate SE
Cithrol DGMO N/E [Croda]—Diethylene glycol monooleate
Cithrol DGMO S/E [Croda]—Diethylene glycol monooleate SE
Cithrol DGMS N/E [Croda Chem. Ltd.]—Diethylene glycol monostearate
Cithrol DGMS S/E [Croda Chem. Ltd.]—Diethylene glycol monostearate SE
Cithrol GMS A/S ES 0743 [Croda]—Glyceryl monostearate
Cithrol GMS Acid Stable, GMS S/E [Croda]—Glyceryl monostearate, self-emulsifying
Cithrol GMS N/E [Croda]—Glyceryl monostearate
Cithrol PGML N/E [Croda Chem. Ltd.]—Propylene glycol monolaurate
Cithrol PGML S/E [Croda Chem. Ltd.]—Propylene glycol monolaurate, self-emulsifying
Cithrol PGMO N/E [Croda Chem. Ltd.]—Propylene glycol monooleate
Cithrol PGMO S/E [Croda Chem. Ltd.]—Propylene glycol monooleate, self-emulsifying
Cithrol PGMS N/E [Croda Chem. Ltd.]—Propylene glycol monostearate
Cithrol PGMS S/E [Croda Chem. Ltd.]—Propylene glycol monostearate, self-emulsifying
Clindrol SDG [Clintwood]—Diethylene glycol monostearate
Collemul H4 [Allied Colloids Ltd]—POE (8) monooleate
Collemul L4 [Allied Colloids Ltd.]—POE (8) monolaurate
Comperlan CD [Henkel/Canada]—Capric acid diethanolamide
Conco AAS Special 3, 3H [Continental]—Isopropylamine dodecylbenzenesulfonate
Conco AAS-35, AAS-40, AAS-45S, AAS-65 [Continental]—Sodium dodecylbenzene-sulfonate
Conco AAS-75S [Continental]—Calcium dodecylbenzene sulfonate
Conco AAS-90 [Continental]—Sodium dodecylbenzenesulfonate
Conco AAS-98S [Continental]—Dodecylbenzene sulfonic acid (linear)
Conco Sulfate 216 [Continental]—Ammonium lauryl ether sulfate
Conco Sulfate A [Continental]—Ammonium lauryl sulfate
Conco Sulfate EP [Continental]—Diethanolamine lauryl sulfate
Conco Sulfate M [Continental]—Magnesium lauryl sulfate
Conco Sulfate P [Continental]—Potassium lauryl sulfate
Conco Sulfate WA, WA Dry, WA Special, WAG, WAS, WB-45, WBS-45 [Continental]—Sodium lauryl sulfate
Conco Sulfate WEM, WM [Continental]—Ammonium lauryl ether sulfate
Conco Sulfate WR, WX [Continental]—Sodium lauryl sulfate
Condanol DLS35 [Dutton & Reinisch]—Diethanolamine lauryl sulfate
Condanol MLS35 [Dutton & Reinisch]—Monoethanolamine lauryl sulfate
Condanol NLS28, NLS30, NLS90 [Dutton & Reinisch]—Sodium lauryl sulfate
Condanol SBF12, SBF-12 Powder [Dutton & Reinisch]—Disodium lauryl sulfosuccinate
Condanol SBFA30, 40% [Dutton & Reinisch]—Disodium lauryl ether sulfosuccinate

Condanol SBL203 [Dutton & Reinisch]—Disodium lauramido MEA-sulfosuccinate
Condasol Sulfonic Acid K [Dutton & Reinisch]—Dodecylbenzene sulfonic acid (linear)
Condensate PM [Continental]—Myristic diethanolamide
CPH-27-N [C.P. Hall]—POE (4) monolaurate
CPH-31-N [C.P. Hall]—Glyceryl monooleate
CPH-39-N [C.P. Hall]—POE (4) monooleate
CPH-41-N [C.P. Hall]—POE (12) monooleate
CPH-43-N [C.P. Hall]—POE (12) monolaurate
CPH-53-N [C.P. Hall]—Glyceryl monostearate
CPH-205-NX [C.P. Hall]—Glyceryl monooleate
CPH-213-N [C.P. Hall]—POE (12) dioleate
CPH-250-SE [C.P. Hall]—Glyceryl monostearate, self-emulsifying
CPH-362-N [C.P. Hall]—Glyceryl monooleate
Cremophor A11 [BASF AG]—POE (11) cetyl/stearyl ether
Cremophor A25 [BASF AG]—POE (25) cetyl/stearyl ether
Crillet 1 [Croda]—POE (20) sorbitan monolaurate
Crillet 2 [Croda]—POE (20) sorbitan monopalmitate
Crillet 3 [Croda]—POE (20) sorbitan monostearate
Crillet 4 [Croda]—POE (20) sorbitan monooleate
Crillet 11 [Croda]—POE (4) sorbitan monolaurate
Crillet 31 [Croda]—POE (4) sorbitan monostearate
Crillet 35 [Croda]—POE (20) sorbitan tristearate
Crillet 45 [Croda]—POE (20) sorbitan trioleate
Crodamet 1.C2 [Croda Chem. Ltd.]—POE (2) coconut amine
Crodamet 1.C5 [Croda Chem. Ltd.]—POE (5) coconut amine
Crodamet 1.C10 [Croda Chem. Ltd.]—POE (10) coconut amine
Crodamet 1.C15 [Croda Chem. Ltd.]—POE (15) coconut amine
Crodamet 1.T15 [Croda Chem. Ltd.]—POE (15) tallow amine
Crodamine 1.O, 1.OD [Croda Universal Ltd.]—Oleyl amine
Crodamine 3.A16D [Croda]—Palmityl dimethyl amine, distilled
Crodamine 3.A18D [Croda Ltd.]—Stearyl dimethyl amine, distilled
Crodazoline O [Croda]—Oleyl imidazoline
Cutina AGS [Henkel]—Ethylene glycol distearate
Cutina E-24 [Henkel Canada]—POE (20) glyceryl monostearate
Cutina GMS, MD, MD-A [Henkel/Canada]—Glyceryl monostearate
Cyclochem EGDS [Cyclo]—Ethylene glycol distearate
Cyclochem GMO [Cyclo]—Glyceryl monooleate
Cyclochem GMS [Cyclo]—Glyceryl monostearate
Cyclochem GMS21, GMS165 [Cyclo]—Glyceryl monostearate, self-emulsifying
Cyclochem PEG 400 DS [Cyclo]—POE (8) distearate
Cyclochem PEG 400 ML [Cyclo]—POE (8) monolaurate
Cyclochem PEG 400 MO [Cyclo]—POE (8) monooleate

Cyclochem PGMS [Cyclo]—Propylene glycol monostearate
Cyclogol EL [Cyclo]—POE (12) lauryl ether
Cyclopol SB-CP [Cyclo]—Disodium coco amide MIPA sulfosuccinate
Cyclopol SBFA30 [Cyclo]—Disodium lauryl ether sulfosuccinate
Cyclopol SBL203 [Cyclo]—Disodium lauramido MEA-sulfosuccinate
Cycloryl 21 [Cyclo]—Sodium lauryl sulfate
Cycloryl 21LS, 21SP [Cyclo]—Sodium lauryl sulfate
Cycloryl ABSA [Cyclo]—Dodecylbenzene sulfonic acid (linear)
Cycloryl DA [Cyclo]—Diethanolamine lauryl sulfate
Cycloryl DDB40 [Cyclo]—Sodium dodecylbenzenesulfonate
Cycloryl MA [Cyclo]—Ammonium lauryl sulfate
Cycloryl MA 330, MA 360 [Cyclo]—Ammonium lauryl ether sulfate
Cycloryl ME361 [Cyclo]—Ammonium myristyl ether sulfate
Cycloryl MG [Cyclo]—Magnesium lauryl sulfate
Cycloryl SA [Cyclo]—Monoethanolamine lauryl sulfate
Cycloteric MV-SF [Cyclo]—Cocoamphopropionate
Cyncal 80% [Hilton-Davis]—Myristyl dimethyl benzyl ammonium chloride
DDBSA 99-b [Monsanto]—Dodecylbenzene sulfonic acid
Dehydol LS-4 [Henkel]—POE (4) lauryl ether
Dehyquart C, C Crystals [Henkel]—Lauryl pyridinium chloride
Dehyquart LT [Henkel KGaA]—Lauryl trimethyl ammonium chloride
DeSonate SA-H [DeSoto]—Dodecylbenzene sulfonic acid (branched)
Dibactol [Hexcel]—Myristyl dimethyl benzyl ammonium chloride (dihydrate)
DMS-33 [Hefti]—Diethylene glycol monostearate
Dobanol 25-3 [Shell]—Pareth-25-3
Dobanol 25-3A/60 [Shell]—Ammonium pareth-25 sulfate
Dobanol 25-7 [Shell]—Pareth-25-7
Drewlene 10 [PVO Int'l.]—Propylene glycol monostearate
Drewmulse 85, GMO [PVO]—Glyceryl monooleate
Drewmulse DGMS [Drew Produtos]—Diethylene glycol monostearate
Drewmulse EGDS [Drew Produtos]—Ethylene glycol distearate
Drewmulse PGML [PVO Int'l.]—Propylene glycol monolaurate
Drewmulse PGMS [Drew Produtos]—Propylene glycol monostearate
Drewmulse POE-SML [PVO Int'l.]—POE (20) sorbitan monolaurate
Drewmulse POE-SMO [PVO Int'l.]—POE (20) sorbitan monooleate
Drewmulse POE-SMS [PVO Int'l.]—POE (20) sorbitan monostearate
Drewmulse POE-STS [PVO Int'l.]—POE (20) sorbitan tristearate
Drewmulse TP, V [PVO Int'l.]—Glyceryl monostearate
Drewmulse V-SE [PVO Int'l.]—Glyceryl monostearate, self-emulsifying
Drewpon 100 [Drew Produtos]—Sodium lauryl sulfate
Drewpon AS, CN [Drew Produtos]—Sodium lauryl sulfate
Drewpon ECM [Drew Produtos]—Monoethanolamine lauryl ether sulfate

Drewpon EKZ [Drew Produtos]—Ammonium lauryl ether sulfate
Drewpon ESG [Drew Produtos]—Magnesium lauryl ether sulfate
Drewpon MEA [Drew Produtos]—Monoethanolamine lauryl sulfate
Drewpon MG [Drew Produtos]—Magnesium lauryl sulfate
Drewpon NH, NHAC [Drew Productos]—Ammonium lauryl sulfate
Drewpone 60 [PVO Int'l.]—POE (20) sorbitan monostearate
Drewpone 65 [PVO Int'l.]—POE (20) sorbitan tristearate
Drewpone 80 [PVO Int'l.]—POE (20) sorbitan monooleate
Duponol C [DuPont]—Sodium lauryl sulfate
Duponol EP [DuPont]—Diethanolamine lauryl sulfate (tech.)
Duponol ME Dry, QC, WA Dry, WA Paste, WAQE [DuPont]—Sodium lauryl sulfate
Dur-Em 114, 204, GMO [Durkee]—Glyceryl monooleate
Dur-Em 117 [Durkee]—Glyceryl monostearate
Dur-Em 207-E [Durkee]—Glyceryl monostearate, self-emulsifying
Durfax 20 [Durkee]—POE (20) sorbitan monolaurate
Durfax 60, 60K [Durkee]—POE (20) sorbitan monostearate
Durfax 65, 65K [Durkee]—POE (20) sorbitan tristearate
Durfax 80, 80K [Durkee]—POE (20) sorbitan monooleate
Durfax EOM, EOM K [Durkee]—POE (20) glyceryl monostearate
Durpeg 400 MO [Durkee]—POE (8) monooleate
Durpro 107-55 [Durkee/SCM Corp.]—Propylene glycol monostearate

Elfan 200 Pulver [Akzo Chemie]—Sodium lauryl sulfate
Elfan 240 [Akzo Chemie]—Sodium lauryl sulfate
Elfan 240M [Akzo Chemie]—Monoethanolamine lauryl sulfate
Elfan 260 S [Akzo Chemie]—Sodium lauryl sulfate
Elfan L310 [Akzo Chemie]—Ethylene glycol distearate
Elfan WA, WA Pulver [Akzo Chemie]—Sodium dodecylbenzenesulfonate
Elfan WA Sulfosäure [Akzo Chemie]—Dodecylbenzene sulfonic acid
Emal AD-25 [Kao Corp.]—Ammonium lauryl sulfate
Emal O, 10 [Kao Corp.]—Sodium lauryl sulfate
Emasol L-120 [Kao Corp.]—POE (20) sorbitan monolaurate
Emasol O-120 [Kao Corp.]—POE (20) sorbitan monooleate
Emasol O-320 [Kao Corp.]—POE (20) sorbitan trioleate
Emasol P-120 [Kao Corp.]—POE (20) sorbitan monopalmitate
Emasol S-120 [Kao Corp.]—POE (20) sorbitan monostearate
Emasol S-320 [Kao Corp.]—POE (20) sorbitan tristearate
Emcol 4400-1 [Witco/Organics]—Disodium lauryl sulfosuccinate
Emcol D 24-25 [Witco]—Calcium sulfate

Emerest 2355 [Emery]—Ethylene glycol distearate
Emerest 2380 [Emery]—Propylene glycol monostearate
Emerest 2381 [Emery]—Propylene glycol monostearate, self-emulsifying
Emerest 2400, 2401 [Emery]—Glyceryl monostearate
Emerest 2407 [Emery]—Glyceryl monostearate, self-emulsifying
Emerest 2421 [Emery]—Glyceryl monooleate
Emerest 2620 [Emery]—POE (4) monolaurate
Emerest 2624 [Emery]—POE (4) monooleate
Emerest 2630 [Emery]—POE (6) monolaurate
Emerest 2640 [Emery]—POE (8) monostearate
Emerest 2642 [Emery]—POE (8) distearate
Emerest 2646 [Emery]—POE (8) monooleate
Emerest 2648 [Emery]—POE (8) dioleate
Emerest 2650 [Emery]—POE (8) monolaurate
Emerest 2660 [Emery]—POE (12) monooleate
Emerest 2661 [Emery]—POE (12) monolaurate
Emerest 2662 [Emery]—POE (12) monostearate
Emerest 2665 [Emery]—POE (12) dioleate
Emerest 2705 [Emery]—POE (8) monolaurate
Emerest 2711 [Emery]—POE (8) monostearate
Emerest 2712 [Emery]—POE (8) distearate
Emerest 2713 [Emery]—POE (20) monostearate
Emerest 2715 [Emery]—POE (40) monostearate
Emersal 6400, 6402, 6403, 6410 [Emery]—Sodium lauryl sulfate
Emersal 6430 [Emery]—Ammonium lauryl sulfate
Emery 5320 [Emery]—Disodium lauryl ether sulfosuccinate
Emid 6544 [Emery]—Capric acid diethanolamide
Empicol 0045 [Albright & Wilson/Detergents]—Sodium lauryl sulfate
Empicol 0185, 0266, 0303 [Albright & Wilson/Detergents]—Sodium lauryl sulfate
Empicol 0919 [Albright & Wilson]—Sodium lauryl sulfate
Empicol AL30 [Albright & Wilson/Australia]—Ammonium lauryl sulfate
Empicol AL30/T [Albright & Wilson]—Ammonium lauryl sulfate
Empicol AL70 [Albright & Wilson/Detergents]—Ammonium lauryl sulfate
Empicol DA, DLS [Albright & Wilson/Marchon]—Diethanolamine lauryl sulfate
Empicol EAB [Albright & Wilson/Marchon]—Ammonium lauryl ether sulfate
Empicol EGB [Albright & Wilson/Detergents]—Magnesium lauryl ether sulfate
Empicol EL [Albright & Wilson/Marchon]—Monoethanolamine lauryl sulfate
Empicol LM [Albright & Wilson/Marchon]—Sodium lauryl sulfate
Empicol LM45 [Albright & Wilson/Marchon]—Sodium lauryl sulfate
Empicol LM/T [Albright & Wilson]—Sodium lauryl sulfate
Empicol LMV [Albright & Wilson/Marchon]—Sodium lauryl sulfate
Empicol LMV/T [Albright & Wilson]—Sodium lauryl sulfate

Empicol LQ33 [Albright & Wilson/Australia]—Monoethanolamine lauryl sulfate
Empicol LQ70 [Albright & Wilson/Detergents]—Monoethanolamine lauryl sulfate
Empicol LS30, LS30B, LS30P [Albright & Wilson]—Sodium lauryl sulfate
Empicol LX, LX28 [Albright & Wilson/Marchon]—Sodium lauryl sulfate
Empicol LXS95 [Albright & Wilson]—Sodium lauryl sulfate
Empicol LXV [Albright & Wilson/Marchon]—Sodium lauryl sulfate
Empicol LY28/S [Albright & Wilson]—Sodium lauryl sulfate
Empicol LZ, LZ34 [Albright & Wilson/Marchon]—Sodium lauryl sulfate
Empicol LZ/D [Albright & Wilson/Marchon]—Sodium lauryl sulfate
Empicol LZ/E, LZG30, LZGV [Albright & Wilson/Marchon]—Sodium lauryl sulfate
Empicol LZGV/C [Albright & Wilson]—Sodium lauryl sulfate
Empicol LZP, LZV [Albright & Wilson/Marchon]—Sodium lauryl sulfate
Empicol LZV/E [Albright & Wilson/Marchon]—Sodium lauryl sulfate
Empicol ML 26 [Albright & Wilson/Detergent]—Magnesium lauryl sulfate
Empicol WA [Albright & Wilson/Marchon]—Sodium lauryl sulfate
Empicol WAK [Albright & Wilson]—Sodium lauryl sulfate
Empicol XT45 [Albright & Wilson/Marchon]—Monoethanolamine lauryl sulfate
Empigen AB [Albright & Wilson/Marchon]—Lauryl dimethyl amine
Empigen BB [Albright & Wilson/Marchon]—Lauryl betaine
Empilan BQ100 [Albright & Wilson/Marchon]—POE (8) monooleate
Empilan GMS LSE 32, GMS SE 32, GMS SE 40 [Albright & Wilson/Marchon]—
    Glyceryl monostearate, self-emulsifying
Empilan GMS NSE 32, GMS NSE 40 [Albright & Wilson/Marchon]—Glyceryl mono-
    stearate
Empilan KL10 [Albright & Wilson/Marchon]—POE (10) oleyl ether
Empilan KL20 [Albright & Wilson/Marchon]—POE (20) oleyl ether
Empimin LR28 [Albright & Wilson]—Sodium lauryl sulfate
Empimin MTT, MTT/A [Albright & Wilson]—Disodium N-oleyl sulfosuccinamate
Emplex [Patco]—Sodium stearoyl lactylate
Emsorb 6900 [Emery]—POE (20) sorbitan monooleate
Emsorb 6903 [Emery]—POE (20) sorbitan trioleate
Emsorb 6905 [Emery]—POE (20) sorbitan monostearate
Emsorb 6906 [Emery]—POE (4) sorbitan monostearate
Emsorb 6907 [Emery]—POE (20) sorbitan tristearate
Emsorb 6910 [Emery]—POE (20) sorbitan monopalmitate
Emsorb 6915 [Emery]—POE (20) sorbitan monolaurate
Emsorb 6916 [Emery]—POE (4) sorbitan monolaurate
Emthox 5882 [Emery]—POE (4) lauryl ether
Emulphogene BC-420 [GAF]—POE (3) tridecyl ether
Emulphogene BC-610 [GAF]—POE (6) tridecyl ether
Emulphor ON-870 [GAF]—POE (20) oleyl ether
Emulsan O [Reilly-Whiteman]—POE (8) monooleate

Emulsifier 99 [Pilot]—Dodecylbenzene sulfonic acid (branched)
Equex S, SP, SW [Procter & Gamble]—Sodium lauryl sulfate
Ethoduomeen T/25 [Akzo Chemie]—POE (15) tallow aminopropylamine
Ethomeen 18/12 [Armak]—POE (2) stearyl amine
Ethomeen 18/15 [Armak]—POE (5) stearyl amine
Ethomeen 18/60 [Armak]—POE (50) stearyl amine
Ethomeen C/12 [Armak]—POE (2) coconut amine
Ethomeen C/15 [Armak]—POE (5) coconut amine
Ethomeen C/20 [Armak]—POE (10) coconut amine
Ethomeen C/25 [Armak]—POE (15) coconut amine
Ethomeen S/15 [Armak]—POE (5) soya amine
Ethomeen S/20 [Armak]—POE (10) soya amine
Ethomeen S/25 [Armak]—POE (15) soya amine
Ethomeen T/12 [Armak]—POE (2) tallow amine
Ethomeen T/15 [Armak]—POE (5) tallow amine
Ethomeen T/25 [Armak]—POE (15) tallow amine
Ethosperse CA-2 [Glyco]—POE (2) cetyl ether
Ethosperse LA-4 [Glyco]—POE (4) lauryl ether
Ethosperse LA-12 [Glyco]—POE (12) lauryl ether
Ethosperse LA-23 [Glyco]—POE (23) lauryl ether
Ethosperse OA-2 [Glyco]—POE (2) oleyl ether
Ethosperse TDA-6 [Glyco]—POE (6) tridecyl ether
Ethylan A2 [Lankro Chem. Ltd.]—POE (4) monooleate
Ethylan A6 [Lankro Chem. Ltd.]—POE (12) monooleate
Eumulgin B3 [Henkel]—POE (30) cetyl/stearyl ether
Eumulgin O10 [Henkel]—POE (10) oleyl ether
Finazoline OA [Finetex]—Oleyl imidazoline
Gardinol WA Paste [Ronsheim & Moore]—Sodium lauryl sulfate
Geleol [Gattefosse]—Glyceryl monostearate
Glycosperse L-20 [Glyco]—POE (20) sorbitan monolaurate
Glycosperse L-20X [Glyco]—POE (20) sorbitan monolaurate (anhyd.)
Glycosperse O-5 [Glyco]—POE (5) sorbitan monooleate
Glycosperse O-20 [Glyco]—POE (20) sorbitan monooleate
Glycosperse O-20 Veg. [Glyco]—POE (20) sorbitan monooleate (vegetable grade)
Glycosperse O-20X [Glyco]—POE (20) sorbitan monooleate (anhyd.)
Glycosperse P-20 [Glyco]—POE (20) sorbitan monopalmitate
Glycosperse S-20 [Glyco]—POE (20) sorbitan monostearate
Glycosperse TO-20 [Glyco]—POE (20) sorbitan trioleate
Glycosperse TS-20 [Glyco]—POE (20) sorbitan tristearate
Graden Glycerol Mono Stearate [Graden]—Glyceryl monostearate
Gradonic 400 ML [Graden]—POE (8) monolaurate
Grindtek FAL1 [Grinsted]—Sodium stearoyl lactylate

Grindtek FAL2 [Grinsted]—Calcium stearoyl lactylate
Grindtek ML90 [Grinsted]—Glyceryl monolaurate
Grindtek MSP 32-6 [Grinsted]—Glyceryl monostearate, self-emulsifying
Grindtek MSP40, MSP40F, MSP52, MSP90 [Grinsted]—Glyceryl monostearate
Grindtek PGMS-90 [Grinsted]—Propylene glycol monostearate
Grocor 2000 [A. Gross]—Glyceryl monooleate
Grocor 5221 SE [A. Gross]—Diethylene glycol monostearate SE
Grocor 5500, 6000, 6000E [A. Gross]—Glyceryl monostearate
Grocor 6000 SE [A. Gross]—Glyceryl monostearate, self-emulsifying
Hartenol LAS-30, LES-60 [Hart]—Sodium lauryl sulfate
Hetoxamate SA-40 (DF) [Heterene]—POE (40) monostearate
Hetoxamine C2 [Heterene]—POE (2) coconut amine
Hetoxamine C5 [Heterene]—POE (5) coconut amine
Hetoxamine C15 [Heterene]—POE (15) coconut amine
Hetoxamine S5 [Heterene]—POE (5) soya amine
Hetoxamine S15 [Heterene]—POE (15) soya amine
Hetoxamine ST2 [Heterene]—POE (2) stearyl amine
Hetoxamine ST5 [Heterene]—POE (5) stearyl amine
Hetoxamine ST50 [Heterene]—POE (50) stearyl amine
Hetoxamine T2 [Heterene]—POE (2) tallow amine
Hetoxamine T5 [Heterene]—POE (5) tallow amine
Hetoxamine T15 [Heterene]—POE (15) tallow amine
Hetoxol CA2 [Heterene]—POE (2) cetyl ether
Hetoxol CA10 [Heterene]—POE (10) cetyl ether
Hetoxol CA20 [Heterene]—POE (20) cetyl ether
Hetoxol CS30 [Heterene]—POE (30) cetyl/stearyl ether
Hetoxol CS50, CS50 Special [Heterene]—POE (50) cetyl/stearyl ether
Hetoxol L4N [Heterene]—POE (4) lauryl ether
Hetoxol L12 [Heterene]—POE (12) lauryl ether
Hetoxol L23N [Heterene]—POE (23) lauryl ether
Hetoxol OA20 Special [Heterene]—POE (20) oleyl ether
Hetoxol OL2 [Heterene]—POE (2) oleyl ether
Hetoxol OL4 [Heterene]—POE (4) oleyl ether
Hetoxol OL10 [Heterene]—POE (10) oleyl ether
Hetoxol OL20 [Heterene]—POE (20) oleyl ether
Hetoxol TD3 [Heterene]—POE (3) tridecyl ether
Hetoxol TD6 [Heterene]—POE (6) tridecyl ether
Hetoxol TD12 [Heterene]—POE (12) tridecyl ether
Hetsorb L-4 [Heterene]—POE (4) sorbitan monolaurate
Hetsorb L-20 [Heterene]—POE (20) sorbitan monolaurate
Hetsorb O-20 [Heterene]—POE (20) sorbitan monooleate
Hetsorb P-20 [Heterene]—POE (20) sorbitan monopalmitate

Hetsorb S-4 [Heterene]—POE (4) sorbitan monostearate
Hetsorb S-20 [Heterene]—POE (20) sorbitan monostearate
Hetsorb TO-20 [Heterene]—POE (20) sorbitan trioleate
Hetsorb TS-20 [Heterene]—POE (20) sorbitan tristearate
Hetsulf 40, 40X, 60S [Heterene]—Sodium dodecylbenzenesulfonate
Hetsulf Acid [Heterene]—Dodecylbenzene sulfonic acid
Hipochem Dispersol SB [High Point]—Butyl oleate, sulfated
Hodag 20-L [Hodag]—POE (4) monolaurate
Hodag 40-S [Hodag]—POE (8) monostearate
Hodag 150-S [Hodag]—POE (1500) monostearate
Hodag DGL [Hodag]—Diethylene glycol monolaurate
Hodag DGO [Hodag]—Diethylene glycol monooleate
Hodag DGS [Hodag]—Diethylene glycol monostearate
Hodag GML [Hodag]—Glyceryl monolaurate
Hodag GMO, GMO-D [Hodag]—Glyceryl monooleate
Hodag GMS [Hodag]—Glyceryl monostearate
Hodag PGML [Hodag]—Propylene glycol monolaurate
Hodag PGMS [Hodag]—Propylene glycol monostearate
Hodag PSML-20 [Hodag]—POE (20) sorbitan monolaurate
Hodag PSMP-20 [Hodag]—POE (20) sorbitan monopalmitate
Hodag PSMS-20 [Hodag]—POE (20) sorbitan monostearate
Hodag PSTO-20 [Hodag]—POE (20) sorbitan trioleate
Hodag PSTS-20 [Hodag]—POE (20) sorbitan tristearate
Homotex PS-90 [Kao Corp.]—Propylene glycol monostearate
Hostacerin O-20 [American Hoechst]—POE (20) oleyl ether
Hybase C-300 [Witco/Sonneborn]—Calcium sulfate
Icomeen 18-50 [BASF Wyandotte]—POE (50) stearyl amine
Icomeen T-15 [BASF Wyandotte]—POE (15) tallow amine
Iconol TDA-3 [BASF Wyandotte]—POE (3) tridecyl ether
Imwitor 191, 900K [Dynamit Nobel]—Glyceryl monostearate
Imwitor 960 [Dynamit Nobel]—Glyceryl monostearate, self-emulsifying
Incronol SLS [Croda Surfactants]—Sodium lauryl sulfate
Incropol CS-30 [Croda Surfactants]—POE (30) cetyl/stearyl ether
Incropol CS-50 [Croda Surfactants]—POE (50) cetyl/stearyl ether
Incropol L-12 [Croda Surfactants]—POE (12) lauryl ether
Incropol L-23 [Croda Surfactants]—POE (23) lauryl ether
Incrosorb S-60 [Croda Surfactants]—POE (20) sorbitan monostearate
Incrosul LMS [Croda Surfactants]—Disodium lauramido MEA-sulfosuccinate
Industrol 400MOT [BASF Wyandotte]—POE (8) monotallate
Industrol COH-25 [BASF Wyandotte]—POE (25) hydrogenated castor oil
Industrol DO-9 [BASF Wyandotte]—POE (8) dioleate
Industrol DO-13 [BASF Wyandotte]—POE (12) dioleate

Industrol L-20-S [BASF Wyandotte]—POE (20) sorbitan monolaurate
Industrol MO-13 [BASF Wyandotte]—POE (12) monooleate
Industrol MS-40 [BASF Wyandotte]—POE (40) monostearate
Industrol O-20-S [BASF Wyandotte]—POE (20) sorbitan monooleate
Industrol S-20-S [BASF Wyandotte]—POE (20) sorbitan monostearate
Industrol STS-20-S [BASF Wyandotte]—POE (20) sorbitan tristearate
Ionet T-20C [Sanyo]—POE (20) sorbitan monolaurate
Jet-Quat 2C-75 [Jetco]—Dicoco dimethyl ammonium chloride
Jordanol SL300 [Jordan]—Sodium lauryl sulfate
Kadif 50 Flakes [Witco Chem. Ltd.]—Sodium dodecylbenzenesulfonate
Katapol PN-430 [GAF]—POE (5) tallow amine
Katapol PN-730 [GAF]—POE (15) tallow amine, distilled
Kemamine P-989 [Humko Sheffield]—Oleyl amine (tech.)
Kemamine Q6502C [Humko Sheffield]—Dicoco dimethyl ammonium chloride
Kemamine T-6502D [Humko Sheffield]—Coco dimethyl amine, distilled
Kemamine T-9902D [Humko Div./Witco]—Stearyl dimethyl amine, distilled
Kemester EGDS [Humko Sheffield]—Ethylene glycol distearate
Kessco Diethylene Glycol Monostearate [Armak]—Diethylene glycol monostearate
Kessco Diglycol Stearate Neutral [Armak]—Diethylene glycol monostearate
Kessco Diglycol Stearate S.E. [Armak]—Diethylene glycol monostearate SE
Kessco Ethylene Glycol Distearate [Armak]—Ethylene glycol distearate
Kessco Glycerol Monolaurate [Armak]—Glyceryl monolaurate
Kessco Glycerol Monooleate [Armak]—Glyceryl monooleate
Kessco Glycerol Monostearate 860, DH-1, Pure [Armak]—Glyceryl monostearate
Kessco Glycerol Monostearate SE [Armak]—Glyceryl monostearate, self-emulsifying
Kessco PEG 200 Dioleate [Armak]—POE (4) dioleate
Kessco PEG 200 Distearate [Armak]—POE (4) distearate
Kessco PEG 200 Monolaurate [Armak]—POE (4) monolaurate
Kessco PEG 200 Monooleate [Armak]—POE (4) monooleate
Kessco PEG 200 Monostearate [Armak]—POE (4) monostearate
Kessco PEG 300 Monolaurate [Armak]—POE (6) monolaurate
Kessco PEG 400 Dioleate [Armak]—POE (8) dioleate
Kessco PEG 400 Distearate [Armak]—POE (8) distearate
Kessco PEG 400 DS-356 [Armak]—POE (8) distearate
Kessco PEG 400 Monolaurate [Armak]—POE (8) monolaurate
Kessco PEG 400 Monooleate [Armak]—POE (8) monooleate
Kessco PEG 400 Monostearate [Armak]—POE (8) monostearate
Kessco PEG 600 Dioleate [Armak]—POE (12) dioleate
Kessco PEG 600 Distearate [Armak]—POE (12) distearate
Kessco PEG 600 Monolaurate [Armak]—POE (12) monolaurate
Kessco PEG 600 Monooleate [Armak]—POE (12) monooleate
Kessco PEG 600 Monostearate [Armak]—POE (12) monostearate

Kessco PEG 1000 Monooleate [Armak]—POE (20) monooleate
Kessco PEG 1000 Monostearate [Armak]—POE (20) monostearate
Kessco PEG 4000 Monostearate [Armak]—POE (75) monostearate
Kessco Propylene Glycol Monolaurate E [Armak]—Propylene glycol monolaurate
Kessco Propylene Glycol Monostearate Pure [Armak]—Propylene glycol monostearate
LA-55-4 [Hefti Ltd.]—POE (4) lauryl ether
LA-55-23 [Hefti Ltd.]—POE (23) lauryl ether
Lakeway 101-10, 101-10P,N, 101-11, 101-19 [Bofors Lakeway]—Sodium lauryl sulfate
Lakeway 101-20 [Bofors Lakeway]—Ammonium lauryl sulfate
Lakeway SA [Bofors Lakeway]—Dodecylbenzene sulfonic acid (linear)
Lamegin CLS [Grunau]—Calcium stearoyl lactylate
Lamegin NSL [Grunau]—Sodium stearoyl lactylate
Lamesoft LMG [Grunau]—Glyceryl monolaurate
Merpasol L47 [Kempen]—Disodium lauramido MEA-sulfosuccinate
Lexemul 55G, 503, 515 [Inolex]—Glyceryl monostearate
Lexemul 530, 561, AR, AS, T [Inolex]—Glyceryl monostearate, self-emulsifying
Lexemul EGDS [Inolex]—Ethylene glycol distearate
Lexemul P [Inolex]—Propylene glycol monostearate, self-emulsifying
Lilamin 172 [Lilachim S.A.]—Oleyl amine
Lilamin 172D [Lilachim S.A.]—Oleyl amine, distilled
Lilamin 312D [Lilachim S.A.]—Lauryl dimethyl amine, distilled
Lilamin 316D [Lilachim S.A.]—Palmityl dimethyl amine, distilled
Lilamin 342D [Lilachim S.A.]—Stearyl dimethyl amine, distilled
Lilamin 367D [Lilachim S.A.]—Coco dimethyl amine, distilled
Lipal 3TD [PVO Int'l.]—POE (3) tridecyl ether
Lipal 4LA [PVO Int'l.]—POE (4) lauryl ether
Lipal 6TD [PVO Int'l.]—POE (6) tridecyl ether
Lipal 12LA [PVO Int'l.]—POE (12) lauryl ether
Lipal 20OA [PVO Int'l.]—POE (20) oleyl ether
Lipal 23LA [PVO Int'l.]—POE (23) lauryl ether
Lipal 39S [PVO Int'l.]—POE (40) monostearate
Lipal 400DS [PVO Int'l.]—POE (8) distearate
Lipal 400S [PVO Int'l.]—POE (8) monostearate
Lipal 600S [PVO Int'l.]—POE (12) monostearate
Lipo DGLS [Lipo]—Diethylene glycol monolaurate, SE
Lipo DGS-SE [Lipo]—Diethylene glycol monostearate SE
Lipo Diglycol Laurate [Lipo]—Diethylene glycol monolaurate
Lipo EGDS [Lipo]—Ethylene glycol distearate
Lipo GMS 450 [Lipo]—Glyceryl monostearate
Lipo GMS 470 [Lipo]—Glyceryl monostearate, self-emulsifying
Lipo PGMS [Lipo]—Propylene glycol monostearate
Lipocol C-2 [Lipo]—POE (2) cetyl ether

Lipocol C-10 [Lipo]—POE (10) cetyl ether
Lipocol C-20 [Lipo]—POE (20) cetyl ether
Lipocol L-1 [Lipo]—PEG-1 lauryl ether
Lipocol L-4 [Lipo]—POE (4) lauryl ether
Lipocol L-12 [Lipo]—POE (12) lauryl ether
Lipocol L-23 [Lipo]—POE (23) lauryl ether
Lipocol O-2 [Lipo]—POE (2) oleyl ether
Lipocol O-10 [Lipo]—POE (10) oleyl ether
Lipocol O-20 [Lipo]—POE (20) oleyl ether
Lipocol TD-6 [Lipo]—POE (6) tridecyl ether
Lipocol TD-12 [Lipo]—POE (12) tridecyl ether
Lipopeg 4-DO [Lipo]—POE (8) dioleate
Lipopeg 4-DS [Lipo]—POE (8) distearate
Lipopeg 4-L [Lipo]—POE (8) monolaurate
Lipopeg 4-O [Lipo]—POE (8) monooleate
Lipopeg 4-S [Lipo]—POE (8) monostearate
Lipopeg 6-L [Lipo]—POE (12) monolaurate
Lipopeg 10-S [Lipo]—POE (20) monostearate
Lipopeg 15-S [Lipo]—POE (1500) monostearate
Lipopeg 39-S [Lipo]—POE (40) monostearate
Liposorb L-20 [Lipo]—POE (20) sorbitan monolaurate
Liposorb O-5 [Lipo]—POE (5) sorbitan monooleate
Liposorb O-20 [Lipo]—POE (20) sorbitan monooleate
Liposorb P-20 [Lipo]—POE (20) sorbitan monopalmitate
Liposorb S-20 [Lipo]—POE (20) sorbitan monostearate
Liposorb TO-20 [Lipo]—POE (20) sorbitan trioleate
Liposorb TS-20 [Lipo]—POE (20) sorbitan tristearate
Lonzest PEG-4-DO [Lonza]—POE (8) dioleate
Lonzest PEG-4-L [Lonza]—POE (8) monolaurate
Lonzest PEG-4-O [Lonza]—POE (8) monooleate
Lonzest SML-20 [Lonza]—POE (20) sorbitan monolaurate
Lonzest SMO [Lonza]—POE (5) sorbitan monooleate
Lonzest SMO-20 [Lonza]—POE (20) sorbitan monooleate
Lonzest SMP-20 [Lonza]—POE (20) sorbitan monopalmitate
Lonzest SMS-20 [Lonza]—POE (20) sorbitan monostearate
Lonzest STO-20 [Lonza]—POE (20) sorbitan trioleate
Lonzest STS-20 [Lonza]—POE (20) sorbitan tristearate
Lonzol LA-300 [Lonza]—Ammonium lauryl sulfate
Lonzol LS-300 [Lonza]—Sodium lauryl sulfate
Lorapon AM-DML [Dutton & Reinisch]—Lauryl betaine
Lorol Liquid NH Sulphonated [Ronsheim & Moore]—Ammonium lauryl sulfate
M-Quat 2475 [Mazer]—Dicoco dimethyl ammonium chloride

Mackam CSF [McIntyre]—Cocoamphopropionate
Mackanate CP [McIntyre]—Disodium coco amide MIPA sulfosuccinate
Mackanate EL, L-1, L-2 [McIntyre]—Disodium lauryl ether sulfosuccinate
Mackanate LM-40 [McIntyre]—Disodium lauramido MEA-sulfosuccinate
Mackanate LO, LO-100 [McIntyre]—Disodium lauryl sulfosuccinate
Macol LA-4 [Mazer]—POE (4) lauryl ether
Macol LA-12 [Mazer]—POE (12) lauryl ether
Macol LA-23 [Mazer]—POE (23) lauryl ether
Macol OA-2 [Mazer]—POE (2) oleyl ether
Macol OA-4 [Mazer]—POE (4) oleyl ether
Macol OA-10 [Mazer]—POE (10) oleyl ether
Macol OA-20 [Mazer]—POE (20) oleyl ether
Macol TD-3 [Mazer]—POE (3) tridecyl ether
Macol TD-6 [Mazer]—POE (6) tridecyl ether
Manro ALES27 [Manro]—Ammonium lauryl ether sulfate
Manro ALS30 [Manro]—Ammonium lauryl sulfate
Manro BA [Manro]—Dodecylbenzene sulfonic acid (linear)
Manro BA25, BA30 [Manro]—Sodium dodecylbenzenesulfonate
Manro HA [Manro]—Dodecylbenzene sulfonic acid (branched)
Manro HCS [Manro]—Isopropylamine dodecylbenzenesulfonate
Manro ML33S [Manro]—Monoethanolamine lauryl sulfate
Manro NA [Manro]—Dodecylbenzene sulfonic acid (linear)
Manro SDBS30 [Manro]—Sodium dodecylbenzenesulfonate
Manro SLS28 [Manro]—Sodium lauryl sulfate
Mapeg 200DO [Mazer]—POE (4) dioleate
Mapeg 200DS [Mazer]—POE (4) distearate
Mapeg 200ML [Mazer]—POE (4) monolaurate
Mapeg 200MO [Mazer]—POE (4) monooleate
Mapeg 200MS [Mazer]—POE (4) monostearate
Mapeg 400DO [Mazer]—POE (8) dioleate
Mapeg 400DS [Mazer]—POE (8) distearate
Mapeg 400ML [Mazer]—POE (8) monolaurate
Mapeg 400MO [Mazer]—POE (8) monooleate
Mapeg 400MOT [Mazer]—POE (8) monotallate
Mapeg 400MS [Mazer]—POE (8) monostearate
Mapeg 600DO [Mazer]—POE (12) dioleate
Mapeg 600DS [Mazer]—POE (12) distearate
Mapeg 600ML [Mazer]—POE (12) monolaurate
Mapeg 600MO [Mazer]—POE (12) monooleate
Mapeg 600MS [Mazer]—POE (12) monostearate
Mapeg 1000MS [Mazer]—POE (20) monostearate
Mapeg 1500MS [Mazer]—POE (1500) monostearate

Mapeg 4000MS [Mazer]—POE (75) monostearate

Mapeg CO-25H [Mazer]—POE (25) hydrogenated castor oil

Mapeg DGLD [Mazer]—Diethylene glycol monolaurate

Mapeg EGDS [Mazer]—Ethylene glycol distearate

Mapeg PGMS [Mazer]—Propylene glycol monostearate

Mapeg S-40 [Mazer]—POE (40) monostearate

Maprofix 563, 563-SD [Onyx/Millmaster]—Sodium lauryl sulfate

Maprofix DLS-35 [Onyx]—Diethanolamine lauryl sulfate

Maprofix LCP, LK USP [Onyx/Millmaster]—Sodium lauryl sulfate

Maprofix LES-60A, MB, MBO [Onyx]—Ammonium lauryl ether sulfate

Maprofix MG [Onyx]—Magnesium lauryl sulfate

Maprofix MM [Onyx/Millmaster]—Sodium lauryl sulfate

Maprofix NH, NH54, NHL, NHL-22 [Onyx]—Ammonium lauryl sulfate

Maprofix WAC, WAC-LA, WAM, WA Paste, WAQ [Onyx/Millmaster]—Sodium
  lauryl sulfate

Marlipal FS [Chem. Werke Huls]—POE (12) dioleate

Marlon A350, A360, A365, A375, A390, A396 [Chem. Werke Huls]—Sodium dodecylbenzenesulfonate

Marlon AFR [Chem. Werke Huls]—Sodium dodecylbenzenesulfonate (modified)

Marlon AS_3 [Chem. Werke Huls]—Dodecylbenzene sulfonic acid (linear)

Mars AMLS [Mars]—Ammonium lauryl sulfate

Mars SA-98S [Mars]—Dodecylbenzene sulfonic acid (linear)

Mars SLS-A95, SLS-30 [Mars]—Sodium lauryl sulfate

Marvanol SBO 60% [Marlowe-Van Loan]—Butyl oleate, sulfated

Mazeen C2 [Mazer]—POE (2) coconut amine

Mazeen C5 [Mazer]—POE (5) coconut amine

Mazeen C10 [Mazer]—POE (10) coconut amine

Mazeen C15 [Mazer]—POE (15) coconut amine

Mazeen S5 [Mazer]—POE (5) soya amine

Mazeen S10 [Mazer]—POE (10) soya amine

Mazeen S15 [Mazer]—POE (15) soya amine

Mazeen T2 [Mazer]—POE (2) tallow amine

Mazeen T5 [Mazer]—POE (5) tallow amine

Mazeen T15 [Mazer]—POE (15) tallow amine

Mazol 80MGK [Mazer]—POE (20) glyceryl monostearate

Mazol 300, GMO [Mazer]—Glyceryl monooleate

Mazol GMS [Mazer]—Glyceryl monostearate

Mazol GMS-D [Mazer]—Glyceryl monostearate, self-emulsifying

Mazol PGMS [Mazer]—Propylene glycol monostearate

Mazoline OA [Mazer]—Oleyl imidazoline

Mercol 15, 25, 30, Special [Capital City]—Sodium dodecylbenzenesulfonate

Merpasol L44 [Kempen]—Disodium lauryl sulfosuccinate

Merpinal LM40, LM50, LS40, LS50 [Kempen]—Sodium lauryl sulfate
Merpisap AE50, AE60, AE70 [Kempen]—Sodium dodecylbenzenesulfonate
Merpisap AP80W, AP85W, AP90, AP90P [Kempen]—Sodium dodecylbenzenesulfonate
Merpisap AS98 [Kempen]—Dodecylbenzene sulfonic acid (linear)
Merpisap KH30 [Kempen]—Sodium dodecylbenzenesulfonate
Michelene LS-90 [M. Michel]—Sodium lauryl sulfate
Miramine OC [Miranol]—Oleyl imidazoline
Miranate LSS [Miranol]—Disodium lauryl sulfosuccinate
Miranol CM-SF Conc. [Miranol]—Cocoamphopropionate
Miranol J2M-SF Conc. [Miranol]—Capryloamphodipropionate
Mirataine LDMB [Miranol]—Lauryl betaine
ML-55-F [Hefti Ltd.]—POE (20) sorbitan monolaurate
ML-55-F-4 [Hefti Ltd.]—POE (4) sorbitan monolaurate
Monamate CPA-100 [Mona]—Disodium coco amide MIPA sulfosuccinate
Monamate LA-100 [Mona]—Disodium lauryl sulfosuccinate
Monamid 150-CW [Mona]—Capric acid diethanolamide
Monamid 150-MW [Mona]—Myristic diethanolamide
Monateric 811 [Mona]—Capryloamphodipropionate
Monateric CA-35% [Mona]—Cocoamphopropionate
Monazoline O [Mona]—Oleyl imidazoline
Monomuls 90-O18 [Grunau]—Glyceryl monooleate
Monomuls 90-L12 [Grunau]—Glyceryl monolaurate
Monosteol [Gattefosse]—Propylene glycol monostearate
Montanox 20 [Seppic]—POE (20) sorbitan monolaurate
Montanox 40 [Seppic]—POE (20) sorbitan monopalmitate
Montanox 60 [Seppic]—POE (20) sorbitan monostearate
Montanox 61 [Seppic]—POE (4) sorbitan monostearate
Montanox 65 [Seppic]—POE (20) sorbitan tristearate
Montanox 80 [Seppic]—POE (20) sorbitan monooleate
Montanox 81 [Seppic]—POE (5) sorbitan monooleate
Montanox 85 [Seppic]—POE (20) sorbitan trioleate
MS-55-F [Hefti Ltd.]—POE (20) sorbitan monostearate
Myrj 45 [ICI United States]—POE (8) monostearate
Myrj 52, 52C, 52S [ICI United States]—POE (40) monostearate
Nacconol 35SL, 40F, 40G [Stepan]—Sodium dodecylbenzenesulfonate
Nacconol 90F, 90G [Stepan]—Sodium dodecylbenzenesulfonate
Nansa 1042, 1042/P [Albright & Wilson/Detergents]—Dodecylbenzene sulfonic acid (linear)
Nansa 1106/P, 1169/P [Albright & Wilson/Detergents]—Sodium dodecylbenzenesulfonate
Nansa EVM 62H, EVM70 [Albright & Wilson/Detergents]—Calcium dodecylbenzene

sulfonate

Nansa HS80 Soft, HS85/S [Albright & Wilson/Marchon]—Sodium dodecylbenzenesulfonate

Nansa HS80/S [Albright & Wilson/Detergents]—Sodium dodecylbenzenesulfonate

Nansa SB25, SB62 [Albright & Wilson/Detergents]—Sodium dodecylbenzenesulfonate

Nansa SBA [Albright & Wilson/Marchon]—Dodecylbenzene sulfonic acid

Nansa SL30 [Albright & Wilson/Marchon]—Sodium dodecylbenzenesulfonate

Nansa SS30, SS60 [Albright & Wilson/Marchon]—Sodium dodecylbenzenesulfonate

Nansa SSA [Albright & Wilson/Marchon]—Dodecylbenzene sulfonic acid

Nansa SSA/P [Albright & Wilson/Detergents]—Dodecylbenzene sulfonic acid (linear)

Nansa YS 94 [Albright & Wilson/Marchon]—Isopropylamine dodecylbenzenesulfonate

Neodol 25-3 [Shell]—Pareth-25-3

Neodol 25-3A [Shell]—Ammonium pareth-25 sulfate

Neodol 25-7 [Shell]—Pareth-25-7

Neodol 25-12 [Shell]—Pareth 25-12

Neopon LS [Nippon Nyukazai]—Sodium lauryl sulfate

Nikkol BC-2 [Nikko]—POE (2) cetyl ether

Nikkol BC-5.5 [Nikko]—POE (6) cetyl ether

Nikkol BC-10TX [Nikko]—POE (10) cetyl ether

Nikkol BC-15TX [Nikko]—POE (15) cetyl ether

Nikkol BC-20TX [Nikko]—POE (20) cetyl ether

Nikkol BC-25TX [Nikko]—POE (25) cetyl ether

Nikkol BC-30TX [Nikko]—POE (30) cetyl ether

Nikkol BO-10TX [Nikko]—POE (10) oleyl ether

Nikkol BO-20TX [Nikko]—POE (20) oleyl ether

Nikkol GBW-25 [Nikko]—POE (6) sorbitol beeswax

Nikkol GBW-125 [Nikko]—POE (20) sorbitol beeswax

Nikkol HCO-5 [Nikko]—POE (5) hydrogenated castor oil

Nikkol MYO-2 [Nikko]—Diethylene glycol monooleate

Nikkol MYS-2 [Nikko]—Diethylene glycol monostearate

Nikkol MYS-4 [Nikko]—POE (4) monostearate

Nikkol MYS-40 [Nikko]—POE (40) monostearate

Nikkol PMS-1C [Nikko]—Propylene glycol monostearate

Nikkol PMS-1CSE [Nikko]—Propylene glycol monostearate, self-emulsifying

Nikkol SLS [Nikko]—Sodium lauryl sulfate

Nikkol TL-10 [Nikko]—POE (20) sorbitan monolaurate

Nikkol TO-10 [Nikko]—POE (20) sorbitan monooleate

Nikkol TO-30 [Nikko]—POE (20) sorbitan trioleate

Nikkol TS-10, TS-10 (FF) [Nikko]—POE (20) sorbitan monostearate

Nikkol TS-30 [Nikko]—POE (20) sorbitan tristearate

Nissan Amine OB [Nippon Oil & Fats]—Oleyl amine

Nissan Cation BB [Nippon Oil & Fats]—Lauryl trimethyl ammonium chloride

Nissan Cation M$_2$-100 [Nippon Oil & Fats]—Myristyl dimethyl benzyl ammonium chloride
Nissan Nonion LT-221 [Nippon Oil & Fats]—POE (20) sorbitan monolaurate
Nissan Nonion ST-221 [Nippon Oil & Fats]—POE (20) sorbitan monostearate
Nissan Tertiary Amine AB [Nippon Oils & Fats]—Stearyl dimethyl amine
Nissan Tertiary Amine BB [Nippon Oil & Fats]—Lauryl dimethyl amine
Nissan Tertiary Amine PB [Nippon Oil & Fats]—Palmityl dimethyl amine
Nonisol 100 [Ciba-Geigy]—POE (8) monolaurate
Nonisol 210 [Ciba-Geigy]—POE (8) dioleate
Nonisol 300 [Ciba-Geigy]—POE (8) monostearate
Nopalcol 1-S [Diamond Shamrock]—Diethylene glycol monostearate
Nopalcol 2-L [Diamond Shamrock]—POE (4) monolaurate
Nopalcol 4-C, 4CH [Diamond Shamrock]—POE (8) monococoate
Nopalcol 4-L [Diamond Shamrock]—POE (8) monolaurate
Nopalcol 4-O [Diamond Shamrock]—POE (8) monooleate
Nopalcol 4-S [Diamond Shamrock]—POE (8) monostearate
Nopalcol 6-L [Diamond Shamrock]—POE (12) monolaurate
Nopalcol 6-O [Diamond Shamrock]—POE (12) monooleate
Nopalcol 6-S [Diamond Shamrock]—POE (12) monostearate
Nopcogen 22-O [Diamond Shamrock]—Oleyl imidazoline
Noram O [Ceca S.A.]—Oleyl amine
Noramac O [Diamond Shamrock Process]—Oleyl amine acetate
Norfox ALS [Norman, Fox & Co.]—Ammonium lauryl sulfate
Norfox MLS [Norman, Fox & Co.]—Magnesium lauryl sulfate
Norfox SLS [Norman, Fox & Co.]—Sodium lauryl sulfate
Nutrapon AL60, KF3846 [Clough]—Ammonium lauryl ether sulfate
Nutrapon HA 3841 [Clough]—Ammonium lauryl sulfate
Nutrapon W, WAC, WAQ [Clough]—Sodium lauryl sulfate
OL-55-F-2 [Hefti Ltd.]—POE (2) oleyl ether
OL-55-F-10 [Hefti Ltd.]—POE (10) oleyl ether
OL-55-F-20 [Hefti Ltd.]—POE (20) oleyl ether
Onamine 12 [Onyx]—Lauryl dimethyl amine
Onamine 16 [Onyx]—Palmityl dimethyl amine
Pationic 122A [Patco]—Sodium capryl lactylate
Pationic 138C [Patco/C.J. Patterson]—Sodium lauroyl lactylate
Pationic 145A [Patco]—Sodium stearoyl lactylate
Pationic CSL [Patco]—Calcium stearoyl lactylate
Pationic SSL [Patco]—Sodium stearoyl lactylate
Peganate COH25 [GAF]—POE (25) hydrogenated castor oil
Pegosperse 50DS [Glyco]—Ethylene glycol distearate
Pegosperse 100L, 100ML [Glyco]—Diethylene glycol monolaurate
Pegosperse 100MR [Glyco]—Diethylene glycol monoricinoleate

Pegosperse 100O [Glyco]—Diethylene glycol monooleate
Pegosperse 100S [Glyco]—Diethylene glycol monostearate
Pegosperse 200ML [Glyco]—POE (4) monolaurate
Pegosperse 400DO [Glyco]—POE (8) dioleate
Pegosperse 400DS [Glyco]—POE (8) distearate
Pegosperse 400MC [Glyco]—POE (8) monococoate
Pegosperse 400ML [Glyco]—POE (8) monolaurate
Pegosperse 400MO [Glyco]—POE (8) monooleate
Pegosperse 400MOT [Glyco]—POE (8) monotallate
Pegosperse 400MS [Glyco]—POE (8) monostearate
Pegosperse 600ML [Glyco]—POE (12) monolaurate
Pegosperse 600MS [Glyco]—POE (12) monostearate
Pegosperse 1000MS [Glyco]—POE (20) monostearate
Pegosperse 1500MS [Glyco]—POE (1500) monostearate
Pegosperse 1750MS [Glyco]—POE (40) monostearate
Pegosperse 4000MS [Glyco]—POE (75) monostearate
PGE-400-DS [Hefti Ltd.]—POE (8) distearate
PGE-400-ML[Hefti Ltd.]—POE (8) monolaurate
PGE-400-MO [Hefti Ltd.]—POE (8) monooleate
PGE-400-MS [Hefti Ltd.]—POE (8) monostearate
PGE-600-DS [Hefti Ltd.]—POE (12) distearate
PGE-600-ML [Hefti Ltd.]—POE (12) monolaurate
PGE-600-MO [Hefti Ltd.]—POE (12) monooleate
PGE-1000-MS [Hefti Ltd.]—POE (20) monostearate
Plurafac A-38 [BASF Wyandotte]—POE (27) cetyl/stearyl ether
Plurafac A-39 [BASF Wyandotte]—POE (55) cetyl/stearyl ether
PMS-33 [Hefti Ltd.]—Propylene glycol monostearate
Polyfac ABS-60C [Westvaco]—Calcium dodecylbenzene sulfonate
Polyfac LAS-97 [Westvaco]—Dodecylbenzene sulfonic acid (linear)
Polyfac MT-610 [Westvaco]—POE (10) monotallate
Polyfac SLS-30, SLS-30 Special [Westvaco]—Sodium lauryl sulfate
Polystep A-17 [Stepan]—Dodecylbenzene sulfonic acid (branched)
Polystep B-3, B-5, B-24 [Stepan]—Sodium lauryl sulfate
Polystep B-7 [Stepan]—Ammonium lauryl sulfate
Polystep B-8 [Stepan]—Diethanolamine lauryl sulfate
Polystep B-9 [Stepan]—Magnesium lauryl sulfate
Polystep B-11 [Stepan Europe]—Ammonium lauryl ether sulfate
Polystep B-21, B-23 [Stepan]—Sodium POE (12) lauryl ether sulfate
Querton 210CL [Lilachim S.A.]—Didecyl dimethyl ammonium chloride
Radiamine 6172 [Oleofina S.A.]—Oleyl amine
Radiamine 6173 [Oleofina S.A.]—Oleyl amine (distilled)
Radiamuls 137 [Oleofina S.A.]—POE (20) sorbitan monolaurate

Radiamuls 141, 341 [Oleofina S.A.]—Glyceryl monostearate, self-emulsifying
Radiamuls 142, 600, 601, 900 [Oleofina S.A.]—Glyceryl monostearate
Radiamuls 147 [Oleofina S.A.]—POE (20) sorbitan monostearate
Radiamuls 152 [Oleofina S.A.]—Glyceryl monooleate
Radiamuls CSL 2980 [Oleofina S.A.]—Calcium stearoyl lactylate
Radiamuls MG2141, MG2142 [Oleofina S.A.]—Glyceryl monostearate
Radiamuls MG2152 [Oleofina S.A.]—Glyceryl monooleate
Radiamuls MG2600, MG2900 [Oleofina S.A.]—Glyceryl monostearate
Radiamuls SORB 2137 [Oleofina S.A.]—POE (20) sorbitan monolaurate
Radiamuls SORB 2147 [Oleofina S.A.]—POE (20) sorbitan monostearate
Radiamuls SSL 2990 [Oleofina S.A.]—Sodium stearoyl lactylate
Radiasurf 7000 [Oleofina S.A.]—POE (20) glyceryl monostearate
Radiasurf 7137 [Oleofina S.A.]—POE (20) sorbitan monolaurate
Radiasurf 7147 [Oleofina S.A.]—POE (20) sorbitan monostearate
Radiasurf 7201 [Oleofina S.A.]—Propylene glycol monostearate
Radiasurf 7206 [Oleofina S.A.]—Propylene glycol monooleate
Radiasurf 7269 [Oleofina S.A.]—Ethylene glycol distearate
Radiasurf 7400 [Oleofina S.A.]—Diethylene glycol monooleate
Radiasurf 7402 [Oleofina S.A.]—POE (4) monooleate
Radiasurf 7403 [Oleofina S.A.]—POE (8) monooleate
Radiasurf 7404 [Oleofina S.A.]—POE (12) monooleate
Radiasurf 7410 [Oleofina S.A.]—Diethylene glycol monostearate
Radiasurf 7411 [Oleofina S.A.]—Diethylene glycol monostearate SE
Radiasurf 7412 [Oleofina S.A.]—POE (4) monostearate
Radiasurf 7413 [Oleofina S.A.]—POE (8) monostearate
Radiasurf 7414 [Oleofina S.A.]—POE (12) monostearate
Radiasurf 7417 [Oleofina S.A.]—POE (1500) monostearate
Radiasurf 7420 [Oleofina S.A.]—Diethylene glycol monolaurate
Radiasurf 7421 [Oleofina S.A.]—Diethylene glycol monolaurate, SE
Radiasurf 7422 [Oleofina S.A.]—POE (4) monolaurate
Radiasurf 7423 [Oleofina S.A.]—POE (8) monolaurate
Radiasurf 7443 [Oleofina S.A.]—POE (8) dioleate
Radiasurf 7453 [Oleofina S.A.]—POE (8) distearate
Radiasurf 7454 [Oleofina S.A.]—POE (12) distearate
Renex 30 [ICI United States]—POE (12) tridecyl ether
Renex 36 [ICI United States]—POE (6) tridecyl ether
Rewomid DLM/SE [Rewo Chem. Werke]—Myristic diethanolamide
Rewomul MG [Rewo Chem. Werke]—Glyceryl monostearate
Rewopal PG280 [Rewo Chem. Werke]—Ethylene glycol distearate
Rewopol MLS30, MLS35 [Rewo Chem. Werke]—Monoethanolamine lauryl sulfate
Rewopol NLS 15/L, NLS 30/L [Rewo]—Sodium lauryl sulfate
Rewopol NLS 28 [Dutton & Reinisch]—Sodium lauryl sulfate

Rewopol NLS 90 [Rewo]—Sodium lauryl sulfate
Rewopol NLS 90-Powder [Dutton & Reinisch]—Sodium lauryl sulfate
Rewopol SBFA30 [Dutton & Reinisch]—Disodium lauryl ether sulfosuccinate
Rewopol SBL203 [Dutton & Reinisch]—Disodium lauramido MEA-sulfosuccinate
Reworyl NKS50, NKS100 [Rewo Chem. Werke]—Sodium dodecylbenzenesulfonate
Reworyl-Sulfonic Acid K [Dutton & Reinisch]—Dodecylbenzene sulfonic acid
Rewoteric AM-DML [Rewo Chem. Werke]—Lauryl betaine
Rewoteric AM-KSF [Rewo Chem. Werke]—Cocoamphopropionate
Richonate 40B, 45B, 45BX, 50BD, 60B [Richardson]—Sodium dodecylbenzene-
    sulfonate
Richonate 1850, 1850U, C-50H [Richardson]—Sodium dodecylbenzenesulfonate
Richonate CS-6021H [Richardson]—Calcium dodecylbenzene sulfonate
Richonate YLA [Richardson]—Isopropylamine dodecylbenzenesulfonate
Richonic Acid B [Richardson]—Dodecylbenzene sulfonic acid
Richonol 2310, 6522, A, A Powder, C [Richardson]—Sodium lauryl sulfate
Richonol AM [Richardson]—Ammonium lauryl sulfate
Richonol Mg [Richardson]—Magnesium lauryl sulfate
Richonol S-1300, S-1300C [Richardson]—Ammonium lauryl ether sulfate
RS-55-40 [Hefti Ltd.]—POE (40) monostearate
Rueterg Sulfonic Acid [Finetex]—Dodecylbenzene sulfonic acid
Sactipon 2A [Lever Industriel]—Ammonium lauryl sulfate
Sactipon 2M [Lever Industriel]—Monoethanolamine lauryl sulfate
Sactipon 2OM [Lever Industriel]—Monoethanolamine lauryl ether sulfate
Sactipon 2OMG [Lever Industriel]—Magnesium lauryl ether sulfate
Sactipon 2S3 [Lever Industriel]—Sodium lauryl sulfate
Sactipon 2OA [Lever Industriel]—Ammonium lauryl ether sulfate
Sactol 2A [Lever Industriel]—Ammonium lauryl sulfate
Sactol 2M [Lever Industriel]—Monoethanolamine lauryl sulfate
Sactol 2OA [Lever Industriel]—Ammonium lauryl ether sulfate
Sactol 2OMG [Lever Industriel]—Magnesium lauryl ether sulfate
Sactol 2S3 [Lever Industriel]—Sodium lauryl sulfate
Sandoz Sulfate EP [Sandoz Colors & Chem.]—Diethanolamine lauryl sulfate
Sandoz Sulfate WA Dry, WAG, WAS, WA Special [Sandoz Colors & Chem.]—Sodium
    lauryl sulfate
Scher PEG 400 Distearate [Scher]—POE (8) distearate
Scher PEG 400 Monolaurate [Scher]—POE (8) monolaurate
Scher PEG 400 Monooleate [Scher]—POE (8) monooleate
Scher PEG 600 Distearate [Scher]—POE (12) distearate
Schercemol DEGMS [Scher]—Diethylene glycol monostearate
Schercemol GMS [Scher]—Glyceryl monostearate
Schercemol PGML [Scher]—Propylene glycol monolaurate
Schercemol PGMS [Scher]—Propylene glycol monostearate

Schercopol CMIS-Na [Scher]—Disodium coco amide MIPA sulfosuccinate
Schercopol LPS [Scher]—Disodium lauryl ether sulfosuccinate
Schercoteric MS-SF [Scher]—Cocoamphopropionate
Schercozoline O [Scher]—Oleyl imidazoline
Serdox NOL2 [Servo B.V.]—POE (2) oleyl ether
Sermul EA129 [Servo B.V.]—Ammonium lauryl sulfate
Sermul EA150 [Servo B.V.]—Sodium lauryl sulfate
Setacin 103 Special [Zschimmer & Schwartz]—Disodium lauryl ether sulfosuccinate
Simulsol 52 [Seppic]—POE (2) cetyl ether
Simulsol 56 [Seppic]—POE (10) cetyl ether
Simulsol 58 [Seppic]—POE (20) cetyl ether
Simulsol 92 [Seppic]—POE (2) oleyl ether
Simulsol 96 [Seppic]—POE (10) oleyl ether
Simulsol 98 [Seppic]—POE (20) oleyl ether
Simulsol M45 [Seppic]—POE (8) monostearate
Simulsol M49 [Seppic]—POE (20) monostearate
Simulsol M52 [Seppic]—POE (40) monostearate
Simulsol P4 [Seppic]—POE (4) lauryl ether
Sipex LCP, SB, SD, UB [Alcolac]—Sodium lauryl sulfate
Sipon 201-20, EA, EA2, EAY [Alcolac]—Ammonium lauryl ether sulfate
Sipon ES-7 [Alcolac]—Sodium POE (7) lauryl ether sulfate
Sipon ES-12 [Alcolac]—Sodium POE (12) lauryl ether sulfate
Sipon L22, L-22HV [Alcolac]—Ammonium lauryl sulfate
Sipon LD [Alcolac]—Diethanolamine lauryl sulfate
Sipon LM, MLS [Alcolac]—Magnesium lauryl sulfate
Sipon LS, LSB [Alcolac]—Sodium lauryl sulfate
Siponate 330 [Alcolac]—Isopropylamine dodecylbenzenesulfonate
Siponate DS-4, DS-10 [Alcolac]—Sodium dodecylbenzenesulfonate
Siponic C-20 [Alcolac]—POE (2) cetyl ether
Siponic E-15 [Alcolac]—POE (30) cetyl/stearyl ether
Siponic L-1 [Alcolac]—PEG-1 lauryl ether
Siponic L-4 [Alcolac]—POE (4) lauryl ether
Siponic L-12 [Alcolac]—POE (12) lauryl ether
Siponic L-25 [Alcolac]—POE (23) lauryl ether
Siponic TD-3 [Alcolac]—POE (3) tridecyl ether
Siponic TD-6 [Alcolac]—POE (6) tridecyl ether
Siponic TD-12 [Alcolac]—POE (12) tridecyl ether
Sole-Onic CDS [Hodag; Swift]—Diethylene glycol monolaurate SE
Sorbax PML-20 [Chemax]—POE (20) sorbitan monolaurate
Sorbax PMO-5 [Chemax]—POE (5) sorbitan monooleate
Sorbax PMO-20 [Chemax]—POE (20) sorbitan monooleate
Sorbax PMP-20 [Chemax]—POE (20) sorbitan monopalmitate

Sorbax PMS-20 [Chemax]—POE (20) sorbitan monostearate
Sorbax PTO-20 [Chemax]—POE (20) sorbitan trioleate
Sorbax PTS-20 [Chemax]—POE (20) sorbitan tristearate
Sorbon T-20 [Toho]—POE (20) sorbitan monolaurate
Sorgen TW20 [Dai-ichi Kogyo Seiyaku Co.]—POE (20) sorbitan monolaurate
Standamid CD [Henkel]—Capric acid diethanolamide
Standamul B-3 [Henkel]—POE (30) cetyl/stearyl ether
Standamul O-10 [Henkel]—POE (10) oleyl ether
Standamul O-20 [Henkel]—POE (20) oleyl ether
Standapol 125-E Conc., 130E [Henkel]—Sodium POE (12) lauryl ether sulfate
Standapol 230E Conc. [Henkel]—Ammonium POE (12) lauryl ether sulfate
Standapol A, AMS-100 [Henkel]—Ammonium lauryl sulfate
Standapol AP-60 [Henkel]—Ammonium pareth-25 sulfate
Standapol DEA [Henkel]—Diethanolamine lauryl sulfate
Standapol EA1, EA2, EA3 [Henkel]—Ammonium lauryl ether sulfate
Standapol EA40 [Henkel]—Ammonium myristyl ether sulfate
Standapol MG [Henkel, Henkel/Canada]—Magnesium lauryl sulfate
Standapol MLS [Henkel]—Monoethanolamine lauryl sulfate
Standapol WA-AC, WAQ-115, WAQ-LC, WAQ-LCX [Henkel]—Sodium lauryl sulfate
Standapol WAQ Special [Henkel/Canada]—Sodium lauryl sulfate
Standapol WAS-100 [Henkel]—Sodium lauryl sulfate
Stearolac C [Paniplus]—Calcium stearoyl lactylate
Steinapol NLS28, NLS90-Powder [Dutton & Reinisch]—Sodium lauryl sulfate
Steinapol SBF12, SBF12-Powder [Dutton & Reinisch]—Disodium lauryl sulfosuccinate
Steinapol SBFA30, 40% [Dutton & Reinisch]—Disodium lauryl ether sulfosuccinate
Steinapol SBL203 [Dutton & Reinisch]—Disodium lauramido MEA-sulfosuccinate
Steinaryl-Sulfonic Acid K [Dutton & Reinisch]—Dodecylbenzene sulfonic acid
Steol CA-460 [Stepan]—Ammonium lauryl ether sulfate
Steol CS-760 [Stepan]—Sodium POE (7) lauryl ether sulfate
Steol CS-760 [Stepan]—Sodium POE (12) lauryl ether sulfate
Stepanol AM, AM-V [Stepan]—Ammonium lauryl sulfate
Stepanol DEA [Stepan]—Diethanolamine lauryl sulfate
Stepanol ME Dry [Stepan]—Sodium lauryl sulfate
Stepanol MG [Stepan]—Magnesium lauryl sulfate
Stepanol WA100, WAC, WA Extra, WA Paste, WAQ, WA Special [Stepan]—Sodium
    lauryl sulfate
Sterling AB-40, AB-80 [Canada Packers]—Sodium dodecylbenzenesulfonate
Sterling AM, AM-HV [Canada Packers]—Ammonium lauryl sulfate
Sterling Emulsifier #3 [Canada Packers]—Pareth-25-3
Sterling ES600 [Canada Packers]—Ammonium lauryl ether sulfate
Sterling LA Acid C, LA Acid Reg. [Canada Packers]—Dodecylbenzene sulfonic acid
    (linear)

Sterling LA Paste 55%, 60% [Canada Packers]—Sodium dodecylbenzenesulfonate
Sterling WA Paste, WA Powder [Canada Packers]—Sodium lauryl sulfate
Sterling WADE [Canada Packers]—Diethanolamine lauryl sulfate
Sterling WAQ-CH, WAQ-Cosmetic, WAQ-LO, WAQ P [Canada Packers]—Sodium lauryl sulfate
Sul-fon-ate AA-9, AA-10 [Tennessee/Cities Service]—Sodium dodecylbenzene-sulfonate
Sul-fon-ate LA-10 [Tennessee]—Sodium dodecylbenzenesulfonate
Sulfatol 33 Pasta, 33 Pasta 13 [Aarhus Olie]—Sodium lauryl sulfate
Sulfatol 33 Paste [Aarhus Olie]—Sodium lauryl sulfate
Sulfatol 33MO [Aarhus Oliefabrik]—Monoethanolamine lauryl sulfate
Sulfopon 101, 101 Special, 102, 103 [Henkel KGaA]—Sodium lauryl sulfate
Sulfopon K-35 [Henkel KGaA]—Sodium lauryl sulfate
Sulfopon WA-3 [Henkel/Canada]—Sodium lauryl sulfate
Sulfopon WA30 [Henkel Argentina]—Sodium lauryl sulfate
Sulfopon WAQLCX, WAQ Special [Henkel/Canada]—Sodium lauryl sulfate
Sulfotex LCX, WA [Henkel]—Sodium lauryl sulfate
Sulfotex OT [Henkel]—Ammonium lauryl ether sulfate
Sulphonated Lorol Liquid MA, Liquid MR [Ronsheim & Moore]—Monoethanolamine lauryl sulfate
Sulphonated Lorol Paste [Ronsheim & Moore]—Sodium lauryl sulfate
Surco 55S, 60S [Onyx]—Sodium dodecylbenzenesulfonate
Surco DDBSA [Onyx]—Dodecylbenzene sulfonic acid (linear)
Surco MG-LS [Onyx]—Magnesium lauryl sulfate
Swanol AM-301 [Nikko]—Lauryl betaine
Swascol 1PC, 3L, 1-P [Swastik]—Sodium lauryl sulfate
T-Maz 20 [Mazer]—POE (20) sorbitan monolaurate
T-Maz 40 [Mazer]—POE (20) sorbitan monopalmitate
T-Maz 60 [Mazer]—POE (20) sorbitan monostearate
T-Maz 61 [Mazer]—POE (4) sorbitan monostearate
T-Maz 65 [Mazer]—POE (20) sorbitan tristearate
T-Maz 80 [Mazer]—POE (20) sorbitan monooleate
T-Maz 81 [Mazer]—POE (5) sorbitan monooleate
T-Maz 85 [Mazer]—POE (20) sorbitan trioleate
Tagat S2 [Goldschmidt]—POE (20) glyceryl monostearate
Tairygent CA-1 [Formosa Chem. & Fibre]—Dodecylbenzene sulfonic acid (linear)
Tairygent CA-2 [Formosa Chem. & Fibre]—Dodecylbenzene sulfonic acid (branched)
Tairygent CB-1, CB-2 [Formosa Chem & Fibre]—Sodium dodecylbenzenesulfonate
Tefose 1500 [Gattefosse]—POE (1500) monostearate
Tegin 90, 90NSE, 515, GRB, M, MAV, NSE [Goldschmidt]—Glyceryl monostearate
Tegin 4480 [Th. Goldschmidt A.G.]—Glyceryl monolaurate
Tegin O Special [Goldschmidt]—Glyceryl monooleate

Tegin P, P-411SE [Goldschmidt]—Propylene glycol monostearate, self-emulsifying

Tegin P412 [Th. Goldschmidt A.G.]—Propylene glycol monostearate

Tegin, Special SE [Goldschmidt]—Glyceryl monostearate, self-emulsifying

Tergitol 25-L-3 [Union Carbide]—Pareth-25-3

Tergitol 25-L-7 [Union Carbide]—Pareth-25-7

Tergitol 25-L-12 [Union Carbide]—Pareth 25-12

Teric 12M2 [ICI Australia Ltd.]—POE (2) coconut amine

Teric 12M5 [ICI Australia Ltd.]—POE (5) coconut amine

Teric 12M15 [ICI Australia Ltd.]—POE (15) coconut amine

Teric 17M15 [ICI Australia Ltd.]—POE (15) tallow amine

Tex-Wet 1197 [Intex Prod.]—Dodecylbenzene sulfonic acid

Texapon A [Henkel/Canada]—Ammonium lauryl sulfate

Texapon A400, Special [Henkel KGaA]—Ammonium lauryl sulfate

Texapon DEA [Henkel/Canada]—Diethanolamine lauryl sulfate

Texapon EA-1, EA-2, EA-3, EA-40 [Henkel/Canada]—Ammonium lauryl ether sulfate

Texapon EA-40 [Henkel/Canada]—Ammonium myristyl ether sulfate

Texapon K-12 [Henkel]—Sodium lauryl sulfate

Texapon K-1294 [Henkel KGaA]—Sodium lauryl sulfate

Texapon K-1296 [Henkel/Canada]—Sodium lauryl sulfate

Texapon L-100 [Henkel]—Sodium lauryl sulfate

Texapon LS Highly Conc. Needles [Henkel KGaA]—Sodium lauryl sulfate

Texapon MG [Henkel KGaA]—Magnesium lauryl ether sulfate

Texapon MLS [Henkel KGaA]—Monoethanolamine lauryl sulfate

Texapon NA [Henkel Argentina]—Ammonium lauryl ether sulfate

Texapon OT Highly Conc. Needles [Henkel KGaA]—Sodium lauryl sulfate

Texapon V Highly Conc. Needles [Henkel]—Sodium lauryl sulfate

Texapon Z Granules [Henkel KGaA]—Sodium lauryl sulfate

Texapon Z Highly Conc. Needles [Henkel]—Sodium lauryl sulfate

Texapon Z Highly Conc. Powder [Henkel]—Sodium lauryl sulfate

Textamine O-1 [Henkel]—Oleyl imidazoline

TO-55-F [Hefti Ltd.]—POE (20) sorbitan trioleate

Trycol LAL-23 [Emery]—POE (23) lauryl ether

Trycol TDA-3 [Emery]—POE (3) tridecyl ether

Trycol TDA-6 [Emery]—POE (6) tridecyl ether

Trylox HCO-5 [Emery]—POE (5) hydrogenated castor oil

Trylox HCO-25 [Emery]—POE (25) hydrogenated castor oil

Trymeen CAM-10 [Emery]—POE (10) coconut amine

Trymeen CAM-15 [Emery]—POE (15) coconut amine

Trymeen SAM-50 [Emery]—POE (50) stearyl amine

Trymeen TAM-15 [Emery]—POE (15) tallow amine

TS-55-F [Hefti Ltd.]—POE (20) sorbitan tristearate

Tween 20 [ICI United States]—POE (20) sorbitan monolaurate

Tween 20SD [Atlas Chem. Ind. N.V.]—POE (20) sorbitan monolaurate (specially deodorized)

Tween 21 [ICI United States]—POE (4) sorbitan monolaurate

Tween 40 [ICI United States]—POE (20) sorbitan monopalmitate

Tween 60 [ICI United States]—POE (20) sorbitan monostearate

Tween 61 [ICI United States]—POE (4) sorbitan monostearate

Tween 65 [ICI United States]—POE (20) sorbitan tristearate

Tween 80 [ICI United States]—POE (20) sorbitan monooleate (USP grade)

Tween 81 [ICI United States]—POE (5) sorbitan monooleate

Tween 85 [ICI United States]—POE (20) sorbitan trioleate

Tylorol A [Thomas Triantaphyllou S.A.]—Ammonium lauryl ether sulfate

Tylorol LM [Thomas Triantaphyllou]—Monoethanolamine lauryl ether sulfate

Tylorol MG [Thomas Triantaphyllou]—Magnesium lauryl ether sulfate

Ultra Sulfate SL-1 [Witco/Organics]—Sodium lauryl sulfate

Unamine O [Lonza]—Oleyl imidazoline

Ungerol AM3-60 [Unger Fabrikker AS]—Ammonium lauryl ether sulfate

Varine O [Sherex]—Oleyl imidazoline

Varonic 400MS [Sherex]—POE (8) monostearate

Varonic 1000MS [Sherex]—POE (20) monostearate

Varonic L205 [Sherex]—POE (5) soya amine

Varonic LI42 [Sherex]—POE (20) glyceryl monostearate

Verv [Patco]—Calcium stearoyl lactylate

Vista SA-597 [Vista]—Dodecylbenzene sulfonic acid (linear)

Volpo 10 [Croda]—POE (10) oleyl ether

Volpo 20 [Croda Inc.]—POE (20) oleyl ether

Volpo L4 [Croda]—POE (4) lauryl ether

Volpo N10 [Croda Chem. Ltd.]—POE (10) oleyl ether, distilled

Volpo N20 [Croda Chem. Ltd.]—POE (20) oleyl ether, distilled

Volpo O10 [Croda Chem. Ltd.]—POE (10) oleyl ether

Volpo O20 [Croda Chem. Ltd.]—POE (20) oleyl ether

Volpo T3 [Croda Chem. Ltd.]—POE (3) tridecyl ether

Witco 942 [Witco/Organics]—Glyceryl monooleate

Witco 1298 Hard Acid, 1298 Soft Acid [Witco/Organics]—Dodecylbenzene sulfonic acid

Witco Acid B [Witco/Organics]—Dodecylbenzene sulfonic acid

Witcolate 6522, A, A Powder [Witco/Organics]—Sodium lauryl sulfate

Witcolate AE-3 [Witco/Organics]—Ammonium pareth-25 sulfate

Witcolate AM [Witco/Organics]—Ammonium lauryl sulfate

Witcolate S1300C [Witco/Organics]—Ammonium lauryl ether sulfate

Witconate 60B, 90 Flakes [Witco/Organics]—Sodium dodecylbenzenesulfonate

Witconate 1250 Slurry [Witco/Organics]—Sodium dodecylbenzenesulfonate

Witconate C-50H [Witco/Organics]—Sodium dodecylbenzenesulfonate

Witconol CA, RHT [Witco/Organics]—Glyceryl monostearate, self-emulsifying
Witconol CAD [Witco/Organics]—Diethylene glycol monostearate
Witconol DOS [Witco/Organics]—Diethylene glycol monooleate SE
Witconol H-31 [Witco/Organics]—POE (8) monotallate
Witconol H-31A [Witco/Organics]—POE (8) monooleate
Witconol H-35A [Witco/Organics]—POE (8) monostearate
Witconol MST [Witco/Organics]—Glyceryl monostearate
Witconol RHP [Witco/Organics]—Propylene glycol monostearate, self-emulsifying
Zoharpon LAA [Zohar Detergent]—Ammonium lauryl sulfate
Zoharpon LAM [Zohar Detergent]—Monoethanolamine lauryl sulfate
Zoharpon LAS, LAS Special [Zohar Detergent]—Sodium lauryl sulfate

GENERIC CHEMICAL SYNONYMS
AND CROSS REFERENCES

Amines, coco alkyl dimethyl. See Coco dimethyl amine

Ammonium laureth sulfate (CTFA). See Ammonium lauryl ether sulfate

Ammonium laureth-5 sulfate (CTFA). See Ammonium POE (5) lauryl ether sulfate

Ammonium laureth-12 sulfate (CTFA). See Ammonium POE (12) lauryl ether sulfate

Ammonium myreth sulfate (CTFA). See Ammonium myristyl ether sulfate

Anhydrite. See Calcium sulfate

Benzenemethanaminium, N,N-dimethyl-N-tetradecyl-, chloride. See Myristyl dimethyl benzyl ammonium chloride

Benzenesulfonic acid, dodecyl-. See Dodecylbenzene sulfonic acid

Benzenesulfonic acid, dodecyl-, compd. with 2-propanamine (1:1). See Isopropylamine dodecylbenzenesulfonate

Benzenesulfonic acid, dodecyl-, sodium salt. See Sodium dodecylbenzenesulfonate

N,N-Bis(2-hydroxyethyl) cocoamine oxide. See Book IV

N,N-Bis (2-hydroxyethyl) decanamide. See Capric acid diethanolamide

N,N-Bis (2-hydroxyethyl) myristamide. See Myristic diethanolamide

Bis (2-hydroxyethyl) tallow amine oxide. See Book IV

N,N-Bis (2-hydroxyethyl) tetradecanamide. See Myristic diethanolamide

Butanedioic acid, sulfo-, 2-cocamido-1-methylethyl, disodium salts. See Disodium coco amide MIPA sulfosuccinate

Butanedioic acid, sulfo-, 1-dodecyl ester, disodium salt. See Disodium lauryl sulfosuccinate

Butanedioic acid, sulfo-, 4-[2-[2-[2-(dodecyloxy) ethoxy] ethoxy] ethyl] ester, disodium salt. See Disodium lauryl ether sulfosuccinate

Butanedioic acid, sulfo-, 1-ester with N-(2-hydroxyethyl) dodecanamide, disodium salt. See Disodium lauramido MEA-sulfosuccinate

Butyl stearate. See Book IV

C_{12-15} alcohol (3EO) ethoxylate. See Pareth-25-3

C_{12-15} alcohol (7EO) ethoxylate. See Pareth-25-7

C_{12-15} alcohol (12EO) ethoxylate. See Pareth-25-12

C_{12-15} alcohol ethoxysulfate, ammonium salt. See Ammonium pareth-25 sulfate

Calcium stearoyl-2-lactylate. See Calcium stearoyl lactylate

Calcium sulfonate. See Calcium sulfate

Capramide DEA (CTFA). See Capric acid diethanolamide

Capric amide DEA. See Capric acid diethanolamide

Capryloamphocarboxypropionate. See Capryloamphodipropionate

Castor oil, sulfated. See Book III

Ceteareth-11 (CTFA). See POE (11) cetyl/stearyl ether

Ceteareth-25 (CTFA). See POE (25) cetyl/stearyl ether

Ceteareth-27 (CTFA). See POE (27) cetyl/stearyl ether

Ceteareth-30 (CTFA). See POE (30) cetyl/stearyl ether
Ceteareth-50 (CTFA). See POE (50) cetyl/stearyl ether
Ceteareth-55 (CTFA). See POE (55) cetyl/stearyl ether
Ceteth-2 (CTFA). See POE (2) cetyl ether
Ceteth-6 (CTFA). See POE (6) cetyl ether
Ceteth-10 (CTFA). See POE (10) cetyl ether
Ceteth-15 (CTFA). See POE (15) cetyl ether
Ceteth-20 (CTFA). See POE (20) cetyl ether
Ceteth-25 (CTFA). See POE (25) cetyl ether
Ceteth-30 (CTFA). See POE (30) cetyl ether
Cetomacrogol 1000 BPC. See POE (20) cetyl ether
Cetomacrogol 1000. See POE (20) cetyl ether
Cetrimonium tosylate (CTFA). See Cetyl trimethyl ammonium *p*-toluene sulfonate
Cetyl alcohol. See Book IV
Cetyl trimethyl ammonium chloride. See Book IV
Cetyl trimethyl ammonium tosylate. See Cetyl trimethyl ammonium *p*-toluene sulfonate
Coco amine + EO (2 moles). See POE (2) coconut amine
Coco amine oxide. See Book IV
Coco betaine. See Book IV
Coconut diethanolamide. See Book III
Coconut monoethanolamide. See Book III
Coconut monoisopropanolamide. See Book III
DDBSA. See Dodecylbenzene sulfonic acid
DEA-lauryl sulfate (CTFA). See Diethanolamine lauryl sulfate
Decaglycerol decaoleate. See Book II
Decaglycerol decastearate. See Book II
Decaglycerol octaoleate. See Book II
Decaglycerol tetraoleate. See Book III
Decanamide, N,N-bis (2-hydroxyethyl)-. See Capric acid diethanolamide
1-Decanaminium, N-decyl-N,N-dimethyl-, chloride. See Didecyl dimethyl ammonium chloride
N-Decyl-N,N-dimethyl-1-decanaminium chloride. See Didecyl dimethyl ammonium chloride
Decylic amide DEA. See Capric acid diethanolamide
Dicocodimonium chloride (CTFA). See Dicoco dimethyl ammonium chloride
Didecyldimonium chloride (CTFA). See Didecyl dimethyl ammonium chloride
Diethanolamine myristic acid condensate. See Myristic diethanolamide
Diethylene glycol monoricinoleate. See Book II
Diethylene glycol oleate. See Diethylene glycol monooleate
Diethylene glycol stearate. See Diethylene glycol monostearate
Diglycol laurate. See Diethylene glycol monolaurate
Diglycol monolaurate. See Diethylene glycol monolaurate

Diglycol monooleate. See Diethylene glycol monooleate
Diglycol monostearate. See Diethylene glycol monostearate
Diglycol oleate. See Diethylene glycol monooleate
Diglycol stearate. See Diethylene glycol monostearate
Dihydrogenated-tallow dimethyl ammonium chloride. See Book IV
2,3-Dihydroxypropyl octadecanoate. See Glyceryl monostearate
Dimethyl cocamine (CTFA). See Coco dimethyl amine
Dimethyl cocoamine. See Coco dimethyl amine
Dimethyl coconut amine. See Coco dimethyl amine
Dimethyl coconut t-amine. See Coco dimethyl amine
Dimethyl dicoco ammonium chloride. See Dicoco dimethyl ammonium chloride
Dimethyl dicoconut ammonium chloride. See Dicoco dimethyl ammonium chloride
Dimethyl didecyl ammonium chloride. See Didecyl dimethyl ammonium chloride
N,N-Dimethyl-1-dodecanamine. See Lauryl dimethyl amine
N,N-Dimethyl-1-hexadecanamine. See Palmityl dimethyl amine
Dimethyl lauramine (CTFA). See Lauryl dimethyl amine
Dimethyl lauryl amine. See Lauryl dimethyl amine
N,N-Dimethyl-1-octadecanamine. See Stearyl dimethyl amine
Dimethyl palmitamine (CTFA). See Palmityl dimethyl amine
Dimethyl palmitylamine. See Palmityl dimethyl amine
Dimethyl stearamine (CTFA). See Stearyl dimethyl amine
Dimethyl stearylamine. See Stearyl dimethyl amine
N,N-Dimethyl-N-tetradecylbenzenemethanaminium chloride. See Myristyl dimethyl benzyl ammonium chloride
Disodium cocamido MIPA-sulfosuccinate (CTFA). See Disodium coco amide MIPA sulfosuccinate
Disodium deceth-6 sulfosuccinate. See Book II
Disodium isodecyl sulfosuccinate. See Book II
Disodium laureth sulfosuccinate. See Disodium lauryl ether sulfosuccinate
Disodium lauryl alcohol polyglycol-ether. See Disodium lauryl ether sulfosuccinate
Disodium lauryl alcohol sulfosuccinate. See Disodium lauryl sulfosuccinate
Disodium stearyl sulfosuccinamate. See Book II
Disodium tallowiminodipropionate. See Book II
1-Dodecanamine, N,N-dimethyl-. See Lauryl dimethyl amine
1-Dodecanaminium N-(carboxymethyl)-N,N-dimethyl-hydroxide, inner salt. See Lauryl betaine
1-Dodecanaminium, N,N,N-trimethyl-, chloride. See Lauryl trimethyl ammonium chloride
Dodecanoic acid, 2,3-dihydroxypropyl ester. See Glyceryl monolaurate
Dodecanoic acid, 2-(2-hydroxyethoxy) ethyl ester. See Diethylene glycol monolaurate
Dodecanoic acid, 2-hydroxypropyl ester. See Propylene glycol monolaurate
Dodecanoic acid, monoester with 1,2-propanediol. See Propylene glycol monolaurate

Dodecanoic acid, monoester with 1,2,3-propanetriol. See Glyceryl monolaurate

1-Dodecanol, hydrogen sulfate, sodium salt. See Sodium lauryl sulfate

Dodecyl ammonium sulfate. See Ammonium lauryl sulfate

Dodecyl dimethyl amine. See Lauryl dimethyl amine

Dodecyl trimethyl ammonium chloride. See Lauryl trimethyl ammonium chloride

Dodecylbenzene isopropylamine sulfonate. See Isopropylamine dodecylbenzene-sulfonate

Dodecylbenzene sodium sulfonate. See Sodium dodecylbenzenesulfonate

Dodecylbenzene sulfonate, calcium salt. See Calcium dodecylbenzene sulfonate

Dodecylbenzene sulfonate, sodium salt. See Sodium dodecylbenzenesulfonate

Dodecylbenzenesulfonic acid, comp. with 2-propanamine (1:1). See Isopropylamine dodecylbenzenesulfonate

Dodecylbenzenesulfonic acid, isopropylamine salt. See Isopropylamine dodecyl-benzenesulfonate

Dodecylbenzenesulfonic acid, sodium salt. See Sodium dodecylbenzenesulfonate

2-(Dodecyloxy) ethanol. See PEG-1 lauryl ether

1-Dodecylpyridinium chloride. See Lauryl pyridinium chloride

Ethanol, 2-(dodecyloxy)-. See PEG-1 lauryl ether

Ethylene glycol monolauryl ether. See PEG-1 lauryl ether

Ethylene glycol monostearate. See Book III

L-Glutamic acid, N-coco acyl derivatives, monosodium salts. See Sodium cocoyl glutamate

Glutamic acid, N-(1-oxododecyl)- monosodium salt. See Sodium lauroyl glutamate

Glycerol monolaurate. See Glyceryl monolaurate

Glycerol monooleate. See Glyceryl monooleate

Glycerol monostearate. See Glyceryl monostearate

Glyceryl laurate (CTFA). See Glyceryl monolaurate

Glyceryl monococoate. See Book III

Glyceryl monoricinoleate. See Book III

Glyceryl oleate (CTFA). See Glyceryl monooleate

Glyceryl stearate (CTFA). See Glyceryl monostearate

Glyceryl triisostearate. See Book IV

Glyceryl trioleate. See Book IV

Glycol distearate (CTFA). See Ethylene glycol distearate

GMS. See Glyceryl monostearate

Gypsum. See Calcium sulfate

2-(8-Heptadecenyl)-4,5-dihydro-1H-imidazole-1-ethanol. See Oleyl imidazoline

1-Hexadecanamine, N,N-dimethyl-. See Palmityl dimethyl amine

1-Hexadecanaminium, N,N,N-trimethyl-, salt with 4-methylbenzenesulfonic acid (1:1). See Cetyl trimethyl ammonium p-toluene sulfonate

Hexadecyl dimethyl amine. See Palmityl dimethyl amine

N-Hexadecyl-N,N,N-trimethyl ammonium p-toluene sulfonate. See Cetyl trimethyl

ammonium p-toluene sulfonate
Hexaglyceryl distearate. See Book II
Hydroxyethyl imidazoline, oleic hydrophobe. See Oleyl imidazoline
1-Hydroxyethyl-2-oleyl imidazoline. See Oleyl imidazoline
2-Hydroxypropyl dodecanoate. See Propylene glycol monolaurate
1H-Imidazole-1-ethanol, 2-(8-heptadecenyl)-4,5-dihydro-. See Oleyl imidazoline
Isopropyl lanolate. See Book IV
Isopropyl myristate. See Book IV
Isostearamidopropyl betaine. See Book IV
Isostearamidopropyl dimethylamine. See Book IV
Isosteareth-2 (CTFA). See POE (2) isostearyl ether
Isosteareth-10 (CTFA). See POE (10) isostearyl ether
Isosteareth-20 (CTFA). See POE (20) isostearyl ether
Isostearyl alcohol. See Book IV
Lanolin. See Book IV
Laureth-1 (CTFA). See PEG-1 lauryl ether
Laureth-4 (CTFA). See POE (4) lauryl ether
Laureth-12 (CTFA). See POE (12) lauryl ether
Laureth-15 (CTFA). See POE (15) lauryl ether
Laureth-23 (CTFA). See POE (23) lauryl ether
Lauric diethanolamide. See Book III
Lauric monoethanolamide. See Book III
Lauric monoisopropanolamide. See Book III
Laurtrimonium chloride (CTFA). See Lauryl trimethyl ammonium chloride
Lauryl dimethyl amine oxide. See Book IV
Lauryl dimethyl glycine. See Lauryl betaine
Lauryl lactate. See Book IV
Lauryl sulfate, diethanolamine salt. See Diethanolamine lauryl sulfate
Lauryl sulfate, monoethanolamine salt. See Monoethanolamine lauryl sulfate
Lecithin. See Book II
Magnesium laureth sulfate (CTFA). See Magnesium lauryl ether sulfate
Magnesium monododecyl sulfate. See Magnesium lauryl sulfate
MEA-laureth sulfate (CTFA). See Monoethanolamine lauryl ether sulfate
MEA-lauryl sulfate (CTFA). See Monoethanolamine lauryl sulfate
Monoolein. See Glyceryl monooleate
Monosodium N-cocoyl-L-glutamate. See Sodium cocoyl glutamate
Monosodium n-hydrogenated-tallowyl-l-glutamate. See Sodium hydrogenated tallow
    glutamate
Monosodium N-lauroyl-L-glutamate. See Sodium lauroyl glutamate
Monostearin. See Glyceryl monostearate
Myristalkonium chloride (CTFA). See Myristyl dimethyl benzyl ammonium chloride
Myristamide DEA (CTFA). See Myristic diethanolamide

Myristic monoethanolamide. See Book III

Myristoyl diethanolamide. See Myristic diethanolamide

Myristyl dimethyl amine oxide. See Book III

1-Octadecanamine, N,N-dimethyl-. See Stearyl dimethyl amine

Octadecanoic acid, 2-(1-carboxyethoxy)-1-methyl-2-oxoethyl ester, calcium salt. See Calcium stearoyl lactylate

Octadecanoic acid, 2-(1-carboxyethoxy)-1-methyl-2-oxoethyl ester, sodium salt. See Sodium stearoyl lactylate

Octadecanoic acid, 2,3-dihydroxypropyl ester. See Glyceryl monostearate

Octadecanoic acid, 1,2-ethanediyl ester. See Ethylene glycol distearate

Octadecanoic acid, 2-(2-hydroxyethoxy) ethyl ester. See Diethylene glycol monostearate

Octadecanoic acid, monoester with 1,2-propanediol. See Propylene glycol monostearate

Octadecanoic acid, monoester with 1,2,3-propanetriol. See Glyceryl monostearate

9-Octadecen-1-amine. See Oleyl amine

9-Octadecenoic acid, 2,3-dihydroxypropyl ester. See Glyceryl monooleate

9-Octadecenoic acid, 2-(2-hydroxyethoxy) ethyl ester. See Diethylene glycol mono-oleate

9-Octadecenoic acid, monoester with 1,2-propanediol. See Propylene glycol monooleate

9-Octadecenoic acid, monoester with 1,2,3-propanetriol. See Glyceryl monooleate

Octadecyl dimethyl amine. See Stearyl dimethyl amine

Oleamidopropyl betaine. See Book III

Oleamine (CTFA). See Oleyl amine

Oleic acid imidazoline. See Oleyl imidazoline

Oleic diethanolamide. See Book III

Oleic fatty acid imidazoline. See Oleyl imidazoline

Oleic hydroxyethyl imidazoline. See Oleyl imidazoline

Oleic imidazoline. See Oleyl imidazoline

Oleic monoisopropanolamide. See Book III

Oleth-2 (CTFA). See POE (2) oleyl ether

Oleth-4 (CTFA). See POE (4) oleyl ether

Oleth-10 (CTFA). See POE (10) oleyl ether

Oleth-20 (CTFA). See POE (20) oleyl ether

Oleth-4 phosphate (CTFA). See POE (4) oleyl ether phosphate

Oleyl alcohol. See Book IV

Oleyl dimethyl benzyl ammonium chloride. See Book IV

Oleyl hydroxyethyl imidazoline. See Oleyl imidazoline

Oleyl polyglycol ether (2 EO). See POE (2) oleyl ether

N-(1-Oxododecyl) glutamic acid, monosodium salt. See Sodium lauroyl glutamate

Palmityl dimethyl tertiary amine. See Palmityl dimethyl amine

PEG-25 castor oil, hydrogenated. See POE (25) hydrogenated castor oil

PEG-2 cetyl ether. See POE (2) cetyl ether

PEG-6 cetyl ether. See POE (6) cetyl ether

PEG-10 cetyl ether. See POE (10) cetyl ether
PEG-15 cetyl ether. See POE (15) cetyl ether
PEG (15) cetyl ether. See POE (15) cetyl ether
PEG-20 cetyl ether. See POE (20) cetyl ether
PEG-25 cetyl ether. See POE (25) cetyl ether
PEG (25) cetyl ether. See POE (25) cetyl ether
PEG-30 cetyl ether. See POE (30) cetyl ether
PEG (30) cetyl ether. See POE (30) cetyl ether
PEG 100 cetyl ether. See POE (2) cetyl ether
PEG 300 cetyl ether. See POE (6) cetyl ether
PEG 500 cetyl ether. See POE (10) cetyl ether
PEG 1000 cetyl ether. See POE (20) cetyl ether
PEG-11 cetyl/stearyl ether. See POE (11) cetyl/stearyl ether
PEG-25 cetyl/stearyl ether. See POE (25) cetyl/stearyl ether
PEG-27 cetyl/stearyl ether. See POE (27) cetyl/stearyl ether
PEG (27) cetyl stearyl ether. See POE (27) cetyl/stearyl ether
PEG-30 cetyl/stearyl ether. See POE (30) cetyl/stearyl ether
PEG (30) cetyl stearyl ether. See POE (30) cetyl/stearyl ether
PEG-50 cetyl/stearyl ether. See POE (50) cetyl/stearyl ether
PEG (50) cetyl stearyl ether. See POE (50) cetyl/stearyl ether
PEG-55 cetyl/stearyl ether. See POE (55) cetyl/stearyl ether
PEG (55) cetyl stearyl ether. See POE (55) cetyl/stearyl ether
PEG-2 cocamine (CTFA). See POE (2) coconut amine
PEG-5 cocamine (CTFA). See POE (5) coconut amine
PEG-10 cocamine (CTFA). See POE (10) coconut amine
PEG-15 cocamine (CTFA). See POE (15) coconut amine
PEG-8 cocoate (CTFA). See POE (8) monococoate
PEG (5) coconut amine. See POE (5) coconut amine
PEG (15) coconut amine. See POE (15) coconut amine
PEG 100 coconut amine. See POE (2) coconut amine
PEG 500 coconut amine. See POE (10) coconut amine
PEG-4 dioleate (CTFA). See POE (4) dioleate
PEG-8 dioleate (CTFA). See POE (8) dioleate
PEG-12 dioleate (CTFA). See POE (12) dioleate
PEG 200 dioleate. See POE (4) dioleate
PEG 400 dioleate. See POE (8) dioleate
PEG 600 dioleate. See POE (12) dioleate
PEG-4 distearate (CTFA). See POE (4) distearate
PEG-8 distearate (CTFA). See POE (8) distearate
PEG-12 distearate (CTFA). See POE (12) distearate
PEG 200 distearate. See POE (4) distearate
PEG 400 distearate. See POE (8) distearate

PEG 600 distearate. See POE (12) distearate
PEG 1000 glyceryl monostearate. See POE (20) glyceryl monostearate
PEG-20 glyceryl stearate (CTFA). See POE (20) glyceryl monostearate
PEG-5 hydrogenated castor oil (CTFA). See POE (5) hydrogenated castor oil
PEG (5) hydrogenated castor oil. See POE (5) hydrogenated castor oil
PEG-7 hydrogenated castor oil (CTFA). See POE (7) hydrogenated castor oil
PEG (7) hydrogenated castor oil. See POE (7) hydrogenated castor oil
PEG-25 hydrogenated castor oil (CTFA). See POE (25) hydrogenated castor oil
PEG (25) hydrogenated castor oil. See POE (25) hydrogenated castor oil
PEG-2 isostearyl ether. See POE (2) isostearyl ether
PEG-10 isostearyl ether. See POE (10) isostearyl ether
PEG-20 isostearyl ether. See POE (20) isostearyl ether
PEG 100 isostearyl ether. See POE (2) isostearyl ether
PEG 500 isostearyl ether. See POE (10) isostearyl ether
PEG 1000 isostearyl ether. See POE (20) isostearyl ether
PEG-2 laurate (CTFA). See Diethylene glycol monolaurate
PEG-4 laurate (CTFA). See POE (4) monolaurate
PEG-6 laurate (CTFA). See POE (6) monolaurate
PEG-8 laurate (CTFA). See POE (8) monolaurate
PEG-12 laurate (CTFA). See POE (12) monolaurate
PEG-75 laurate (CTFA). See POE (75) monostearate
PEG-4 lauryl ether. See POE (4) lauryl ether
PEG-12 lauryl ether. See POE (12) lauryl ether
PEG-15 lauryl ether. See POE (15) lauryl ether
PEG (15) lauryl ether. See POE (15) lauryl ether
PEG-23 lauryl ether. See POE (23) lauryl ether
PEG (23) lauryl ether. See POE (23) lauryl ether
PEG 200 lauryl ether. See POE (4) lauryl ether
PEG 600 lauryl ether. See POE (12) lauryl ether
PEG 400 monococoate. See POE (8) monococoate
PEG 100 monolaurate. See Diethylene glycol monolaurate
PEG 200 monolaurate. See POE (4) monolaurate
PEG 300 monolaurate. See POE (6) monolaurate
PEG 400 monolaurate. See POE (8) monolaurate
PEG 600 monolaurate. See POE (12) monolaurate
PEG 100 monooleate. See Diethylene glycol monooleate
PEG 200 monooleate. See POE (4) monooleate
PEG 400 monooleate. See POE (8) monooleate
PEG 600 monooleate. See POE (12) monooleate
PEG 1000 monooleate. See POE (20) monooleate
PEG 100 monostearate. See Diethylene glycol monostearate
PEG 200 monostearate. See POE (4) monostearate

PEG 400 monostearate. See POE (8) monostearate
PEG 600 monostearate. See POE (12) monostearate
PEG 1500 monostearate. See POE (1500) monostearate
PEG 2000 monostearate. See POE (40) monostearate
PEG 4000 monostearate. See POE (75) monostearate
PEG 400 monotallate. See POE (8) monotallate
PEG 500 monotallate. See POE (10) monotallate
PEG-1 nonyl phenyl ether. See Book II
PEG-1 octyl phenyl ether. See Book II
PEG-2 oleate (CTFA). See Diethylene glycol monooleate
PEG-4 oleate (CTFA). See POE (4) monooleate
PEG-8 oleate (CTFA). See POE (8) monooleate
PEG-12 oleate (CTFA). See POE (12) monooleate
PEG-20 oleate (CTFA). See POE (20) monooleate
PEG 200 oleate. See POE (4) monooleate
PEG 600 oleate. See POE (12) monooleate
PEG-2 oleyl ether. See POE (2) oleyl ether
PEG-4 oleyl ether. See POE (4) oleyl ether
PEG-10 oleyl ether. See POE (10) oleyl ether
PEG-20 oleyl ether. See POE (20) oleyl ether
PEG 100 oleyl ether. See POE (2) oleyl ether
PEG 200 oleyl ether. See POE (4) oleyl ether
PEG 500 oleyl ether. See POE (10) oleyl ether
PEG 1000 oleyl ether. See POE (20) oleyl ether
PEG-4 oleyl ether phosphate. See POE (4) oleyl ether phosphate
PEG 200 oleyl ether phosphate. See POE (4) oleyl ether phosphate
PEG-25 propylene glycol stearate (CTFA). See POE (25) propylene glycol monostearate
PEG-6 sorbitan beeswax (CTFA). See POE (6) sorbitol beeswax
PEG-20 sorbitan beeswax (CTFA). See POE (20) sorbitol beeswax
PEG 300 sorbitan beeswax. See POE (6) sorbitol beeswax
PEG 1000 sorbitan beeswax. See POE (20) sorbitol beeswax
PEG 4 sorbitan monostearate. See POE (4) sorbitan monostearate
PEG (5) soya amine. See POE (5) soya amine
PEG (15) soya amine. See POE (15) soya amine
PEG 500 soya amine. See POE (10) soya amine
PEG-5 soyamine (CTFA). See POE (5) soya amine
PEG-10 soyamine (CTFA). See POE (10) soya amine
PEG-15 soyamine (CTFA). See POE (15) soya amine
PEG-2 stearamine (CTFA). See POE (2) stearyl amine
PEG-5 stearamine (CTFA). See POE (5) stearyl amine
PEG-50 stearamine (CTFA). See POE (50) stearyl amine
PEG-2 stearate (CTFA). See Diethylene glycol monostearate

PEG-4 stearate (CTFA). See POE (4) monostearate
PEG-6-32 stearate (CTFA). See POE (1500) monostearate
PEG-8 stearate (CTFA). See POE (8) monostearate
PEG-12 stearate (CTFA). See POE (12) monostearate
PEG-40 stearate (CTFA). See POE (40) monostearate
PEG (5) stearyl amine. See POE (5) stearyl amine
PEG (50) stearyl amine. See POE (50) stearyl amine
PEG 100 stearyl amine. See POE (2) stearyl amine
PEG-8 tallate (CTFA). See POE (8) monotallate
PEG-10 tallate (CTFA). See POE (10) monotallate
PEG-2 tallow amine (CTFA). See POE (2) tallow amine
PEG-5 tallow amine (CTFA). See POE (5) tallow amine
PEG (5) tallow amine. See POE (5) tallow amine
PEG-15 tallow amine (CTFA). See POE (15) tallow amine
PEG (15) tallow amine. See POE (15) tallow amine
PEG 100 tallow amine. See POE (2) tallow amine
PEG-15 tallow aminopropylamine (CTFA). See POE (15) tallow aminopropylamine
PEG (15) tallow aminopropylamine. See POE (15) tallow aminopropylamine
PEG-3 tridecyl ether. See POE (3) tridecyl ether
PEG (3) tridecyl ether. See POE (3) tridecyl ether
PEG-6 tridecyl ether. See POE (6) tridecyl ether
PEG-12 tridecyl ether. See POE (12) tridecyl ether
PEG 300 tridecyl ether. See POE (6) tridecyl ether
PEG 500 tridecyl ether. See POE (12) tridecyl ether
Plaster of Paris. See Calcium sulfate
POE (5) castor oil. See Book II
POE (30) castor oil. See Book II
POE (40) castor oil. See Book II
POE (2) cetyl alcohol. See POE (2) cetyl ether
POE (10) cetyl alcohol. See POE (10) cetyl ether
POE (20) cetyl alcohol. See POE (20) cetyl ether
POE (30) cetyl/stearyl alcohol. See POE (30) cetyl/stearyl ether
POE (50) cetyl/stearyl alcohol. See POE (50) cetyl/stearyl ether
POE (2) cetyl/stearyl ether. See Book II
POE (3) cetyl/stearyl ether. See Book II
POE (4) cetyl/stearyl ether. See Book II
POE (6) cetyl/stearyl ether. See Book II
POE (10) cetyl/stearyl ether. See Book II
POE (12) cetyl/stearyl ether. See Book II
POE (15) cetyl/stearyl ether. See Book II
POE (20) cetyl/stearyl ether. See Book II
POE (2) coco amine. See POE (2) coconut amine

POE (4) dilaurate. See Book II
POE (6) dilaurate. See Book II
POE (8) dilaurate. See Book II
POE (12) dilaurate. See Book II
POE (20) dilaurate. See Book III
POE (32) dilaurate. See Book III
POE (75) dilaurate. See Book III
POE (150) dilaurate. See Book III
POE (150) dinonyl phenyl ether. See Book II
POE (10) dinonyl phenyl ether phosphate. See Book IV
POE (150) distearate. See Book III
POE (20) glycerol monostearate. See POE (20) glyceryl monostearate
POE (7) glyceryl monococoate. See Book IV
POE (40) hydrogenated castor oil. See Book II
POE (60) hydrogenated castor oil. See Book II
POE (100) hydrogenated castor oil. See Book II
POE (200) hydrogenated castor oil. See Book II
POE (20) isocetyl ether. See Book II
POE (23) lauryl alcohol. See POE (23) lauryl ether
POE (2) monolaurate. See Diethylene glycol monolaurate
POE (2) monooleate. See Diethylene glycol monooleate
POE (2) monostearate. See Diethylene glycol monostearate
POE (2) nonyl phenyl ether. See Book II
POE (4) nonyl phenyl ether. See Book II
POE (5) nonyl phenyl ether. See Book II
POE (6) nonyl phenyl ether. See Book II
POE (8) nonyl phenyl ether. See Book II
POE (9) nonyl phenyl ether. See Book II
POE (3) octyl phenyl ether. See Book II
POE (5) octyl phenyl ether. See Book II
POE (40) octyl phenyl ether. See Book II
POE (2) oleyl alcohol. See POE (2) oleyl ether
POE (10) oleyl alcohol. See POE (10) oleyl ether
POE (3) oleyl ether. See Book IV
POE (20) sorbitan monoisostearate. See Book II
POE (10) sorbitan monolaurate. See Book IV
POE (80) sorbitan monolaurate. See Book III
POE (40) sorbitan peroleate. See Book II
POE (2) soya amine. See Book IV
POE 8 stearate. See POE (8) monostearate
N,N´, N´-POE (15)-N-tallow-1,3-diaminopropane. See POE (15) tallow aminopropyl-
   amine

POE (3) tridecyl alcohol. See POE (3) tridecyl ether

POE (6) tridecyl alcohol. See POE (6) tridecyl ether

Poly (oxy-1,2-ethanediyl), α-(3-carboxy-1-oxo-3-sulfopropyl)-ω-(dodecyloxy)-, disodium salt. See Disodium lauryl ether sulfosuccinate

Poly (oxy-1,2-ethanediyl), α-sulfo-ω-(dodecyloxy)-, ammonium salt. See Ammonium lauryl ether sulfate

Poly (oxy-1,2-ethanediyl), α-sulfo-ω-(tetradecyloxy)-, ammonium salt. See Ammonium myristyl ether sulfate

Polyoxyethylene (9) polyoxypropylene (3) isostearyl ether. See PPG-3-isosteareth-9

Polyoxyl 8 stearate. See POE (8) monostearate

Polyoxyl 40 stearate. See POE (40) monostearate

Polyoxypropylene (3) polyoxyethylene (9) isostearyl ether. See PPG-3-isosteareth-9

Polysorbate 20 (CTFA). See POE (20) sorbitan monolaurate

Polysorbate 21 (CTFA). See POE (4) sorbitan monolaurate

Polysorbate 40 (CTFA). See POE (20) sorbitan monopalmitate

Polysorbate 60 (CTFA). See POE (20) sorbitan monostearate

Polysorbate 61 (CTFA). See POE (4) sorbitan monostearate

Polysorbate 65 (CTFA). See POE (20) sorbitan tristearate

Polysorbate 80 (CTFA). See POE (20) sorbitan monooleate

Polysorbate 81 (CTFA). See POE (5) sorbitan monooleate

Polysorbate 85 (CTFA). See POE (20) sorbitan trioleate

Propylene glycol laurate (CTFA). See Propylene glycol monolaurate

Propylene glycol monoricinoleate. See Book III

Propylene glycol oleate (CTFA). See Propylene glycol monooleate

Propylene glycol stearate (CTFA). See Propylene glycol monostearate

Pyridinium, 1-dodecyl-, chloride. See Lauryl pyridinium chloride

Quaternary ammonium compounds, dicoco alkyl dimethyl, chlorides. See Dicoco dimethyl ammonium chloride

Quaternium-34. See Dicoco dimethyl ammonium chloride

Quaternium-52. See Book IV

Sodium dicyclohexyl sulfosuccinate. See Book II

Sodium dioctyl sulfosuccinate. See Book II

Sodium laureth-7 sulfate (CTFA). See Sodium POE (7) lauryl ether sulfate

Sodium laureth-12 sulfate (CTFA). See Sodium POE (12) lauryl ether sulfate

Sodium lauryl benzene sulfonate. See Sodium dodecylbenzenesulfonate

Sodium olefin sulfonate. See Book III

Sodium PEG (7) lauryl ether sulfate. See Sodium POE (7) lauryl ether sulfate

Sodium PEG 600 lauryl ether sulfate. See Sodium POE (12) lauryl ether sulfate

Sodium POE (4) lauryl ether phosphate. See Book IV

Sodium salt of coconut acid amide of glutamic acid. See Sodium cocoyl glutamate

Sodium salt of hydrogenated tallow acid amide of glutamic acid. See Sodium hydrogenated tallow glutamate

GENERIC CHEMICAL SYNONYMS AND CROSS REFERENCES

Sodium salt of the lauric acid amide of glutamic acid. See Sodium lauroyl glutamate
Sodium stearyl-2-lactylate. See Sodium stearoyl lactylate
Sodium xylene sulfonate. See Book II
Sorbimacrogol laurate 300. See POE (20) sorbitan monolaurate
Sorbimacrogol oleate 300. See POE (20) sorbitan monooleate
Sorbimacrogol palmitate 300. See POE (20) sorbitan monopalmitate
Sorbimacrogol stearate 300. See POE (20) sorbitan monostearate
Sorbimacrogol trioleate 300. See POE (20) sorbitan trioleate
Sorbimacrogol tristearate 300. See POE (20) sorbitan tristearate
Sorbitan monolaurate. See Book III
Sorbitan monooleate. See Book IV
Sorbitan monopalmitate. See Book II
Sorbitan POE (20) monolaurate. See POE (20) sorbitan monolaurate
Sorbitan POE (20) monopalmitate. See POE (20) sorbitan monopalmitate
Sorbitan POE (20) monostearate. See POE (20) sorbitan monostearate
Sorbitan POE (20) tristearate. See POE (20) sorbitan tristearate
Sorbitan sesquioleate. See Book IV
Sorbitan trioleate. See Book III
Sorbitan tristearate. See Book III
Sorbitan, monododecanoate, poly (oxy-1-2-ethanediyl) derivatives. See POE (20) sorbitan monolaurate
Sorbitan, monohexadecanoate, poly (oxy-1,2-ethanediyl) derivatives. See POE (20) sorbitan monopalmitate
Sorbitan, monooctadecanoate, poly (oxy-1,2-ethanediyl) derivatives. See POE (20) sorbitan monostearate
Sorbitan, mono-9-octadecenoate, poly (oxy-1,2-ethanediyl) derivatives. See POE (20) sorbitan monooleate
Sorbitan, trioctadecanoate, poly (oxy-1,2-ethanediyl) derivatives. See POE (20) sorbitan tristearate
Sorbitan, tri-9-octadecenoate, poly (oxy-1,2-ethanediyl) derivatives. See POE (20) sorbitan trioleate
Stearethate 40. See POE (40) monostearate
Stearyl alcohol. See Book III
Stearyl dimethyl amine oxide. See Book IV
Stearyl dimethyl benzyl ammonium chloride. See Book IV
Substituted imidazoline of oleic acid. See Oleyl imidazoline
Succinic acid, sulfo-, 1-ester with N-(2-hydroxyethyl) dodecanamide, disodium salt. See Disodium lauramido MEA-sulfosuccinate
Succinic acid, sulfo-, monododecyl ester, disodium salt. See Disodium lauryl sulfosuccinate
Sulfobutanedioic acid, 2-cocamido-1-methylethyl, disodium salts. See Disodium coco amide MIPA sulfosuccinate

391

Sulfobutanedioic acid, 1-dodecyl ester, disodium salt. See Disodium lauryl sulfosuccinate

Sulfobutanedioic acid, 4-[2-[2-[2-(dodecyloxy) ethoxy] ethoxy] ethyl] ester, disodium salt. See Disodium lauryl ether sulfosuccinate

Sulfobutanedioic acid, 1-ester with N-(2-hydroxyethyl) dodecanamide, disodium salt. See Disodium lauramido MEA-sulfosuccinate

Sulfuric acid, calcium salt (1:1) dihydrate. See Calcium sulfate

Sulfuric acid, monododecyl ester, ammonium salt. See Ammonium lauryl sulfate

Sulfuric acid, monododecyl ester, compd. with 2-aminoethanol (1:1). See Monoethanolamine lauryl sulfate

Sulfuric acid, monododecyl ester, compd. with 2,2'-iminodiethanol (1:1). See Diethanolamine lauryl sulfate

Sulfuric acid, monododecyl ester, magnesium salt. See Magnesium lauryl sulfate

Sulfuric acid, monododecyl ester, potassium salt. See Potassium lauryl sulfate

Sulfuric acid, monododecyl ester, sodium salt. See Sodium lauryl sulfate

Tallow dimethyl amine oxide. See Book IV

Tetradecamide, N,N-bis. See Myristic diethanolamide

Tetradecyl dimethyl benzyl ammonium chloride. See Myristyl dimethyl benzyl ammonium chloride

Tetrasodium dicarboxyethyl stearyl sulfosuccinamate. See Book II

Trideceth-3 (CTFA). See POE (3) tridecyl ether

Trideceth-6 (CTFA). See POE (6) tridecyl ether

Trideceth-12 (CTFA). See POE (12) tridecyl ether

Tridecyl alcohol, ethoxylated (3 EO). See POE (3) tridecyl ether

Tridecyloxypoly (ethyleneoxy) ethanol (3 EO). See POE (3) tridecyl ether

N,N,N-Trimethyl-1-dodecanaminium chloride. See Lauryl trimethyl ammonium chloride

Trimethyl dodecyl ammonium chloride. See Lauryl trimethyl ammonium chloride

N,N,N-Trimethyl-1-hexadecanaminium salt with 4-methylbenzenesulfonic acid. See Cetyl trimethyl ammonium p-toluene sulfonate

TRADENAME PRODUCT MANUFACTURERS

Aarhus Oliefabrik A/S
DK-1800
Aarhus C. Denmark

Ajinomoto USA, Inc.
9 West 57th St., #4625
New York, NY 10019

Akzo Chemie America
300 S. Wacker Dr.
Chicago, IL 60606

Akzo Chemie B.V.
Stationsstraat 48, PO Box 247
3800 AE-Amersfoort, Netherlands

Albright & Wilson (Australia) Ltd.
610 St. Kilda Rd., PO Box 4544
Melbourne 3001, Australia

Albright & Wilson/
Detergents Div., Marchon Works
PO Box 15, Whitehaven
Cumbria CA28 9QQ, UK

Albright & Wilson Inc.
PO Box 26229
Richmond, VA 23260

Alcolac Inc.
3440 Fairfield Rd.
Baltimore, MD 21226

Alkaril Chem. Inc.
Industrial Pkwy., PO Box 1010
Winder, GA 30680

Allied Colloids Ltd.
PO Box 38, Low Moor, Bradford
Yorkshire BD12 OJZ, UK

Amerchol Corp.
PO Box 4051, 136 Talmadge Rd.
Edison, NJ 08818

American Hoechst
Rt. 202–206 North
Somerville, NJ 08876

Arco Chem. Co./
Div. Atlantic Richfield Co.
1500 Market St.
Philadelphia, PA 19101

Ardmore Chem. Co.
29 Riverside Ave., Bldg. #14
Newark, NJ 07104

Arjay Inc.
PO Box 45045
Houston, TX 77045

Armak Co.
8401 West 47th St.
McCook, IL 60525

Armak Ind. Chem. /
Div. Akzo Chemie America
300 S. Wacker Dr.
Chicago, IL 60606

Armstrong Chem. Co., Inc.
1530 South Jackson St.
Janesville, WI 53545

Atlas Chem. Ind. N.V.
Imperial Chemical House
Milbank, London, SW1P 3JF, UK

393

BASF AG
Carl-Bosch-Strasse 38
D-6700 Ludwigshafen, F.R.G.

BASF Wyandotte Corp.
100 Cherry Hill Rd.
Parsippany, NJ 07054

Berol Kemi AB
Box 851, S-444 01
Stenungsund 1, Sweden

Bio-Lab Inc.
627 E. College Ave.
Decatur, GA 30030

Bofors Lakeway
5025 Evanston Ave.
PO Box 328
Muskegon, MI 49443

Canada Packers Ltd.
5100 Timberlea Blvd.
Mississauga, Ontario L4W 2S5, Canada

Capital City Products/
Div. of Stokely-Van Camp
PO Box 569
Columbus, OH 43216

Carson Chemicals, Inc.
2779 East El Presidio
Long Beach, CA 90810

Catawba-Charlab, Inc.
5046 Pineville Rd., PO Box 240497
Charlotte, NC 28224

Ceca S.A.
12/16 Ave. des Vosges
La Defense 5-Cedex 54-92062
Paris-La Defense, France

Chem-Y Fabriek Van Chemische
    Producten B.V.
PO Box 50
2410 AB, Bodegraven, Netherlands

Chemax, Inc.
PO Box 6067, Highway 25 South
Greenville, SC 29606

Chemical Products Corp.
125 Main Ave., PO Box 360
Elmwood Park, NJ 07407

Ciba-Geigy Corp.
PO Box 18300
Greensboro, NC 27419

Cindet Chemicals Inc.
Greensboro, NC

Clintwood Chem. Co.
4342 S. Wolcott Ave.
Chicago, IL 60609

Clough Chemical Co., Ltd.
178 St. Pierre
St.-Jean, Quebec J3B 7B5, Canada

Continental Chem. Co.
270 Clifton Blvd.
Clifton, NJ 07015

Croda Chem. Ltd.
Cowick Hall, Snaith Goole
North Humberside DN14 9AA, UK

Croda Inc.
51 Madison Ave.
New York, NY 10010

Croda Surfactants Inc.
183 Madison Ave.
New York, NY 10016

TRADENAME PRODUCT MANUFACTURERS

Croda Universal Ltd.
Oak Rd., Clough Rd.
Hull, UK

Cyclo Chem. Corp.
7500 N.W. 66th St.
Miami, FL 33166

Dai-ichi Kogyo Seiyaku Co., Ltd.
Miki Building, 3-12-1, Nihombashi,
Chuo-ku
Tokyo 103, Japan

DeSoto Inc.
PO Box 23523
Harahan, LA 70183

Diamond Shamrock/Process Chem. Div.
350 Mt. Kemble Ave.
Morristown, NJ 07960

Domtar Inc./CDC Div.
1136 Matheson Blvd.
Mississauga, Ontario L4W 2V4, Canada

Dow Chem. Co.
1703 S. Saginaw Rd.
Midland, MI 48640

Drew Produtos Quimicos Ltds.
Rua Sampaio Viana, 425
04004, Sao Paulo, Brazil-CP4885

DuPont de Nemours, E.I. & Co.
Nemours Bldg.
Wilmington, DE 19898

Durkee Industrial Foods/SCM Corp.
900 Union Commerce Bldg.
Cleveland, OH 44115

Dutton & Reinisch Ltd.
Crown House, London Rd., Morden
Surrey SM45DU, UK

Dynamit Nobel
10 Link Dr.
Rockleigh, NJ 07647

Emery Industries Inc.
1501 W. Elizabeth Ave.
Linden, NJ 07036

Emulsion Systems Inc.
215 Kent Ave.
Brooklyn, NY 11211

Ethyl Corp.
451 Flordia Blvd.
Baton Rouge, LA 70801

Finetex Inc.
418 Falmouth Ave.
Elmwood Park, NJ 07407

Formosa Chem. & Fibre Corp.
1 Nanking East Road, Sec. 2
Taipei, Taiwan, China

GAF Corp.
1361 Alps Rd.
Wayne, NJ 07470

Gattefosse Ets.
36 Chemin de Genas, 69800 Saint Priest
92100-Boulogne, France

Geronazzo Ind. Chim. SpA, Mario
78, Ospiate Di Bollate
Milano, Italy 20021

Glyco Chemicals Inc.
PO Box 700, 51 Weaver St.
Greenwich, CT 06830

Goldschmidt AG, Th.
Goldschmidtstr. 100
4300 Essen 1, Postfach 101461
West Germany

Goldschmidt Chem. Corp.
914 Randolph Rd., Box 1299
Hopewell, VA 23860

Grace, W.R. & Co.
55 Hayden Ave.
Lexington, MA 02173

Graden Chem. Co., Inc.
426 Bryan St.
Havertown, PA 19083

Grinsted Products Inc.
201 Industrial Pkwy., PO Box 26
Industrial Airport, KS 66031

Gross, A. & Co./
Div. of Millmaster Onyx Corp.
652 Doremus Ave., PO Box 818
Newark, NJ 07101

Chemische Fabrik Grunau GmbH
Robert-Hansen Str. 1, Postfach 1063
D-7918 Jllertissen
Bavaria, West Germany

Hall, C.P. Co.
7300 South Central Ave.
Chicago, IL 60638

Harrison, A.
PO Box 494
Pawtucket, RI 02862

Hart Chem. Ltd.
256 Victoria Rd. South
Guelph, Ontario N1H 6K8, Canada

Hart Products Corp.
173 Sussex St.
Jersey City, NJ 07302

Hefti Ltd.
PO Box 1623, CH-8048
Zurich, Switzerland

Henkel Argentina S.A.
Avda. E. Madero Piso 14
1106 Capital Federal
Argentina

Henkel Chem. (Canada) Ltd.
9550 Ray Lawson Blvd.
Ville d'Anjou, Quebec H1J 1L3, Canada

Henkel Inc.
480 Alfred Ave.
Teaneck, NJ 07666

Henkel KGaA
Postfach 1100, D-4000
Dusseldorf 1, West Germany

Heterene Chem. Co., Inc.
POB 247, 792 21 Ave.
Paterson, NJ 07513

Hexcel Chem. Products
7 Century Dr.
Parsippany, NJ 07054

High Point Chem. Corp.
601 Taylor St., PO Box 2316
High Point, NC 27261

TRADENAME PRODUCT MANUFACTURERS

Hilton-Davis Chem. Co./
Div. of Sterling Drug, Inc.
2235 Langdon Farm Rd.
Cincinnati, OH 45237

Hodag Chem. Corp.
7247 N. Central Park Ave.
Skokie, IL 60076

Hüls AG, Chemische Werke
Postfach 1320 D-4370
Marl 1, West Germany

Humko Chem./Div. Witco Chem.
PO Box 125, 755 Crossover Lane
Memphis, TN 38101

Humko Sheffield Chem./Div. Kraft Inc.
PO Box 398
Memphis, TN 38101

ICI Australia Ltd.
ICI House, 1 Nicholson St.
Melbourne 3000, Australia

ICI United States Inc.
New Murphy Rd. & Concord Pike
Wilmington, DE 19897

Inolex Chem.Co.
4221 S. Western Blvd.
Chicago, IL 60609

Intex Products Inc.
PO Box 6648
Greenville, SC 29606

Jetco Chem.
PO Box 1898
Corsicana, TX 75110

Jordan Chem. Co.
1830 Columbia Ave.
Folcroft, PA 19032

Kao Corp.
14-10 Nihonbashi, Kayabacho 1-chome,
Chuo-ku
Tokyo 103, Japan

Elektrochemische Fabrik Kempen GmbH
Postfach 100 260
D-4152 Kempen 1, West Germany

Lankro Chem. Ltd.
PO Box 1, Eccles
Manchester M30 0BH, UK

La Tassilchimica
Bergamo, Italy

Lever Industriel
103 Rue de Paris
9300 Bobigny, France

Lilachim S.A.
37-33 Rue de la Loi
1040 Brussels, Belgium

Lipo Chemicals Inc.
207 19th Ave.
Paterson, NJ 07504

Lonza Inc.
22-10 Route 208
Fair Lawn, NJ 07410

Lyndal Chem. Co.
PO Box 1740
Dalton, GA 30720

Magna Corp./
Arjay Chem. Div.
PO Box 33387
Houston, TX 77233

Manro Products Ltd.
Bridge St., Stalybridge
Cheshire SK15 1PH, UK

Marlowe-Van Loan Corp.
PO Box 1851
High Point, NC 27261

Mars Chem. Corp.
762 Marietta Blvd. N.W.
Atlanta, GA 30318

Mazer Chem. Inc.
3938 Porett Dr.
Gurnee, IL 60031

McIntyre Chem. Co., Ltd.
4851 S. St. Louis Ave.
Chicago, IL 60632

Michel & Co., M.
90 Broad St.
New York, NY 10004

Miranol Chem. Corp.
68 Culver Rd., PO Box 411
Dayton, NJ 08810

Mona Industries, Inc.
PO Box 425, 76 E. 24th St
Paterson, NJ 07544

Monsanto Co.
800 N. Lindbergh Blvd.
St. Louis, MO 63167

Nikko Chem. Co., Ltd.
1-4-8 Nihonbashi-Bakurocho, Chuo-ku
Tokyo 103, Japan

Nippon Nyukazai Co., Ltd.
19-9, 3-chome, Ginza, Chuo-ku
Tokyo 104, Japan

Nippon Oil & Fats Co., Ltd.
5-1 chome, Yurakucho, Chiyoda-ku
Tokyo, Japan

Norman, Fox & Co.
5511 S. Boyle Ave.
PO Box 58727
Vernon, CA 90058

Oleofina S.A.
Rue de Science 37 Wetenschapsstraat
1040 Brussels, Belgium

Onyx Chem. Co./
Millmaster Onyx Group
190 Warren St.
Jersey City, NJ 07302

Paniplus
100 Paniplus Rdwy.
Olathe, KS 66061

Patco Products/
Div. C.J. Patterson Co.
3947 Broadway
Kansas City, MO 64111

Pilot Chem. Co.
11756 Burke St.
Santa Fe Springs, CA 90670

PPF International
Lindtsedjik 8, 3336 LE
Zwijndrecht, The Netherlands

TRADENAME PRODUCT MANUFACTURERS

Procter & Gamble Co.
301 E. 6th St., PO Box 599
Cincinnati, OH 45201

PVO International Inc.
416 Division St.
Boonton, NJ 07005

Reilly-Whiteman Inc.
Washington & Righter Sts.
Conshohocken, PA 19428

Rewo Chemicals Inc.
107B Allen Blvd.
E. Farmingdale, NY 11735

Rewo Chemische Werke GmbH
Postfach 1160, Industriegebiet West
D-6497 Steinau an der Strasse
West Germany

Richardson Co.
2400 Devon Ave.
Des Plaines, IL 60018

Ronsheim & Moore Ltd.
Ings Lane, Castleford
Yorkshire WF10 2JT, UK

Sandoz Colors & Chem.
Route 10
East Hanover, NJ 07936

Sanyo Chem. Industries Ltd.
11-1 Ikkyo Nomoto-cho Higashiyama-ku
Kyoto 605, Japan

Scher Chem. Inc.
Industrial West & Styertowne Rd.
Clifton, NJ 07012

Seppic
70 Champs Elysees
75008 Paris, France

Servo Chemische Fabriek B.V.
PO Box 1, 7490 AA
Delden, Holland

Shell Chem. Co.
One Shell Plaza
Houston, TX 77001

Sherex Chem. Co.
5777 Frantz Rd., PO Box 646
Dublin, OH 43017

Stauffer Chem. Co.
Nyala Farm Rd.
Westport, CT 06880

Stepan Co.
Edens & Winnetka Rds.
Northfield, IL 60093

Stepan Europe
BP127
38340 Voreppe, France

Stokely-Van Camp Inc.
PO Box 569
Columbus, OH 43216

Swastik Household & Industrial Products
    Ltd.
Shahibag House, PB 362
13 Walchand Hirachand Marg
Ballard Estate
Bombay 400 038, India

Swift Chem. Co.
1211 West 22nd St.
Oak Brook, IL 60521

Tennessee Chem. Co.
3475 Lenox Rd. N.E., Suite 670
Atlanta, GA 30326

Tennessee/Cities Service
ICD, Cities Service Bldg., 3445
Peachtree Rd. N.E.
Atlanta, GA 30326

Toho Chem. Industry Co., Ltd.
14-9, 1-chome, Kakigara-cho
Nihonbashi, Chuo-ku
Tokyo 103, Japan

Triantaphyllou S.A., Thomas
405 Tatoiou Av., TK 136 71, Acharnes
Athens, Greece

Unger Fabrikker AS
N-1601
Fredrikstad, Norway

Union Carbide Corp.
39 Old Ridgebury Rd.
Danbury, CT 06817

Van Dyk & Co., Inc.
11 Williams St.
Belleville, NJ 07109

Vanderbilt & Co., Inc., R.T.
30 Winfield St.
Norwalk, CT 06855

Vista Chem. Co.
15990 N. Barkers Landing Rd.
PO Box 19029
Houston, TX 77079

Westvaco Chem. Div.
PO Box 70848
Charleston Hts, SC 29415

Witco Chem. Ltd.
Union Lane, Droitwich
Worcester WR9 9BB, UK

Witco/Organics Div. & Sonneborn Div.
277 Park Ave.
New York, NY 10017

Zohar Detergent Factory
Kibbutz Dalia, Israel 18920

Zschimmer & Schwartz
Postfach 2179, 4-5 Max-Schwarz-Strasse
D-5420 Lahnstein, West Germany

www.ingramcontent.com/pod-product-compliance
Lightning Source LLC
Chambersburg PA
CBHW021026210326
41598CB00016B/920